# The Man Who Flattened the Earth

Portrait of Maupertuis by Robert Levrac-Tournières, 1740.
By permission of Musée de Saint-Malo.

# The Man Who Flattened the Earth

## MAUPERTUIS and the Sciences in the Enlightenment

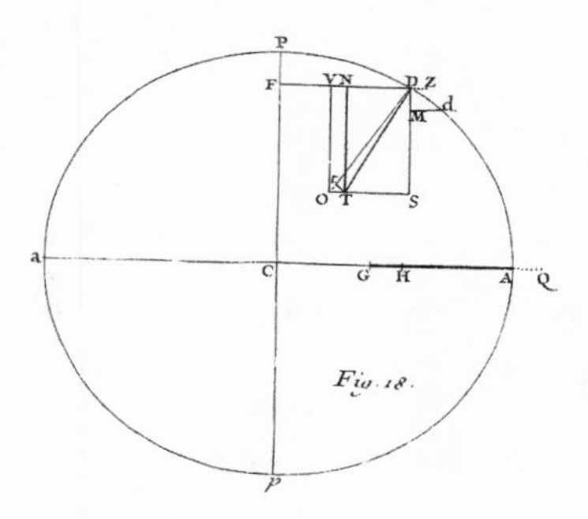


Mary Terrall

THE UNIVERSITY OF CHICAGO PRESS

*Chicago and London*

The University of Chicago Press, Chicago 60637
The University of Chicago Press, Ltd., London

Paperback edition 2006
Printed in the United States of America
11 10 09 08 07 06 2 3 4 5

ISBN 0-226-79360-5 (cloth)
ISBN 0-226-79361-3 (paperback)

Library of Congress Cataloging-in-Publication Data

Terrall, Mary.
The man who flattened the earth : Maupertuis and the sciences in the enlightenment / Mary Terrall.
p. cm.
Includes bibliographical references and index.
ISBN 0-226-79360-5 (hardcover : alk. paper)
1. Maupertuis, 1698–1759. 2. Scientists—France—Biography. I. Title.
Q143.M28 T47 2002
509.2—dc21
2002006778

# CONTENTS

# ILLUSTRATIONS

# ACKNOWLEDGMENTS

I HAVE MANY PEOPLE TO THANK for many different kinds of assistance and support. For reading and commenting helpfully on all or part of the manuscript, I am grateful to Ken Alder, Tom Broman, Jan Golinski, Pamela Smith, Michael Dettelbach, Amir Alexander, Mi Gyung Kim, and Ted Porter. For valuable conversations over the years, in addition to those just mentioned, I thank Peter Reill (in whose seminar I first encountered Maupertuis), Simon Schaffer, Paula Findlen, Jim Evans, Bob Brain, Olivier Courcelle, John Greenberg, Bob Frank, Shirley Roe, Emma Spary, and Norton Wise. For generously sharing research materials, I thank Elisabeth Badinter, Olivier Courcelle, Tom Hankins, Andrea Rusnock, and J. B. Shank. I would also like to thank Ann Blair for help translating a Latin epigraph; Ludmilla Jordanova for art historical insights about scientific portraiture; Jörg Sacher for his expert knowledge of Berlin Academy prize questions, and his patient investigation of Prussian genealogies; Richard Sorrenson and Sara Schechner for answers to technical questions about astronomical instruments; and Matthew Jones for his invaluable expertise on seventeenth- and eighteenth-century mathematics. I had crucial research assistance at various stages of this project from Michael Gordin, Donyne Choo, Michael Sauter, and Corey Hollis. On the personal front, I thank my family, and especially David Politzer, for patience and support over the years it has taken to complete this book. The younger generation did not always believe it would ever be finished.

The research for this book took me into the collections of libraries and archives too numerous to mention. I would like to give special thanks to a few of the many librarians and archivists who made my work possible. For sympathetic assistance in finding, using, and reproducing materials in their collections: Christine Demeulenaere-Douyère and Claudine Pouret of the Archives at the Académie des Sciences in Paris; Wolfgang Knobloch of the Archives of the Akademie der Wissenschaften in Berlin; Fritz Nagel and Martin Mattmüller of the Bernoulli-Edition in Basel; Bruce Whiteman of the Clark Library at UCLA; Katherine Donahue of the History Division of the UCLA Biomedical Library.

I also acknowledge financial support from the National Science Foundation; the Committee on Research of the UCLA Academic Senate; and the Harvard University Joseph H. Clark Fund.

Parts of Chapters 5 and 8 have appeared in different form in *Isis* and *History of Science*, respectively.

# ABBREVIATIONS

| | |
|---|---|
| AdW | Akademie der Wissenschaften, Archives (Berlin) |
| AS p-v | Procès-verbaux, Académie Royale des Sciences, Paris |
| AS | Archives de l'Académie Royale des Sciences, Paris |
| *BAS* | *Histoire de l'Académie Royale des Sciences et Belles-Lettres,* Berlin (includes the *Mémoires*) |
| BEB | Bernoulli-Edition Basel, Universitätsbibliothek Basel |
| BJ | Biblioteka Jagiellonska, Cracow |
| BN | Bibliothèque nationale de France, Paris |
| *Correspondance* | *Correspondance de Madame de Graffigny.* |
| FII | Frederick II of Prussia |
| *HAS* | *Histoire de l'Académie Royale des Sciences* (Paris) |
| JB | Johann I Bernoulli |
| JBII | Johann II Bernoulli |
| Koser, *Briefwechsel* | Reinhold Koser, ed. *Briefwechsel Friedrichs des Grossen mit Grumbkow und Maupertuis* |
| *Lettres* | T. Besterman, ed., *Les lettres de la marquise du Châtelet* |
| LS | Achille Le Sueur, *Maupertuis et ses correspondants* |
| M | Maupertuis |
| *MAS* | *Mémoires de l'Académie Royale des Sciences* (Paris) (published with the *Histoire,* separately paginated) |
| MS Obs. | Bibliothèque de l'Observatoire de Paris, manuscript collection |
| *Oeuvres* | Maupertuis, *Oeuvres,* 4 vols., 2nd ed. Lyon 1768 |
| RS | Royal Society of London |
| SBB | Staatsbibliothek, Preussischer Kulturbesitz, Berlin |
| SM | Archives municipales de Saint-Malo, Manuscript collection, ms. ii.24 |
| Volt. *Corr.* | T. Besterman, ed., *The Complete Works of Voltaire: Correspondence* |

# A NOTE ON TRANSLATIONS

All translations in the text are my own unless otherwise indicated. In some cases, especially when citing unpublished letters, quotations appear in the footnotes in the original language, preserving original spelling and punctuation.

# 1

# Portrait of a Man of Science

In 1739, at the peak of his scientific career, Pierre-Louis Moreau de Maupertuis sat for his portrait (see frontispiece). The finished picture represented this rather eccentric man of science to the cosmopolitan world of letters and to the smaller world of his immediate friends, admirers, and enemies. The painting was publicly displayed in the Salon exhibition in the Louvre palace in 1741 and discussed in the press. Copies were commissioned, in oils and in copperplate engravings, and sent to friends and patrons; subsequently a version of the same image appeared as the frontispiece to Maupertuis's collected works. The subject of the painting was a senior member of the Paris Academy of Sciences and the author of numerous technical and polemical books and papers, connected socially with prominent families in the capital and with men and women of letters across Europe. He worked with the artist, Robert Tournières, to imbue the portrait with the marks of a carefully constructed persona as a mathematician, an explorer and adventurer, a man of action and wit. His central role in a tortuous dispute in the Academy of Sciences about the shape of the earth had recently enhanced his fame—or notoriety. The painting contributed to a polemic that had engrossed the public and divided the Academy. Painted at a key moment in a controversial and publicly visible career, the image gives us some insight into the way its subject wished to be seen.[1]

The face directs its confident and self-assured gaze outward to the viewer. Its untroubled and slightly bemused expression defies the observer to find it ridiculous. The face belongs to a man dressed in fur hat and reindeer-skin robe, sumptuous with gilt-embroidered red trimming. The costume, ostensibly the native dress of the northern regions of Lapland, marks him as the triumphant survivor of a journey above the Arctic Circle, where he made his most famous and controversial measurements.[2] The landscape of those frozen mountains and rivers is just

1. Born Robert Levrac, the artist was known as Levrac-Tournières, or Tournières, after his birthplace (Bénézit, *Dictionnaire critique et documentaire des peintres, sculpteurs, dessinateurs et graveurs*). See also chapter 5 below.

2. There is no reason to believe in the authenticity of the apparel. Another northern scientific traveler, Carl Linnaeus, also had himself depicted in the dress of the native Sami people, in the

visible in the background, vaguely indicated by a conventional pointing finger. The outlandishly dressed mathematician presses firmly downward on the north pole of a terrestrial globe, marked with lines of latitude and longitude, deforming it into a slightly oblate shape, effectively claiming ownership of the flattened earth. The pose asks the viewer to agree that the calculator and voyager is also somehow responsible for flattening the earth itself. By advertising the truth about the earth's shape, Maupertuis also asserts his own power. This strength derives partly from the feat of having traveled so far, and partly from the mathematics indicated in the half-visible diagrams spilling over the windowsill. A contemporary viewer would have recognized something we cannot see directly, that this strength is also grounded in the power of the French king, who magnanimously authorized (and financed) the expedition. The mathematician supplied the knowledge and instruments, along with physical stamina and courage, but without the state's resources the expedition would never have left Paris.

The picture also domesticates the difficult and dangerous work involved in making astronomical observations in remote corners of the world, by dressing up the result in luxury. No frostbite or ravaged lungs are evident, nor are any of the many other participants in the expedition, including his closest collaborators. The painting effectively distills the meaning of the expedition into the person of one man. With its references to the far north, it appeals to a taste for the exotic in both learned and fashionable circles. Although the pose of the mathematician denotes strength and mastery, it is also slightly coquettish. The man of science presents himself as a denizen of high society; he is courting the same audience that frequented theaters and salons, performing a role much as his aristocratic friends performed their own theatricals. Maupertuis's reputation as mathematician was matched by a reputation as a conversationalist and man-about-town, even a libertine, at home in the salons and boudoirs of aristocratic women.[3] The portrait, then, graphically demonstrates the way these audiences, the learned and the fashionable, overlapped and interacted. This overlap is emblematic of the practice of science in a setting where validation and rewards, as well as challenges and attacks, came from different sources simultaneously.

By sitting for his portrait, Maupertuis consolidated his public image. Painted in the aftermath of a painful and sometimes embarrassing controversy, the calm-

---

frontispiece to his *Flora lapponica* (1737) and in a more formal portrait. On Linnaeus as "Lapp," see Koerner, *Linnaeus,* 64–67.

3. Maupertuis's movements in this world can be traced in the voluminous correspondence of Françoise de Graffigny, who developed a friendship with him when she was living with the duchesse de Richelieu in the late 1730s (see Graffigny, *Correspondance*).

ness and command of the portrait belie the contention surrounding the expedition by representing Maupertuis's controverted claim (the flattened globe) as a fait accompli. The controversy was not far from the minds of viewers, however. Jean-François Nollet described the portrait sardonically as the latest sally of a Don Quixote "defending his Dulcinea." Some years later, the journalist Charles Collé used the portrait as evidence that Maupertuis was "devoured by jealousy and thirst for reputation," since his collaborators were nowhere to be seen.[4] Nothing was more unstable than reputation in this sociable world of privilege; Maupertuis's authority was controversial, however secure he appeared in his portrait.

The portrait crystallizes a moment in the convoluted story of how Maupertuis rose to prominence in a society where such arcane practices as precise astronomical measurements could translate into a variety of rewards, including status and honor. This book follows the movements and choices he made in his bid for fame and reputation, through the careful crafting of his mathematical and literary work, as well as through his sociability and conversation. His quest for knowledge was also a quest for a persona that would incorporate the posture of the portrait with an intellectual commitment to the rational investigation of nature. If we had only the portrait as evidence of this career, it would be a story of adventure and resounding acclaim. With the addition of his published works on diverse subjects and in many genres, the record of his activities in scientific institutions, and personal letters that document his social and intellectual networks, the story becomes more nuanced, more contested, and less directed toward a triumphant climax. The portrait can be read as one strategy among many—of mixed success as the reactions of his contemporaries attest—for the representation of the enlightened man of science. Maupertuis's biography, then, illuminates the place of science in the cosmopolitan Republic of Letters, as well as the role of science in the making of his own identity.

## Sociability

In the middle decades of the eighteenth century, the loose confederation of writers and readers who thought of themselves as enlightened identified enlightenment with what they called "sociability." As the element of human nature responsible for the social bond, this term carried connotations of citizenship as well as conviviality. Maupertuis participated in many of the forms of sociability, public and private, that have elicited scholarly interest in the wake of Jürgen

4. Nollet to Jallabert, in Benguigui, ed., *Théories électriques du XVIIIe siècle.* For more of Nollet's account, see chapter 4 below. Collé, *Journal historique,* August 1759, in Bonhomme, ed., *Journal et mémoires,* 2:296–99.

Habermas's work on the public sphere.[5] Cafés, public lectures, salon gatherings, public gardens, scientific and literary academies, and even royal courts provided the settings for the sociable exchange that characterized production of knowledge in the Enlightenment. As Roger Chartier has shown, this more or less public sociability depended on private reading and writing.[6] These interlocking practices informed the strategies deployed by Maupertuis in building his persona and in solidifying his reputation as a member of a new elite that drew its status from scientific credentials, a status whose value was still being negotiated over the course of Maupertuis's life.

Enlightened prescriptions for the pursuit of natural knowledge placed it definitively in society, linking knowledge to social utility, but also to sociable behavior. The true philosopher, Denis Diderot told readers of the *Encyclopédie*, "knows how to divide his time between solitude and social intercourse." Unlike his less enlightened brethren, such a man recognizes the drawbacks of isolated reflection and rigid systematic thinking. "Man is not a monster who should live only in the depths of the sea or the farthest reaches of the forest. . . . Reason demands that he know and study the qualities of sociability and endeavor to acquire them."[7] If philosophy was a sociable pursuit in self-consciously enlightened circles in mid-eighteenth-century Europe, it also required an awareness of the human mind's limitations and a willingness to give up constricting and all-encompassing theoretical "systems." Diderot's philosopher articulated his claims in dialogue with nature, much as the man of letters engaged in witty conversation with companions or wrote books for enlightened readers of both sexes.

Diderot was only one among many of his contemporaries who defined reason, wit (*esprit*), and good sense so as to put them to work in the service of a socially engaged philosophy that would challenge dogmatism and superstition. This definition was of course polemical, with enlightened or true philosophy standing opposed to the outmoded, rigid (and often-caricatured) "spirit of system" associated with Descartes and his followers.[8] Diderot's portrayal of the philosopher exemplified the self-conscious impulse to articulate a new kind of identity, claiming a spot for "philosophy" in the society of men and women worthy of "this enlightened century." The trend, epitomized in the encyclopedia project of Diderot and d'Alembert, to ground knowledge in sensory experience and to derive both social

5. Habermas, *The Structural Transformation of the Public Sphere*. For definitions of sociablility, see Diderot, *Encyclopédie* s.v. "Sociabilité" and "Naturelle."

6. Chartier, *Forms and Meanings;* idem, *Cultural Uses of Print.*

7. Diderot, *Encyclopédie*, s.v. "Philosophe."

8. For the canonical Enlightenment distinction between *esprit de système* and *esprit systématique*, see d'Alembert, *Preliminary Discourse*, 22–23.

utility and pleasure from it, made science sociable. The *Encyclopédie* itself, one of the most elaborate and controversial publishing ventures of its time, made manifest the connection of print to sociability.[9] These convivial and reasoned forms of interaction permeated not only conversation and letter writing, but books and journals as well. Traces of dialogue and exchange abound in printed works, in footnotes, prefaces, dialogues, and critical reviews; this literary angle was essential to the connection between science and sociability. Reading might seem a solitary and unsociable activity, but discussion and debate about books dominated many social gatherings and epistolary exchanges. To be sociable meant, among other things, to converse and correspond about books, their authors, their attackers, their supporters, and any attendant scandal. Writers interested in making a reputation in this world attuned themselves to these discussions, and fostered them. Maupertuis meticulously positioned himself and his books in this complex web of audiences, not just in order to be admired but to provoke discussion and even controversy. Louis-Bertrand Castel, a Jesuit mathematician and journalist, recalled hearing this advice from Maupertuis: "He who causes himself to be often spoken of is always discussed, and that is everything. It was [some] years ago that the illustrious president M[aupertuis] said to me, 'Publish small works often, and you will dominate all of literature.'"[10]

Much historical attention has been paid to the political consequences of a wide range of forms of sociability, for articulating challenges to the regime and for contributing to a newly significant public opinion. Enlightened *salonnières* consciously promoted a code of sociability for their guests to distinguish their conversations and letters from the courtly civility of the previous century, a form of civility they branded as superficial. Salons were only one of many types of gathering where literature, philosophy, and politics were discussed and debated. The exchange between "institutions of sociability" (including male-dominated cafés and clubs as well as female-dominated salons) points to the multiplicity of venues where men and women pursued and promoted science, along with other forms of edification and amusement.[11] Maupertuis's trajectory through the world of salons, cafés, and academies shows that science also belonged in the realm of conversation and wit. Discourse about natural knowledge—all the topics that might be grouped under the rubric of "the sciences"—took place in and among many kinds of social groups, including mixed-gender salons and public lectures, but also

9. Darnton, *Business of Enlightenment;* Proust, *Diderot et l'Encyclopédie.*

10. Castel to La Condamine, [30 July 1751], excerpted in Thierry Bodin, Auction Catalogue (autograph manuscripts), July 1992. (The current location of the original letter is unknown.)

11. On sociability in the immediate pre-Revolution period in France, see Goodman, *Republic of Letters;* Gordon, *Citizens without Sovereignty.*

informal all-male gatherings in private homes or public cafés, aristocratic house parties, clubs, and a variety of organized institutions. The boundaries between these settings were never entirely rigid, though different rules of comportment and performance obtained in different settings. In fact, Maupertuis's career demonstrates the permeability of these boundaries, as he adapted his rhetoric and behavior to the appropriate context and exploited his ability to pass from one setting to another.

The Academy of Sciences was the most exclusive venue for scientific investigation, but it was integrated into the Republic of Letters, and many of its members wrote for that unstructured, cosmopolitan, and often contentious wider community of readers. Maupertuis himself was one of a small number of members of the science academy who was also elected to the elite literary academy, the Académie française, which in turn was closely linked to the salons of powerful aristocratic hostesses. One anecdote about the early part of his career is telling in this regard. Although he later broke with Fontenelle, in the 1730s the patronage of the star of salon society served Maupertuis well. As the story goes, Maupertuis "wished to be admitted *chez* Mme Lambert, who assembled men of letters at her home. Fontenelle, in introducing him, said, 'I have the honor of presenting M. de Maupertuis, who is a great mathematician and who nevertheless is not a fool [*sot*].'"[12]

## Crossing Intellectual Boundaries

The idealized sociability of reasoned and witty conversations coexisted with vigorous and often vitriolic contention in the letters and sciences of the eighteenth century. Maupertuis lived the life of the sociable enquiring philosopher described by Diderot, embroiled in intellectual controversy and intrigue, and exploiting the medium of print to further his goals and enhance his reputation. His career exemplifies the way in which the various scientific disciplines were interleaved with literature and philosophy in mid-eighteenth-century Europe. Although he did not write for the *Encyclopédie,* he belonged to the diffuse community of "gens de lettres" from which it drew its authors and its readers. His many works, primarily academic papers and short books, range across an encyclopedic variety of topics, belying anachronistic notions of specialization or expertise. Reviews and abstracts of his books appeared in the pages of all the leading literary and scientific journals of the day. Citations and summaries are sprinkled throughout the volumes of the *Encyclopédie* in articles on mathematics, physics, life science, reproduction, naviga-

12. Abbé Raynal, "Nouvelles littéraires," in Grimm, *Correspondance littéraire,* 10, 1:114. Mme. Lambert died in 1733; this meeting, if it is not apocryphal, would have taken place in the late 1720s or early 1730s.

tion, astronomy, epistemology and language. All of this made him a visible figure, admired by some and reviled by others, in the world of enlightened science and letters.

Maupertuis set out in the 1720s to make his way as a mathematician; he learned Leibnizian analysis from Johann Bernoulli and applied it to problems addressed by Newton in the *Principia.* This was a way to stake a claim at the cutting edge of mathematical physics, where methods and concepts gave rise to rancorous debate. He went on to write about geodesy (the shape of the earth) and mechanics, both mathematical disciplines, but then ventured into life science, cosmology, metaphysics, and philosophy of language. Maupertuis has been remembered, in the history of science and in physics and biology, for certain of his ideas that gained significance later by virtue of some relation, genealogical or not, to more modern concepts like evolution or the principle of least action.[13] Most often, in the historical literature, his works have been treated piecemeal, in isolation from their context in the social and political environment of old-regime France and Prussia. Even when they have been examined as the product of a single thinker, as in studies by Pierre Brunet and David Beeson, the theories and arguments have not been read in light of what it meant to do science and be a man of science in the eighteenth century.[14] This was above all a literary world, where books and periodicals provided not only reading matter, but stimulated conversation and gossip as well. In this world, hierarchical structures and power relations coexisted with the cosmopolitan and egalitarian ideals of the Republic of Letters.[15] These features of the social location of science informed the way Maupertuis formulated his ideas, his presentation of them to his contemporaries, and the reactions they provoked. He maintained friendships and collegial relations with men of science and aristocratic women across Europe.[16] He frequented the fashionable intellectual salons of Paris, as well as Louis XV's court, and the intimate circle of Frederick II of Prussia. His success in these venues aroused jealousy, pique, and ridicule, but also admiration and flattery.

Reputation was crucially important in this world of gossip, performance, and reading, and the sciences could foster reputation in a sphere that extended beyond the limited specialist elite of the academy. As a man of science and a man of letters,

13. Glass, *Forerunners of Darwin;* Guéroult, *Dynamique et métaphysique leibniziennes;* Brunet, *Etude historique.*

14. Brunet, *Maupertuis;* Beeson, *Maupertuis.*

15. Daston, "The Ideal and Reality of the Republic of Letters in the Enlightenment."

16. A partial list of his correspondents includes Montesquieu, Martin Folkes, Cromwell Mortimer, Voltaire, the duchesse d'Aiguillon, the duchesse de Chaulnes, the comtesse de Bentinck, the comtesse de Verteillac, Denis Diderot, Pieter von Musschenbroek, Willem 'sGravesande, Jean

Maupertuis systematically crafted his public identity by building relations with a variety of constituencies and patrons, and by writing for several overlapping audiences. He first established his reputation in the cafés and salons of Paris, then in the Paris Academy of Sciences, and went on to dominate the Francophone scientific world of Prussia under Frederick the Great, in the Berlin Academy of Sciences. His books, in form as well as in content, projected an image of the author as adventurer, wit, and philosopher, equally comfortable in salon and academy, fluent in the language of mathematics and astronomical measurement but also the master of an elegant literary style. He wrote for an elite readership avid for books on provocative subjects, as well as for academicians versed in technical evidence and arguments. To call him a popularizer would be to miss the point of his concerted efforts to build a career and reputation among the social and intellectual elite, including technically adept men of science. Along the way, he aspired to asserting the aristocratic status of science, compatible with nobility as well as with the freewheeling and even freethinking world of the *philosophes.*

Although his reputation as a brilliant innovator was largely eclipsed by the end of the century, in his day Maupertuis operated at the center of European scientific and literary life. In practice, this center was not localized in geographical space, and he sought fame and fortune by moving around a good deal. In 1728, he traveled to London as a budding mathematician in search of connections to the Newtonian establishment. Shortly after returning from his trip across the Channel, he journeyed eastward to Basel to cultivate the good graces of Johann Bernoulli, the foremost living mathematician of the previous generation. The intellectual patronage of the older man, crucial for Maupertuis's reputation as a mathematician in Paris, became the foundation for lifelong personal ties to the Bernoulli family. Once established in the Paris Academy of Sciences in the early 1730s, he traveled even farther afield, as the leader of a high-profile expedition, under the banner of the Academy and the crown. The objective was astronomical and geodetical measurements in the Arctic, in what was then the Swedish territory known as Lapland (now part of Finland). This expedition, the product of a dispute about the Newtonian theory of gravitation and the proper methods for testing it, gave Maupertuis the opportunity to expand his reputation outward from the specialist environment of the academy and the mathematics community to the genteel public who followed such disputes in the press and in conversations in elevated social circles. He wrote up the expedition as an adventure, embedding the technical results in an engaging narrative, widely read and admired. The portrait of Mauper-

LeRond d'Alembert, Georges-Louis Leclerc de Buffon, Etienne Bonnot de Condillac, Charles-Marie de La Condamine, Francesco Algarotti, Emilie du Châtelet, the Johann Bernoullis (father and son), Daniel Bernoulli, and Leonhard Euler.

tuis as an earth flattener dates from this period. In the aftermath of the Lapland expedition, he became a frequent guest at the best houses in Paris, not to mention the royal court. "Heavens!," one of his well-placed female friends exclaimed, "Maupertuis is a flea. Is he ever in one place?"[17]

A flea might seem a strange emblem for a man of letters, but the image is peculiarly apt. There were certainly some who found him as annoying as a biting insect, especially because of his persistent self-aggrandizement, and a goodly number felt the caustic sting of his bite over the years. But it was rather his frenetic ubiquity that his friend was identifying (although the nickname, coined by Françoise de Graffigny, was meant affectionately, if ironically, and only used in private).[18] Maupertuis's energy and eclecticism took him not only into the social arena, where he was famous for amusing anecdotes and witty repartee, but into new intellectual and literary ventures as well. He took the opportunity afforded by the appearance of a comet in 1742 to try his hand at an elegantly styled dialogue about cosmology, a kind of update of Fontenelle's *Conversations on the Plurality of Worlds.*[19] Published anonymously, Maupertuis's short explication of comets, and their place in Newtonian physics, took his conversational talents into print. Not long after this literary experiment, he went considerably further with another anonymous book, *Vénus physique,* this time about the fraught question of the generation of organisms.[20] Although its theory was unabashedly speculative, and its style provocative, this was nevertheless a serious examination of an intransigent theoretical problem, examined in light of extensive empirical evidence from anatomy, microscopy, embryology, and animal breeding. He came back to the topics of generation and heredity repeatedly over the years. As with the book on comets and the account of the Lapland expedition, his writings on this subject reinforced his image as a man of letters, and as someone willing to take risks, intellectual as well as physical.

Acclaim for his book about the Lapland expedition brought Maupertuis to the attention of Frederick, then crown prince of Prussia, who subsequently invited him to preside over the Berlin Academy of Sciences. It took several years for the circumstances to line up to make this offer attractive, but he did finally decamp to Berlin in 1745. The expedition to Lapland was the first major turning point in his career; taking over the Berlin Academy was the next. Both were carefully

17. Graffigny to Devaux, 27 June [1739], in Graffigny, *Correspondance,* 2:19.

18. Graffigny gave nicknames to almost everyone of her acquaintance; some of these nicknames were charming and some satirical.

19. Maupertuis, *Lettre sur la comète.*

20. [Maupertuis], *Vénus physique,* Paris 1745. A shorter version appeared in 1744, also anonymously, under the title *Dissertation physique sur le nègre blanc.* See chapter 7 below.

considered strategic actions, not without risk. The move to Prussia, to an official position in a foreign monarchy, was highly unusual among French intellectuals, especially those who were not under threat of police action. It widened the scope of his reputation and his personal power, but also meant splitting personal and political loyalties. He left behind not only Paris, where he had made his name, but his native Brittany, which continued to exert a nostalgic pull throughout the years of voluntary exile (1745–1756). Berlin, although the capital city of an increasingly powerful ruler, was culturally a far cry from Paris, and not an easy place for a Frenchman to feel at home, notwithstanding Frederick's unrelenting Francophilia. Philosophical and personal conflicts only heightened the contrast to the remembered peace and familiarity of his childhood home—which nevertheless was too provincial to support the grand scale of his desire for power and position.

While ruling over the scientific scene in Berlin, Maupertuis also moved into the exclusive circle of Frederick's court at Potsdam, where a number of French philosophers and writers took refuge from censorship at home. Although his position in Berlin gave him a high profile in the Republic of Letters, he was frustrated in his efforts to transform the Prussian Academy to his own specifications, and he spent the last three years of his life in a liminal status induced by the Seven Years' War—caught between allegiance to France, his fatherland and spiritual home, and to Frederick, his most illustrious patron. Over the course of a career dominated by controversy and contention, he crossed and recrossed geographical and political boundaries, as well as the more fluid boundaries that defined categories of readers and audiences: learned and polite, male and female, admirers and detractors.

Crafting a life as a man of science meant navigating through complex webs of allegiances and animosities, personal and institutional and political. Most of Maupertuis's scientific accomplishments were contested, to a greater or lesser extent; he was an original thinker and an energetic controversialist. To examine these controversies and contentions is to see into the mechanics of struggles over intellectual territory. His historical importance lies not in any list of particular "great discoveries," although his work had an impact on many scientific fields, especially rational mechanics, observational astronomy, and life sciences. Beyond these particular intellectual and organizational contributions, his story illuminates the pitfalls and rewards of forging a career in science at a time when there was not yet a prescribed route to such a career, or even a clear consensus about what it meant to be a man of science. A close look at someone who moved so widely across the scientific map, both geographically and intellectually, provides insights into the development of scientific ideas and practices in the Enlightenment and into the place of science in the wider culture.

In the 1720s and 1730s, when Maupertuis was establishing himself in the Paris Academy of Sciences, traditional forms of personal patronage mingled with newly articulated standards of merit to create an elite, exclusive body of men whose practical and intellectual work contributed to the eminence and prosperity of the monarchy.[21] Senior academicians were working to entrench their institution firmly as the arbiter of technical knowledge, legitimated by ties to the crown and the government. The expedition to Lapland played a role in this effort, affirming the capabilities of men of science to promote the interests and the glory of the king and his ministers, as well as to display their own perspicacity and fortitude. In Berlin, Frederick looked to Paris for the model for his Academy of Sciences and Belles-Lettres. He imagined the Prussian institution as an emblem of his enlightened rule and also of his personal commitment to letters and philosophy. As the king's representative in the Academy, Maupertuis had to ensure that the work of his academicians lived up to royal expectations. His ventures into print in this period were intended to enhance his own reputation, but also to contribute to the public image of the Berlin Academy, and of the Prussian monarch as a patron of enlightened thinking.

The dynamic balance between personal patronage and institutional power, and the flexible social strategies of intellectuals, undergirded the substance of eighteenth-century science and philosophy. To tell the story of Maupertuis's career, I integrate analysis of private correspondence and archival records with an examination of his published works, which were filled with eclectic and contentious ideas and arguments. My attention to questions of "self-fashioning," strategy, and rhetoric runs in parallel to analysis of ideas, arguments, calculations, and observations.[22] Ultimately, I am assessing not the psychological state or the personality of my subject, but the contours of his thought and the way his works (and especially the published works) contributed to an evolving identity as enlightened man of science. This individual persona, with its particularities and idiosyncrasies, also fed into and drew upon institutional and group identities. Pursuing his own glory, Maupertuis took seriously his service to the monarchy (in both France and Prussia), and contributed to defining the value of science for the emerging state.[23] The possibilities were more open-ended in this period than they had been in the seventeenth century, when the "new science" was just getting established. Much valuable historical work has addressed the function of social codes such as courtly

21. Smith, *Culture of Merit.*

22. On science as consummate strategy, see Biagioli, *Galileo Courtier.*

23. In Maupertuis's lifetime, this was not yet the bureaucratic state. On science at the end of the old regime, see Gillispie, *Science and Polity;* Outram, "Politics and Vocation"; idem, *Georges Cuvier.*

etiquette, gentlemanly civility, and Jesuit self-discipline in framing the observational, experimental, and literary practices that came to define the scientific revolution in its various aspects.[24] Given the consolidation of political power in the absolutist states where Maupertuis operated, it is perhaps surprising that the rules for determining status and assigning value to various activities were more fluid in the eighteenth century than they had been earlier.

A man of letters, of course, is known through his works. Maupertuis, like many of his contemporaries, was attuned to the strategic uses of print, in all its forms. He thought carefully about how to publish, distribute, and revise his writings, as well as how to formulate them stylistically. The physical attributes of the books, the content and layout of prefaces and title pages, plans for distribution, and revisions for subsequent editions were all weighed by the author. He took his own early advice to Castel to publish frequently but not at great length, keeping himself in the public eye quite deliberately, even when publishing anonymously. The development of his thought and his shifting interests can be traced through these works; in addition to this intellectual history, I examine how contemporaries read them, why he wrote them in one form rather than another, and how they played into debates about the practice of science and the workings of nature. Intellectual history runs the risk of separating texts from their authors' lives and circumstances, but it need not do so. Maupertuis's books were not best-sellers of the type analyzed by Robert Darnton; nor were they subsequently canonized as essential texts of the Enlightenment. They were, however, at the center of scientific discourse in many different areas, and many of them also exemplify the enlightened literary natural philosophy that circulated through the Republic of Letters. As subsequent chapters will show, Maupertuis explored the boundaries of genres and played with textual forms, as did many of his friends and associates.

The authorial identity that Maupertuis projected in his books shifted over the course of his career. Early on, after his return from Lapland, he played the man of action and the adventurer, as well as the intrepid observer and calculator. In his satirical attacks on his enemies, he played the literary wit. When he wrote the *Letter on the Comet,* he adapted Fontenelle's worldly didactic posture to a newly fashionable Newtonian cosmology, a posture he pushed further in *Vénus physique.* Subsequently, he wrote a number of books for the same audience, although he backed off from the slightly salacious style that lent *Vénus physique* its notoriety. By the time he published his *Essay on Cosmology,* also anonymously, he had adopted a tone of philosophical equanimity, to undermine natural theology and to promote his version of mechanics. Shortly thereafter, in a *Letter on the Progress of the Sciences,* he

24. Westman, "Astronomer's Role"; Shapin and Schaffer, *Leviathan;* Biagioli, *Galileo Courtier;* Shapin, *Social History of Truth;* Dear, "A Mechanical Microcosm."

combined the speculative attitude of many of these works with his mature persona as statesman of science, secure in his position at the seat of power. Reading this quite disparate corpus of work, with an eye to the way it was presented and the audiences to whom it appealed, we see how a man of science could present himself in many possible guises.

Positioning themselves with respect to the social and cultural constructs in which they live, individuals draw on resources of all kinds, but they also affect the people and institutions and ideas around them, altering the resources available to others. The story of one man's life opens up the story of the institutions and practices that shaped that life.[25] In the case of an early-modern mathematician and philosopher of nature, this means looking at the situation and status of science, defined not as the sum of knowledge about particular subjects at a given time, but as the process of producing and evaluating and representing knowledge. Maupertuis's formal institutional location in the early part of his career was the Paris Academy of Sciences, but he also sought the approval of audiences and readers outside the narrow confines of the Academy. The relatively young institution was still finding its footing on a shifting ground of cultural politics and political culture. As Maupertuis and his colleagues jockeyed for position inside the Academy, they also molded the institution and its interactions with the crown, with the wider scientific community, with the genteel public in Paris, and with the readers of books and periodicals across Europe. Similarly, the Prussian science academy's reorganization as the francophone arbiter of intellectual life in Berlin raised questions about the place of Berlin relative to German universities and to the rest of Europe. The cultural politics at work in Frederick's Berlin impinged on at least some of the papers published by the Academy or the prize questions set for its competitions. The story of Maupertuis's rise to prominence is thus also a story about institutions and the ways that science became part of the expanding flow of ideas avidly consumed by an increasingly self-conscious public in the Enlightenment.

I have organized my study of Maupertuis along roughly chronological lines, without doing violence to thematic coherence. At several times in his life, he worked simultaneously on distinct areas; in these cases, I have grouped my discussion of the works in question according to subject matter. Maupertuis's family origins in the adventurous corsair culture of Brittany contributed to his sense of his own place, socially and geographically, and to the opportunities available to him. He was not an introspective person, at least as far as surviving sources indicate, but even the outlines of the story of his early life and education, which emerge in

25. Shapin, *The Social History of Truth*; idem, "Personal Development and Intellectual Biography."

the next chapter, illuminate his ambitious transition from province to capital and into the Paris Academy of Sciences as a young man. In chapter 3, I give a detailed account of his mathematical work, based on an extensive exchange of letters with Johann Bernoulli. The Bernoullis kept both sides of all their correspondences, so that this is a rare case (for Maupertuis) of a complete correspondence, in which technical material is integrated with personal reflections. The letters reveal how Maupertuis used his connection with Bernoulli, not only to secure a place in the Paris Academy, but also to acquire the skills necessary to read Newton's *Principia.* This led him into the problem of the shape of the earth, where his efforts were in part inspired by the possibility of making Newton intelligible to his French colleagues. Chapter 4 concerns the Academy-sponsored expedition to Lapland and examines the literary means used to represent the experience to urbane readers and spectators. The scientific results of the expedition met with rather more skepticism from the Paris astronomers than the travelers expected, and the protracted polemics between the two sides are examined in chapter 5. The polemic was partly disciplinary, with the astronomers defending their record and their techniques, and the mathematician-observers of the Arctic expedition arguing for the strength of their imported English instruments and their new methods of calculation. The dispute became a literary dispute, albeit rather one-sided, as Maupertuis explored the uses of satire along with calculations and diagrams.

By the early 1740s, Maupertuis was known throughout Europe as the champion of Newtonian physics. In chapter 6, the narrative moves to his efforts to establish a name for himself as more than a Newtonian, as an original thinker in his own right. No longer working directly with Bernoulli, or explicating Newton's difficult demonstrations, he turned to statics, optics, and finally dynamics, and argued for a new general principle, which he dubbed the principle of least action. As he deployed it, this economy principle functioned both metaphysically and mechanically, encompassing the accepted laws of motion. However, this work initially elicited very little reaction from the Paris Academy. As it happened, this was the moment when Frederick called Maupertuis to Berlin, and the remainder of chapter 6 unravels the tangled strands of motivations and interests at play in the negotiations that finally resulted in his move to Prussia.

Chapter 7 interrupts the biographical narrative to analyze Maupertuis's excursion into the life sciences, with his anonymous *Vénus physique.* The next chapter explores the ideological and political bases of the Berlin Academy of Sciences, and the ways in which Maupertuis adapted to the cultural setting in Berlin, quite different from that of Paris. Alongside his role of philosopher-courtier, Maupertuis pursued his program for a metaphysical mechanics, with the able assistance of Leonhard Euler, the most accomplished mathematician at the revived Berlin Academy. Chapter 9 picks up with Maupertuis's mature formulation of a teleological mechanics as the basis for a rationalist theology and a proof for God's existence. The key text here is *Essay on Cosmology* (1750), which embroiled its author in yet another controversy, this time in the context of the Berlin Academy, when his originality was challenged by Johann Samuel König, a follower of the German philosopher Christian Wolff. When Voltaire rather perversely entered the polemic in defense of König, the dispute became a full-blown literary quarrel, involving the Prussian king as well as the Berlin Academy. At stake in this apparent priority dispute were honor and reputation, certainly, but also the credibility of mechanics based on the principle of least action. In the aftermath of this bitter controversy, Maupertuis returned to the problems of generation and heredity, extending his earlier speculations on active matter and organization. This work is the subject of chapter 10.

Throughout his career, Maupertuis's engagement with scientific problems was accompanied by a persistent concern, sometimes bordering on obsession, with his reputation. He wished to be known as a man of action, a man of letters, and a man of science. His cultural setting, whose agenda he helped to set, encompassed the overlapping spaces of café, academy, salon, and court. He took advantage of institutional possibilities and the evolving position of science to fashion an idiosyncratic identity for himself. Although much of his work centered on arguments accessible only to a few technically proficient individuals, he often reformulated these arguments for other audiences, sometimes in the pages of the same book. For Maupertuis himself, the adventurer was never far from the man of science. The geographical trajectory of his life—from Saint-Malo to Paris, and then back and forth between France and London, Basel, Stockholm, Lapland, and finally Berlin—graphically represents the fluidity of Enlightenment cosmopolitanism. He made scientific knowledge the key to moving around in this philosophical and geographical landscape.

# 2

## From Saint-Malo to Paris

Maupertuis was born and raised in the prosperous port city of Saint-Malo in Brittany. His social and geographical origins informed not only his character but also his subsequent successes in the flourishing intellectual and social scene of Paris in the 1720s and 1730s. The interests, values, and idiosyncrasies nurtured in the maritime and commercial setting of Saint-Malo stood him in good stead in the lively and sociable urban culture of the French capital. In a world where science occupied no more than a tiny niche, he exploited a series of opportunities to prepare the way for his initiation into the realm of philosophy and letters. His campaign to make a place for himself in this realm traversed a wide array of formal institutions and informal social groupings: literary and scientific academies, cafés, salons, and royal courts.

Saint-Malo, at the time of Maupertuis's birth in 1698, carried on a thriving overseas trade with French colonies in the Americas, China, India (Pondicherry), and North Africa.[1] Ship's captains and their financial backers made their fortunes from this commerce and used their new wealth to take over some of the social and political prerogatives once reserved for the old aristocracy. These upwardly mobile merchant families consolidated their wealth and local power on the model of the aristocracy, through intermarriage and involvement in provincial politics.[2] The trend was typical of the expansion of commercial capitalism more generally, but the Malouin sea captains distinguished themselves from merchants in other parts of the realm through their privateering ventures against the shipping of other nations. In the wake of the virtual collapse of the French navy in the 1690s, Secretary of the Navy Jérôme Pontchartrain promoted privateering by changing the regulations governing the financing of such voyages and the division of their booty. Responding to the latest fiscal crisis, he encouraged investment in armed corsair ships by nobles, courtiers, and government ministers. Maupertuis's contemporary Charles Duclos recalled the Saint-Malo of their youth: "Everyone was a merchant or a corsair, and often both at once. In the midst of the misfortunes of

1. Dahlgren, "Jérôme de Pontchartrain," 225–27.

2. Meyer, *La noblesse bretonne.* For biographical summaries that demonstrate the extent of intermarriage among merchant-privateer families, see Cunat, *Saint-Malo illustré.*

a war that desolated and ruined France, the Malouin *armateurs,* and those associated with them, saw their enterprises thrive on all the seas."[3] The *guerre de course* (privateering war) fulfilled the goals of both economic and military strategy, as Pontchartrain recognized the value of attacking enemy commerce as well as mobilizing private wealth to finance armaments that the royal treasury could ill afford.[4] The government saw these ventures as sufficiently successful to warrant substantial rewards for the privateers in the form of royal offices and other honors.[5]

Maupertuis's father, René Moreau, made his fortune as a merchant and privateer in the heyday of the *guerre de course* (1688–1697), arming and commanding a series of vessels that attacked English shipping. When he married Jeanne-Eugénie Baudran, daughter of another prosperous corsair family, he forged the kind of alliance that strengthened the interests of the Breton merchant class. His maritime exploits led not only to substantial profits, but also to a prestigious and potentially remunerative commission in a squadron headed for America in 1696. Moreau's privateering voyages ended with the Peace of Ryswick (1697), before the birth of his first son, Pierre-Louis, the following year. By this time, Moreau *père* could comfortably retire from seafaring to invest in the voyages of other captains and to become the proprietor of a profitable nearby fishery.[6] His fellow merchants elected him deputy to the Estates of Brittany, and then to represent their city on the Council of Commerce in Paris.

The Council of Commerce had been set up in 1700 to bring merchants from commercial centers across the realm into the royal power structure. The deputies deliberated on all sorts of trade questions and advised the royal ministers on policy matters.[7] To take up his post, with its required weekly meetings, Moreau moved to Paris in 1706, leaving his wife and three children in Brittany. Two years later, he was ennobled and awarded the coveted Order of Saint Michel. The king bestowed this honor and the hereditary title of "sieur de Maupertuys" in recognition of Moreau's privateering. The letter of ennoblement applauded his sizable investment in arms and his capture of enemy ships: "He has commanded his own vessels armed for war with 40 to 50 cannon, which he mounted himself, and with which he made many expeditions and raids on the enemies of our State. . . . He has known . . . how to reconcile the merit and prudence of a good merchant with

3. Duclos, *Mémoires,* 2.

4. Geoffrey Symcox argues that the corsairs contributed crucially to the French war effort, especially since they were fighting a war of attrition (*The Crisis of French Sea-power,* 225–26).

5. See generally Meyer, *La noblesse bretonne.*

6. Morel, "La guerre de course à St. Malo"; Meyer, *La noblesse bretonne,* 338–39. Moreau armed several medium-sized ships characteristic of the St-Malo trade, carrying forty to fifty cannon (Velluz, *Maupertuis,* 9, n.1; Kerviler, *La Bretagne à l'Académie française*).

7. See generally Schaeper, *The French Council of Commerce.*

Fig. 1. Maupertuis Family by Robert Levrac-Tournières, 1715 (?). There is some doubt about the date and the identity of the family, since René Moreau had three children. The painting is titled "La famille Maupertuis" in the catalogue of a private collection acquired by the Musée des Beaux-Arts in Nantes in 1837. By permission of Musée des Beaux-Arts de Nantes.

the valor and intrepidity of a soldier [*homme de guerre*]." The king reiterated the crown's vital interest in the investments responsible for Moreau's success, stressing that he should not construe his new noble status as incompatible with commercial activity. "We ennoble him and his descendants, without on this account obliging him to cease his business, which we have on the contrary expressly enjoined him to continue."[8]

The exploits of Moreau's privateering days opened up new social vistas, centered in Paris. No longer maneuvering on the high seas, he nevertheless maintained his financial and personal ties to his native city and to overseas commerce.[9] He exemplified the burgeoning aristocracy of merit in a nobility that became increasingly diverse in the course of the eighteenth century.[10] As merchants insinuated themselves into the status hierarchy of the old regime, exploits such as privateering were useful sources of legitimacy for the new beneficiaries of royal patronage. A seat on the Council of Commerce further solidified the honor attendant on such service by integrating it directly into the government of the kingdom. Once in Paris, the merchants on the Council had the opportunity to protect their investments and to identify new ones, since they were now insiders to the regulation of commerce.

Thus, the family fortunes were thriving in the years of Maupertuis's youth (fig. 1). The very few surviving accounts of his early life, though written in light of his later accomplishments, give some sense of the options available to a young man of his social background in the early eighteenth century. In the long run, his choices were necessarily atypical, since a career in the Academy of Sciences was not possible for more than a handful of individuals. His father's recent rise in wealth and status certainly opened up possibilities, however, as well as providing a model for emulation. With his father's world of politics and economic policy in the middle distance, and the seafaring life of a bustling port close at hand, Maupertuis had the kind of childhood that later fosters nostalgia for lost peace and contentment.[11] Saint-Malo itself retained a vital importance as well, not only

8. "Lettres d'annoblissement octroyées par Louis XIV à René Moreau, 'sieur de Maupertuys,' Septembre 1708," Registres des mandements de la Chambre des Comptes de Bretagne, Archives départmentales de Loire-Atlantique; copy in AS, dossier Maupertuis.

9. Moreau became a director of the Compagnie de l'Occident and continued to invest in overseas trade (Cunat, *Saint-Malo illustré,* 125). He also continued to be involved in the illegal, but very profitable, trade in the South Seas (Schaeper, *The French Council of Commerce,* 86).

10. See generally J. Smith, *The Culture of Merit;* Chaussinand-Nogaret, *The French Nobility;* idem, *Histoire des élites.*

11. He remembered the family of his childhood tutor in his will; see Testament, AS Fonds Maupertuis; Formey, *Souvenirs d'un citoyen;* Moreau de la Primerais, "Ecrits biographiques," AS Fonds Maupertuis.

as the family home and the source of his fortune, but as a counterpoint to the string of controversies he was to face in Paris and Berlin. Over the years, he traveled back to his native city as often as possible, to breathe the *air natale* as an all-purpose restorative.

When he articulated his public image in later years he drew on heroic tropes reminiscent of the corsair culture of Saint-Malo. His origins were well known to his contemporaries, and he modeled his scientific adventures on the successes of his privateer father and uncles, casting the quest for knowledge as adventure and action.[12] This was not the only way to construe knowledge, of course, but it resonated with attempts to represent the enterprise of science as an essential service to the king and to the nation, and it made for good anecdotes. As we will see, in the case of Maupertuis's geodetical expedition, the king's authority conveyed status to men designated by the Academy for the job, while they dedicated themselves to serving the crown's interests. The privateer's arrangement with the crown was not so different, as he risked his life and his capital to benefit himself as well as the royal treasury. When he had his portrait painted in fancy "Lapp" dress, Maupertuis was also linking his own accomplishments with the taking of trophies by the Malouin privateers.

He must have been conscious at an early age of the wider world of the metropolis and the distant destinations of the ships in the busy harbor, places his father had seen on his own travels. Even before his ennoblement, René Moreau had good enough connections to get an audience at Versailles for his family in 1704. This episode loomed large in the family memory, as it involved attending the king's supper, where the six-year-old eldest son was fussed over by the duchesse de Bourgogne.[13] When Maupertuis reached the age of sixteen, his father insisted that he continue his education in Paris. He duly enrolled at the College de la Marche, one of the many small colleges affiliated with the University of Paris. Here he studied philosophy for two years with a Cartesian professor, LeBlond.[14] At this point, the question of what was to become of the young gentleman was still open. Among the wealthy Malouin families of his social circle, many sons followed their fathers or uncles into the lucrative business of overseas commerce or went to sea themselves. Whatever penchant Maupertuis may have felt for joining them, and several

12. He was rumored to have completed the unfinished memoirs of one of his most famous compatriots, the Malouin corsair DuGuay-Trouin. Voltaire found the rumor plausible: "C'est un homme comme vous, unique en son genre." Voltaire to M, 29 June 1740, Volt., *Corr.,* 91:226–27; Graffigny to Devaux, 19 May [1739], *Correspondance,* 1:497.

13. Moreau de la Primerais, "Ecrits biographiques," AS Fonds Maupertuis; La Beaumelle, *Vie de Maupertuis,* 6.

14. Tuilier, *Histoire de l'Université de Paris,* 2:103–110. Moreau de la Primerais gives the name of the professor, "Ecrits biographiques," AS Fonds Maupertuis.

sources refer to his predilection for the sea, he never entered the world of commerce.[15] Some accounts attribute this choice to his mother's anxiety about the dangers of seafaring. It seems rather more plausible that Maupertuis *père* had taken enough steps up the social and economic ladder and that, away from the provinces, he saw opportunities for his son to profit from his own success by moving into the more genteel, even aristocratic, circles now open to him.

Making a place in the privileged urban aristocracy did not require denying the commercial roots of the family's good fortune. René Moreau was able to use his wealth and position to make certain choices for his son. Having set the young man up at a secular college, he looked around for a mathematician to teach him geometry. As a navigator, Moreau had learned geometry himself: "recognizing the utility of this science for most any profession one could embrace, he wished [his son] to join the study of geometry to that of philosophy."[16] The story of Maupertuis's early mathematics education later came to be told as the flowering of a "passion" for geometry, following a logical path from tutor to academy. The progression worked so well, not simply because of Maupertuis's native mathematical abilities, but also because his father undertook to hire a teacher of the best reputation. The tutor chosen was Nicolas Guisnée, a former protégé of the marquis de l'Hôpital and an associate member of the Academy of Sciences, who made a living by teaching geometry and the rudiments of algebra to wealthy young men.[17]

After he completed the philosophy course, Maupertuis continued to study mathematics and took up music as well, studying with Bernier, the court composer. His father remained intent on finding a profession for his son and, instead of sending him to sea, arranged a position for him in a company of the king's guards (the *mousquetaires gris*), housed in Paris. Here he spent a comfortable two years, combining the profession of arms with his mathematical studies, in the manner of a young nobleman. The only contemporary source concerning his conduct in this period refers to his involvement in a mysterious, but apparently philosophical, dispute that nearly led to a duel.[18] He owed his standing in this high-living society to his father, who used his considerable credit within the mil-

15. Moreau de la Primerais, "Ecrits biographiques"; La Beaumelle, *Vie de Maupertuis,* 9. Maupertuis recalled that his mother favored him over her two other children and indulged his every wish (Formey, "Eloge de Maupertuis," *BAS,* 1759, 465). According to Formey, Maupertuis blamed his intolerance for being contradicted on his overindulgent mother. He had one brother, Louis-Malo, born in 1700 or 1701, and a younger sister, Marie.

16. Moreau de la Primerais, "Ecrits biographiques."

17. Guisnée had written an elementary mathematics textbook (1704); he was not pensioned at the academy and willingly accepted the patronage of the young and wealthy Maupertuis.

18. D'Aguesseau to M. de Fresne, 4 July 1718, in Rives, ed., *Lettres inédites du chancelier d'Aguesseau,* 81–82. Maupertuis and Fresne, the son of d'Aguesseau, were on the same side of a "querelle

itary patronage system to acquire a captain's commission for his son in a cavalry company based in Lille. Such coveted commissions were primarily held by aristocrats and entailed no great risk or time commitment during this rare interlude of peace. In the three years that he held the commission, Maupertuis wintered in Paris, kept up with mathematics, and pursued a vigorous social life. Ultimately, the life of the fashionable intellectual proved irresistible, and he sold his military commission and moved to the capital.

## The Café Scene

By this time, Maupertuis was on cordial terms with several mathematicians in the Academy of Sciences, as well as a number of other men and women of letters. He cultivated these connections in cafés, where intellectuals of varied social backgrounds conversed, unconstrained by the hierarchies of military, academic, or courtly life. By one count, there were 300 cafés in Paris in 1716, a number that had grown to 380 just seven years later.[19] A substantial cross-section of city dwellers frequented these cafés. By the 1720s, some proprietors had invested in luxurious furnishings, such as mirrors and crystal chandeliers, in an effort to appeal to prosperous *honnête gens* (including aristocrats) and to distinguish their premises from lower-class haunts.[20] Intellectuals cultivated the habit of drinking coffee, contrasting the sharp wits of a coffee-drinker to the irrationality of a drunkard. In the cafés, men developed reputations for wit (*esprit*), that elusive but much vaunted quality of intellectual life in the old regime.

Historians have drawn attention to the importance of these public places as centers of discussion and dissemination of political news and other controversial questions. *Nouvellistes,* who earned a livelihood by collecting and exchanging news and gossip, circulated through cafés and other public places.[21] In addition to political news, some cafés functioned as centers of literary gossip and discussion. Politics and literary discussions went hand in hand, as one contemporary observer noted of the cafés he visited: "One spoke of everything there, of moral philosophy,

philosophique" that their adversaries wanted to turn into a "guerre civile." I owe this reference to Olivier Courcelle.

19. Fosca, *Histoire des cafés,* 14; Brennan, *Public Drinking,* 85–87. These figures come from records of the guild of *limonadiers,* licensed to serve coffee and brandy.

20. Fosca, *Histoire des cafés,* 8; Brennan, *Public Drinking,* 129–32. Using inventories of furnishings, Brennan shows the range of decoration and by extension the range in clientele frequenting taverns, cafés, and other drinking establishments. The literary and iconographic images of Enlightenment cafés portray a decidedly upper-class clientele. For the recording of political gossip in cafés, see generally Farge, *Subversive Words.*

21. Funck-Brentano, *Les nouvellistes;* Farge, *Subversive Words.* See also Habermas, *Structural Transformation,* for the role of such cultural settings in the development of public opinion and political

of physics, of medicine, of politics, of history, of theology, of jurisprudence, of anatomy, of mathematics, of literature."[22] Not all discussion was learned, certainly, but many subjects that look arcane to the modern eye, like mathematics and jurisprudence, were acceptable topics of conversation. In the more literary cafés, discussions of books, plays, and operas became part of the experience of reading literature.

By dint of hard work and perhaps some native ability, some men carved out a niche for themselves in this malleable scene and made themselves into men of letters. Some were able to use conversational skill to make the connections that could lead to a more formal and stable station in life. Duclos, for example, the son of a Breton merchant, quite consciously put his facile wit and penchant for conviviality to good use when he sought an entrée into high society.

> Thus, having no more position than that of a student who isn't studying, . . . I was on the threshold of being received into society of a rank superior to my own. This only happens in Paris, for men, provided that they are from respectable [*honnêtes*] families and are not in a dependent position. They can live with those who have a more elevated status, if the same tastes bind them to each other.[23]

Duclos fared well, becoming a successful author of essays and fiction, and eventually serving as secretary of the Académie française. Maupertuis started from a stronger position, as the eldest son of a wealthy family with the trappings of influence and even nobility already in place. The coffeehouse turned out to be a useful place for such young men from the provinces, regardless of wealth, to establish reputations as men of wit and as bons vivants. This in turn made them desirable guests for fashionable hosts and hostesses who eventually became their readers and perhaps their patrons. The cafés could also open the doors of the Academy of Sciences.

Maupertuis was often seen at two establishments frequented by men of letters, Café Procope and Café Gradot. He found a congenial place in the circle around the critic and poet Antoine Houdar de La Motte, an aging but still vibrant intellectual mentor figure who was a fixture at the Gradot.[24] La Motte's fame dated

---

consciousness; Farge, *Subversive Words.* For analogous developments in England, see Stewart, *The Rise of Public Science.*

22. Léger, cited by Fosca, *Histoire des cafés,* 19.

23. Duclos, *Mémoires,* 36.

24. Brunet mentions only the Procope, and most subsequent biographies follow him (e.g., Beeson). The Gradot was Maupertuis's more habitual circle (Maupertuis–Châtelet correspondence; Duclos, *Mémoires*). Maupertuis's initial introduction into this group may have been through his fellow Malouin, the Abbé Trublet, himself a journalist and man of letters. See Jacquart, *L'Abbé Trublet,* 48.

back to the celebrated dispute of the ancients and the moderns. His "modern" approach to reading literature elevated individual aesthetic and moral responses above timeless critical standards, leaving room for readers to judge "by their own lights." He had also reflected on the role of the reading public in such disputes.[25] These crucial issues for Enlightenment criticism suggest the sorts of discussions that occupied the café clientele. Maupertuis would later rank La Motte on a par with Newton and Leibniz, indicating the powerful impression the older man made on him.[26]

The circles of regulars *chez* Procope or Gradot were exclusive, but claims to reputation in café society did not depend on the traditional social distinctions of birth or wealth, as Duclos had discovered. When Maupertuis's protégé, L. A. de la Beaumelle, came to write his biography in the 1760s, he evoked the halcyon days of the intellectual cafés, populated by "savants and beaux esprits." Although La Beaumelle's book is not always reliable, the section on café life was substantially written by Charles-Marie de la Condamine, an intimate friend of Maupertuis from the 1730s to the end of his life, and is closer to a firsthand account than other parts of the biography. La Condamine provides a few telling insights into the formative stages of Maupertuis's intellectual career.

> The celebrated la Motte-Houdard [*sic*] of the Académie française, blind and infirm from the age of thirty, knew hardly any pleasure on leaving his study other than conversation, and no one made it more pleasing than he. . . . As soon as they spotted his bearers, the crowd opened to make a passage for him to the place he had chosen. . . . Bel esprit, likeable and of sweet character, he had gradually become the perpetual dictator of a republic whose principal citizens were people like La Faye, Saurin, Terrasson, Fréret, Melon, Nicole, Marivaux. . . . Everyone conversed, held forth, disputed vivaciously, but without bitterness. The stranger, the provincial, the newly arrived were sure of being listened to from the first time there, and they soon knew whether they should speak a second time.[27]

25. In 1715, La Motte sparred with the classicist Anne Dacier over the principles of literary criticism, specifically with respect to judging the works of the ancients. This was the second round of the literary "battle of the ancients and moderns." The first round had coincided with the early years of the Academy of Sciences with Fontenelle as a key player. See generally DeJean, *Ancients Against Moderns.*

26. "Un de nos plus grands géomètres [Maupertuis] m'a dit plusieurs fois qu'il y avait dans M. de la Motte de quoi faire un Newton, un Leibniz" (Trublet, "Lettre à Madame T.D.L.F.," in Houdar de la Motte, *Oeuvres,* x–xi).

27. "Le célebre la Motte-Houdard [*sic*] de l'Academie francaise, aveugle et infirme dès l'age de trente ans, ne connaissait guère d'autre plaisir au sortir de son cabinet que celui de la conversation, et personne n'y mettait plus d'agrément que lui. . . . Dès qu'on apercevait ses porteurs, la foule s'ouvrait pour lui faire un passage à la place qu'il s'était choisie. . . . Bel esprit, aimable et d'un caractère

This nostalgic portrait of a lost paradise, probably based on Maupertuis's recollections as well as those of La Condamine, stresses the ideal of republican egalitarianism in La Motte's domain, where performance determined status. The café promoted a meritocracy of wit, judged by a kind of philosopher-prince, whose bodily weakness only enhanced the brilliance of his mind. Those welcomed into the circle found themselves drawn into a lively exchange orchestrated by the master conversationalist. Maupertuis apparently fit in without too much trouble: "because of his frank and open manners he was forgiven for the superiority he displayed too quickly."[28]

In many ways, Maupertuis modeled himself on La Motte in the years to come. As he worked to establish a name as a mathematician, he kept up his regular attendance at the Gradot. After La Motte died in 1731, La Condamine recalled that Maupertuis took his place as center of this group.

> Since he moved very little in high society at that time, one only saw M. de Maupertuis at the homes of his personal friends or at his café. You were sure to find him there at midday; almost every evening he spent two hours there after a promenade or the theater. Nowhere else was he so much at his ease. He asked one of his friends very seriously what he would do in the great houses [of the rich]. His taste for independence, his distaste for any kind of trouble . . . was an unbreakable habit, and even when he was sought out by the most brilliant society, with his choice of the best company, he always preferred the free and perhaps blunt tone of his little republic to those grand polite considerations, those formal attentions that resemble flattery and that sometimes lead to falseness.[29]

---

doux, il était insensiblement devenu le dictateur perpétuel d'une république dont les principaux citoyens étaient les la Faye, les Saurin, les Terrasson, les Freret, les Melon, les Nicole, les Marivaux. . . . On conversait, on dissertait, on disputait avec vivacité, mais sans aigreur. L'étranger, le provincial, le nouveau venu étaient sûrs d'être écoutés dès la première fois, et bientôt ils savaient s'ils devaient parler une seconde" (La Beaumelle, *Vie de Maupertuis,* 11–12). The biography remained unpublished until 1856, when the manuscript was found by a descendant of the author. La Condamine had been part of the café scene in the 1730s and worked closely with La Beaumelle on the manuscript of Maupertuis's biography, correcting errors and checking details. See La Condamine to JBII, 25 June 1769, BEB, where he claims authorship of the section on the café Gradot.

28. La Beaumelle, *Vie de Maupertuis,* 12.

29. "Peu répandu jusqu'alors dans le grand monde, on ne voyait M. de Maupertuis que chez ses amis particuliers ou à son café. On était sûr de l'y trouver à midi; presque tous les soirs il y passait deux heures au retour de la promenade ou des spectacles. Nulle part il ne se trouvait si fort à son aise. Il demandait très sérieusement à un de ses amis ce qu'il allait faire dans les maisons. Son goût pour l'indépendance, son éloignement pour toute sorte de gêne . . . était une habitude insurmontable, et lorsque recherché dans le monde le plus brillant il eut le choix de la meilleure compagnie, il préféra toujours le ton libre et peut-être brusque de sa petite république à ces grands égards polis, à ces

This description, no doubt idealizing Maupertuis's independence from the corrupting influences of the glittering world of the aristocracy, divides the high society in private houses from the comfortable ambiance of the café, which was free of the constraints of decorum. The masculine independence exhibited in the café contrasts with the realm of flattery and superficial brilliance associated with women's drawing rooms and salons. The contrast between republican brusqueness and aristocratic politeness is a telling one, though La Condamine may well have exaggerated Maupertuis's distaste for high society. Later, he would happily frequent gatherings of clever women and great ladies, as well as the royal court at Fontainebleau. He continued to cultivate his "taste for independence," but he also learned how to sparkle in feminine society, thriving on the "flattery that sometimes leans toward falseness." As he worked his way up in the world, he came to realize that high society had its pleasures too, and he sought them avidly. If we are to believe La Condamine, fashionable sociability was an acquired taste, and in his first years in Paris not yet part of his persona.

## Salon and Court

At this point, Maupertuis's reputation flourished in a relatively small but sociable set of intellectuals of various stripes. He was well known to the denizens of the literary cafés, but had not yet courted a wider public by venturing into print. He was also beginning to be introduced into some of the "great houses" where women received regular guests for conversation and reading aloud of new literary works. Salons run by women have attracted a good deal of attention from scholars investigating alternatives to the canonical venues for the origins of the French Revolution and the Enlightenment.[30] Some of these hostesses held regularly scheduled gatherings, at which they set the agenda for reading, performance, and discussion, modeling their activities on the famous salons of the seventeenth-century *précieuses*.[31] There were also many other more informal social gatherings in private homes, where women presided, and where similar matters were discussed in mixed groups for amusement and edification. Guests exchanged personal, political, and literary gossip alongside card games, letter reading, and discussions of current controversies. These comfortable social interactions, often in the context of large supper parties or extended visits at aristocratic country houses, make a

---

attentions compassées qui ressemblent assez à la flatterie et qui tiennent quelquefois de la fausseté" (ibid., 12–13).

30. Goodman, *The Republic of Letters;* idem, "Enlightenment Salons"; idem, "Governing the Republic of Letters"; and idem, "Public Sphere and Private Life"; Gordon, *Citizens Without Sovereignty;* Craveri, *Madame du Deffand et son monde.*

31. Lougee, *"Le paradis des femmes."* Dena Goodman, *The Republic of Letters,* argues that serious intellectual work was done in these feminine spaces.

nice contrast to the institutionalized forms of intellectual exchange in the academies.[32] Certainly not all such gatherings were devoted to the "serious" philosophical and political work described by Dena Goodman in her discussion of the salons of Marie-Thérèse Geoffrin, Julie de Lespinasse, and Suzanne Necker. For the purposes of understanding Maupertuis's audiences and networks of alliances, I use the shorthand term "salon" loosely to designate this fluid realm of social relationships, witty conversations, and sexual liaisons. It stands in contrast to the Academy, with its formal rules and exclusive elections, and to the all-male "republics" in the cafés.

In Maupertuis's day, this form of sociability was decidedly aristocratic, but more freewheeling than that of the protocol-driven royal court. Some noblewomen cultivated men of letters, and the salons were an important node in the French cultural matrix, where alliances formed, dissolved, and re-formed under the pressures of influence transmitted through hierarchies of power. Many of the individuals, women as well as men, who frequented these circles had substantial pull with government ministers, and some spent time at court as well. Several of the most prominent hostesses had strong connections to the academies, and especially the literary academy, the Académie française.[33] Maupertuis's connections to salon hostesses and other women who regularly entertained the literary and political elite can be traced through the letters of Mme. Geoffrin, Mme. du Deffand, Mme. de Graffigny, the marquise du Châtelet, and the duchesses of Saint-Pierre, Richelieu, Chaulnes, and Aiguillon.[34] For someone concerned to make his mark in society, as well as in the Republic of Letters, these connections were crucial. He relished the attentions of the titled and powerful, and turned them to his advantage whenever possible. Still, though he was a welcome guest in many drawing rooms and boasted of his ties to ministers and his adventures at Fontainebleau, he was never far from the Academy of Sciences.

## The Academy of Sciences

When Maupertuis returned to Paris in 1723, he abandoned his military career with the understanding that his election to the Academy of Sciences was virtually assured. Several academicians frequented La Motte's circle, among them the mathematicians Jean-Baptiste Terrasson (who also wrote literary criticism), Joseph

32. Many such occasions are described in Graffigny, *Correspondance.*

33. Over the course of the century, the center of gravity in some salons shifted away from the aristocracy, and some hostesses welcomed a broader mix of guests into their circles. For connections to academy elections, see Craveri, *Madame du Deffand et son monde.*

34. Châtelet, *Lettres;* Graffigny, *Correspondance;* letters to Maupertuis from various noblewomen, BN, n.a.f. 10398.

Saurin (who was also a journalist), and François Nicole. These connections secured Maupertuis an adjunct position at the end of 1723. No longer a captain of the cavalry, he found ways to incorporate the military officer's glamour into his new persona. The Academy became his base of operations and the primary reference point for conceiving and maintaining his role as a man of science. He moved rapidly up the hierarchy by virtue of his productivity, his ability to forge alliances with his seniors, and his sociability. As part of the institutional apparatus of the absolutist state, the royal academies could not have been more different from the fluid social and intellectual space in the cafés. Yet quite a number of the men who honed their reputations in cafés eventually found positions in the highly regulated spaces of the literary and scientific academies. Membership in the Academy of Sciences would become the primary touchstone of Maupertuis's identity, and the institution itself the core venue for his intellectual activity over many years. This is not to say that once he gained admission he spoke only to his *confrères* in the specialized language of mathematics—far from it. But his new academic identity informed his other activities, his writing, his exchanges with friends and enemies, and his strategies for self-promotion. Thus, the structure and functioning of the institution set some of the basic parameters of his ambition and accomplishments.

The founding of royal academies in the 1660s had signaled a shift in cultural policy under Colbert's administration, replacing the personal patronage ties of prince or noble and client with patronage mediated by institutions and controlled by ministers of the government.[35] Although still clients of the king, academicians entered into this relationship in their capacity as members of a quasi-autonomous body, rather than as individuals. In return for the crown's "protection" and privileges granted to the institution, the academicians consulted on practical matters like navigation, ballistics, or waterworks. Even when they were not asked to attack specific problems, they often chose subjects of at least hypothetical utility or cast their investigations in terms of future practical benefits.[36] The Academy also represented its accomplishments as contributions to the sovereign's glory, in the manner of clients offering the fruits of their labors to the patron. But the Academy's usefulness to the crown went beyond producing practical or glorious results. The institution represented the crown's power to shape knowledge and culture in the French realm, and sometimes even beyond its borders.

35. Harth, *Ideology and Culture*. For the early history of the Academy of Sciences, see generally Hahn, *Anatomy of a Scientific Institution;* Alice Stroup, *A Company of Scientists;* McClellan, *Science Reorganized;* Maindron, *L'Académie des sciences*.

36. On utility in the Academy of Sciences, see Briggs, "The Académie Royale des Sciences." Maupertuis, "Balistique arithmétique" (1731); idem, "Courbes de poursuite" (1732) (charting courses for intercepting enemy ships).

From 1699, when it was given formal statutes, the Paris Academy of Sciences operated as a corporation under the protection and supervision of the king and his ministers.[37] In modeling the Academy of Sciences along traditional corporate lines, the royal ministers adapted old forms to new practices, intending to keep those practices under the oversight of the crown. Originating at a time when the state was consolidating its own power, the Academy, in fact, had less statutory autonomy than many older corporations like universities and occupational guilds.[38] The regulations structured the institution hierarchically; in this way, it resembled virtually all the formalized institutions or social divisions of the old regime: the church, the nobility, the government, and the commercial and artisanal guilds. The academic hierarchy, divided into six classes by subject matter, descended from *honoraires* to senior *pensionnaires* and on down to *associés* and *adjoints*.[39] Each rank had carefully defined privileges and obligations. The honorary category consisted of amateurs or patrons of science drawn from the high clergy, nobility, or the upper echelons of government. They did not attend meetings regularly but made themselves visible at the semiannual public meetings. These men were not just figureheads, however; they voted on policy matters and, more importantly, on elections. In this regard, they had more power than the associates and adjuncts, who had no voice at all in choosing new members. Every adjunct member aimed at working his way up the hierarchy to a pensioned slot. Because of the limited number of places, however, and the requirements for productivity, adjuncts could not count on ever receiving a pension. (Of the sixty working academicians, only twenty collected pensions at any given time.)[40]

The statutes of an occupational *corps* defined its domain and restricted nonmembers from intruding upon it. Only members of the Academy could attend the biweekly meetings, and since the number of places was very limited, these few individuals held the exclusive right to judge excellence, whether for prizes, patents, or entrance and advancement within the institution. Although the two top ranks voted on candidates for vacant slots, all choices ultimately rested with the king. In order to be proposed for election, an aspiring candidate had to offer proof of

37. The reorganization formalized procedures and regularized the number of spaces and pensions. The Academy's letters-patent were formally issued by the crown and registered by the Paris Parlement only in 1713 (Hahn, *Anatomy of a Scientific Institution,* 5). For all documents pertaining to structure of Academy, see Maindron, *L'Académie des sciences.*

38. For example, the regulations stated that the Academy "will receive its orders by that Secretary of State to whom it pleases His Majesty to give the responsibility" (Reg. I, Règlement du 26 janvier 1699, in Maindron, *L'Académie des sciences,* 18).

39. Originally labeled *élèves* and attached to senior pensioned members as apprentices to masters, this category was changed to *adjoints* in 1716.

40. Aucoc, *L'Institut de France,* for pension information. The average pension was 2000 *livres.*

his accomplishments, just as apprentices had to produce a masterwork to advance into a guild.[41] In principle, the academician was required to continue producing and to report on his work to his colleagues. Knowledge thus originated with individuals but was validated by the authority of the corporate body. The institution policed its own borders from within and passed judgment on other knowledge of potential interest to the crown. The most visible form of judgment came in the awarding of prizes for questions set by the Academy, starting in 1720. These competitions drew submissions from all over France and the rest of Europe, including England.

The esoteric knowledge of specialists and the political power of the crown reinforced each other. The patronage relationship that bound crown to Academy was the key to this reinforcement. Election to the Academy gave an intellectual not only a place to do his work, but a status that could be vitally important in establishing a social identity, in Paris and in the wider Republic of Letters. The value of a position in the Academy cannot be measured solely by the monetary value attached to it, nor can a pension of this sort be equated with salary. Even as an unremunerated adjunct, a young man gained an identity as an insider in an elite body, with the prospect of advancing up the ladder. Academicians aspired to elevated status in the cultural hierarchy, where institutional autonomy was always balanced against control by the state. They did not see this situation as hampering their experiments, observations, or calculations; they willingly participated in the vertical relations of power that authenticated their cultural position. The pensions awarded to the most productive members were tokens of their ties to the crown and evidence of the king's liberality.[42]

The social and cultural value of academic rank depended on the status of the institution itself. In its early years, the Academy occupied only a small corner of the French administrative structure. The establishment of the institution in 1666 hardly guaranteed its success, since it was not immediately clear how far the "experimental philosophy" of its founders would go. While the Academy's hierarchical organization and corporate privileges later came under fire from reformers

41. For the language of corporations, see Sewell, "*Etat, Corps,* and *Ordre,*" 54. Hahn refers to the corporate nature of the Academy of Sciences (*Anatomy of a Scientific Institution,* 73). The candidate had to be "known by some considerable printed work, by outstanding success in his studies, by some machine he invented or by some particular discovery" (Reg. 13, Règlement de l'Académie, in Maindron, *L'Académie des sciences,* 19).

42. I take issue with Roger Hahn's depiction of academicians as civil servants, paid to perform services for the crown. "Like other bureaucrats, academicians were given a salary tied to their specific function rather than to their literary production" (Hahn, *Anatomy of a Scientific Institution,* 51). Salary for services rendered does not adequately represent the intangible values exchanged between Academy and crown.

and critics, both inside and outside the institution, in the first decades of the eighteenth century, the Academy worked to solidify its status within the parameters of its new statutes. Following a reorganization in 1699 and especially after the death of Louis XIV in 1715, with the political situation in flux, academicians tried to stabilize their institution's position, not only with respect to the government, but also with respect to an elite public that followed the competitions, elections, disputes, and self-proclaimed triumphs within the Academy. In fact, the official statutes prescribed a public function for the Academy that went beyond consulting on technical problems to providing a kind of spectacle to interested onlookers.[43] The Academy carefully orchestrated this spectacle to reflect the glory of the protector as well as the accomplishments of the academicians. The elite specialists never lost sight of the importance of displaying their work to a curious public, including women.[44] In the first instance, this was a local Parisian audience. But by means of the press and its own official publications, the Academy also asserted its preeminence before a cosmopolitan audience across Europe. Interested readers could turn to the annual *Histoire* (issued with the more technical *Mémoires*), for a summary of the year's accomplishments shorn of arcane technical language. The public in Paris could also attend semiannual meetings of the Academy, which were covered in the periodical press as noteworthy events, often with long abstracts of the papers and eulogies.

Much of the scientific work of the Academy was undertaken and presented internally, by members for the edification of their fellows. Most of these men were also quite conscious of their relations with the outside world, of how their activities appeared to the uninitiated. Many of them moved easily from the meeting room of the Academy to the cafés, gardens, salons, and theaters where they mingled with a public of literary men and women, aristocrats, government ministers, and writers. The audience and readers lumped together in academic discourse as "the public" can be differentiated into several different, if overlapping, publics. These included, at any given time, some or all of the following: the Academy's patrons in the government; the populace at large who might benefit from discoveries and technological improvements; women who invited people to their homes to discuss natural phenomena and philosophical issues; buyers of books of philosophy and natural history; people of all classes curious about strange events and phenomena; men of science in other academies and other countries; spectators at the public sessions; men and women who followed the news and gossip of the Republic of Letters in periodicals. Partitioning of these publics into separate sociological groups

43. Reg. 35, Règlement de l'Académie, in Maindron, *L'Académie des sciences*, 22.

44. On the relations of the Academy with a feminine audience, see Terrall, "Gendered Spaces, Gendered Audiences."

would be a fruitless exercise. Nevertheless, the notion of multiple publics helps clarify the status of science in Enlightenment Europe. The appetites of these various constituencies for different forms of scientific knowledge sometimes provoked tensions between a commitment to esoteric knowledge, which would perforce remain the province of a specialist elite, and amusing or useful knowledge that opened up that province to curious eyes.

## Mathematics and Curiosity

When Maupertuis gave up his military career to devote himself to more intellectual pursuits, he saw mathematics as a possible route to an established position in Parisian intellectual circles. In no way precocious or brilliant as a mathematician (as were Alexis Clairaut and Jean Le Rond d'Alembert, for example), he worked hard to define problems in the calculus that would yield to his more modest abilities. He did have a remarkable talent for self-promotion and for sociability, and an eye for choosing material that would distinguish him from his fellows. Unlike many of the habitués of the literary and academic world, Maupertuis had no pressing need for a paying position—though that came in good time—and so could afford to cultivate friendships and build a reputation in the cafés while continuing his studies informally with Nicole, one of the mathematicians who frequented La Motte's table at the Gradot.[45]

By all accounts, Maupertuis's easy gaiety and cleverness in conversation endeared him to the older men who set the tone in these circles. When he submitted two papers for the approbation of the Academy in 1723, as a formal introduction to that body, he chose natural history, rather than mathematics for his subject. It seems likely that his supporters, who could well have included the minister in charge of the Academy, the comte de Maurepas, as well as the mathematicians who frequented the cafés, knew of an impending move to expel an unproductive adjunct member, opening up a spot. The vacancy was in the mechanics section; Maupertuis was proposed as one of the candidates to fill it, even though he had not yet done any mechanics whatsoever. A few days later, he was elected to the Academy, and his position was almost immediately redefined as adjunct geometer.[46] He did not impress his future colleagues as exceptionally accomplished, but he convinced

45. Formey, "Eloge de Maupertuis." Maupertuis's first mathematics teacher, Guisnée, had died in 1718. On the training of academic mathematicians, see Paul, *Science and Immortality*, 75–77.

46. See AS p-v, 4 December 1723, for mention of natural history papers. A committee was assigned to evaluate them; the committee reported on 7 December; on 11 December, Maupertuis's name was put forward for the vacant mechanics position. The slot changed categories when Beaufort asked to switch from geometry to mechanics, leaving the opening in geometry "qui luy conviendra mieux que celle de Méchanicien, pour laquelle il avoit été proposé" (AS p-v, 22 December 1723). Maurepas was also minister of the navy, and well known to Maupertuis's father.

them that he had the potential to "become capable to an eminent degree."[47] Although he filled a mathematics slot on the roster, "he did not rush to show himself to the Academy as a geometer," as La Beaumelle noted.[48] In fact, he gave his first mathematical paper only in 1726, after he had advanced to the rank of associate, one step up the hierarchy.

Maupertuis impressed and charmed his seniors in the Academy as he had in the café. Rather unusually for a new member, he was selected to deliver his first formal paper to the public session of the Academy in November 1724, on the shape of stringed instruments. This paper, an elegantly-written reflection on the design of instruments played in drawing rooms, may seem hardly worth pausing over. Without treating the question mathematically, it turned a philosophical gaze on the guitars, lutes, and harps familiar to the elite audience at the public assembly. The paper generalized from many particular examples, and discoursed easily on the relation of theory to practice and art to reason. It bore the mark of a neophyte who had not yet found his place, but who was not tentative or apologetic either. A guitar player himself, Maupertuis had studied music quite seriously, which many in his audience would have known.[49] By collecting and comparing stringed instruments and showing that they shared the property of resonating at a full range of frequencies, he joined the "enjoyable [*agréable*]" and the rational. The instruments themselves appear in the illustration as partially abstracted (they have no strings or embellishments), but they are not fully reduced to geometrical diagrams. Artisans adopted certain shapes for musical instruments through trial and error, but physics can show why these shapes function so well, the paper suggested. "Testing by trial and error [*tatonnement*] is often a very long route, but it is almost always the most reliable. We shall see that time has given instruments the shape that physics prescribed to them."[50] This paper did not set out to define a research program; it was more of an entertainment to affirm the place of the Academy in elite culture and to display the facility of its author with the language of physics and the language of pleasure. The *Mercure de France* gave it an appreciative extract: "M de M. gives a detailed account of all the stringed instruments with great exactitude. One is surprised to see that their shapes, so various, so bizarre and seemingly only the result of chance, should all be founded on this necessary property of providing fibers of all tones [i.e., of all lengths]."[51] This first public effort was

47. La Beaumelle, *Vie de Maupertuis,* 14, quoting Terrasson's assessment of Maupertuis's promise.

48. "Il ne se pressa pas de se montrer à l'Académie comme géometre" (La Beaumelle, *Vie de Maupertuis,* 15).

49. Moreau de la Primerais, "Ecrits biographiques," AS Fonds Maupertuis.

50. Maupertuis, "Sur la forme des instrumens de musique," *MAS* 1724, 215–26.

51. *Mercure de France* (November 1724): 2429.

a perfect exemplar of Fontenelle's style of witty philosophy worthy of the admiration of people of taste and discernment.

Maupertuis had undoubtedly been encouraged to develop his mathematical talents by his café associates, several of whom combined literary pursuits with mathematics. His first performance was a success, but did not entail any sophisticated mathematics. Eventually, however, he began to produce a steady stream of mathematical papers that took him into the realm of the Leibnizian calculus. As he became more ambitious about choosing problems, Maupertuis made strategic decisions to maximize his own chances of success at something that never came easily to him. Mathematics was something he purposefully cultivated, seeking out mentors wherever he could find them, at a time when Parisian mathematics was hardly flourishing. The Academy's mathematical work in the early 1720s was less focused on a program of research than it had been earlier, in part because of the absence of any dominant individuals to give it direction. In this sense, it was ripe for new blood, and new insights.

3

# Mathematics and Mechanics in the Paris Academy of Sciences

Around the turn of the eighteenth century, European mathematical practice had begun to shift away from geometry and toward algebraic analysis.[1] In this period, G.-W. Leibniz, Jacob and Johann Bernoulli, Jacob Hermann, and Pierre Varignon, among others, worked out techniques of integration and applied them to numerous problems concerning the properties of curves and the motions of bodies constrained by forces. Disregarding, for the most part, the logical and conceptual problems associated with manipulating infinitesimals, these mathematicians exploited the power of algebraic notation to solve problems analytically, without attempting Euclidean synthetic proofs. Instead, they solved many canonical problems over and over again in different ways to display improvements in techniques; in other cases, they considered familiar problems under restrictive conditions for a profusion of variant solutions. The scholarly journals published these solutions alongside the empirical results of chemists, astronomers, and anatomists. The posing and solving of new variants of geometrical and mechanical problems were analogous to the collection of observational and experimental data; mathematicians used their techniques as instruments for obtaining results, often with the goal of increasing elegance or generality.[2]

When Maupertuis entered the Paris Academy as *adjoint géometre*, French mathematics was in a lull between the vigorous debates about the viability of the calculus at the turn of the century and the explosion of interest in analysis in the 1730s. Varignon, the most accomplished French mathematician of the previous generation, had died in 1722, just before a new generation of mathematicians

1. Grabiner, *Cauchy's Rigorous Calculus*, 28. See also Blay, *La naissance de la mécanique analytique*. In France, this shift coincided with the literary "battle of the ancients and the moderns" that had been crucial to the establishment of the literary and scientific academies. See generally DeJean, *Ancients Against Moderns*.

2. As Greenberg has shown, French mathematicians were slow to do innovative work with integral calculus. Varignon and l'Hôpital tried; so did Montmort and Maupertuis's mentor Nicole. Montmort to Taylor, 5 August, 1718: "It is disgraceful for France that we have no one capable of entering the lists with the English and the Germans" (cited in Greenberg, *The Problem of the Earth's Shape*, 243). See also Shank, "Before Voltaire," on Parisian analytical mechanics at the turn of the century.

gradually made its way into the Academy. The most innovative mathematics on the Continent came out of Basel and Saint Petersburg; in England, mathematicians developed a different set of techniques based on Newton's method of fluxions. As Maupertuis cast about for subjects for his academic papers, he was searching for an angle to claim as his own. It took him about five years to find this angle; in the meantime, he worked at developing his mathematical skills. His early papers represent the possibilities opened up by the calculus.[3] "On a question of maxima and minima," (1726) is as much an exploration of the power of the method to discover unexpected results as it is the solution of a specific problem. The problem is a simple one from plane geometry: to find trapezoids of greatest and least area, given certain conditions for the lengths of the sides.[4] As Nicole had taught him to do, he solved the problem analytically simply by rewriting the geometrical problem in algebraic form, then differentiating the expression for the area. This approach gives four different solutions, including two that are not trapezoids. These unanticipated solutions, artifacts of the method, reveal facts about a class of plane figures that share a given relationship between the lengths of their sides. "Nothing shows better the advantage of algebra over geometry in the solution of problems than this abundance with which it gives not only what we had meant to ask of it, but also everything depending on the same conditions and that we did not think of asking it. The calculation answers not only the question we put to it, but it also teaches us that the question has a broader meaning than we had thought, and it answers the whole thing."[5] This marks the beginning of Maupertuis's interest in the calculus as a problem-solving device. He spent the next few years studying the properties of curves, finding quadratures (areas) and rectifications (lengths), and investigating relationships among families of curves. Following the tradition established in the Academy by Varignon and l'Hôpital, and continued by Maupertuis's own mentor Nicole, these workmanlike but relatively uninspired papers allowed him to practice techniques of applying differential methods to curves.[6]

Like the experimental and observational sciences that were also investigated in the Academy, mathematics was both controversial and sociable. In spite of a rhetoric of shared standards of truth and demonstration, mathematicians across

3. Greenberg, "Mathematical Physics," 66–68.

4. Maupertuis, "Sur une question de maximis et minimis," *MAS* 1726, 84–94.

5. Ibid., 84, 86.

6. Maupertuis, "Quadrature et rectification des figures . . . ," *MAS* 1727; idem, "Nouvelle manière de développer les courbes," *MAS* 1727; idem, "Sur toutes les développées . . . ," *MAS* 1728; idem, "Sur quelques affection des courbes," *MAS* 1729. For the state of French mathematics in this period, see Greenberg, "Mathematical Physics"; idem, *The Problem of the Earth's Shape,* 243–49.

Europe were divided over such questions as how to evaluate generality, how to understand infinitesimals, or how to apply analysis to mechanics. English and Continental mathematicians continued to disagree about how best to apply the calculus in the aftermath of the Newton-Leibniz priority dispute, and about how to interpret their results.[7] The community was a small one, crisscrossed by ties of intellectual patronage and mentorship. These personal ties colored not just relations among individuals but research directions more generally. Problems and results circulated through the scholarly journals that moved across national and linguistic borders, but also through private correspondence. In fact, personal communication through letters exchanged directly or indirectly drove much of the production of knowledge in this small and specialized corner of the Republic of Letters, and many of the most fruitful mathematical correspondences came out of mentor relationships.

## The *Vis Viva* Controversy

Officially, the Academy frowned on contention and on dogmatism. Thus, Fontenelle, the permanent secretary and public voice of the Academy, insisted that the institution would promote "no general system [at all], out of fear of falling into the disadvantage of [promoting] rash systems."[8] In practice, not surprisingly, these ideals were not so easily upheld, and the secretary found himself managing disputes that divided the Academy internally and connected it to the outside world. Before turning to Maupertuis's role in one of these disputes, the so-called *vis viva* quarrel, we need to pause briefly over the question of Cartesianism. The Academy has sometimes been represented as the home of retrograde Cartesians, unwilling to accept the truths of experiment or of Newtonian natural philosophy.[9] Like most simplistic contrasts, this cannot hold up to scrutiny, but the various associations of Cartesian physics and philosophy in the Academy repay attention, if only to illuminate the many ways in which the label was used and misused. The content and consistency of Descartes's own writings are not at issue here. When someone was called a Cartesian in early eighteenth-century France, it might imply any or all of the following: belief in a material plenum with vortices of subtle matter carrying the celestial bodies; commitment to "system," which usually involved

7. On this dispute, see generally Bertoloni Meli, *Equivalence and Priority.*

8. Fontenelle, *HAS* 1699, 11–12; reprinted 1733. For a careful analysis of the rhetoric of system in the first half of the eighteenth century, see Loveland, *Rhetoric and Natural History,* 102–14.

9. These representations started in the eighteenth century with Voltaire's *Lettres philosophiques* (1734). D'Alembert, in the *Preliminary Discourse* to the *Encyclopédie,* articulated another overly schematic distinction, between *esprit de systême* and *esprit systématique,* the former linked to Descartes. The distinction is echoed in Brunet, *Introduction des théories de Newton* and Beeson, *Maupertuis.*

hypothesizing about causal mechanisms beyond the reach of observation; matter defined by extension, with no added fundamental properties; "rationalism" rather than "empiricism"; conservation of quantity of motion ($mv$) in the universe; measuring the force of bodies by their quantity of motion rather than by their "living force" ($mv^2$); refusal to accept forces acting across empty space; equating such forces with "occult qualities" of scholastic philosophy; observable effects of gravity caused by pressure of matter in celestial vortices. Many of these terms were rhetorically laden, and many of the concepts became conflated together. The Academy did have some members who identified themselves as Cartesians, the most visible being Fontenelle, who had made his name with the *Conversations on the Plurality of Worlds* (1st ed. 1686) with its engaging picture of the qualitative physics of Descartes's cosmic plenum. This does not mean that Fontenelle ascribed to all the tenets listed above. Year after year, his summary of the Academy's work in the annual *Histoire* stressed the collection of indisputable matters of fact and discouraged attempts at system building or ad hoc hypothesizing. Although some academicians devoted considerable attention to trying to make Cartesian planetary vortices work, there was nothing like a prescribed academic orthodoxy on philosophical or physical questions. There was, however, contention.

In the 1720s, while Maupertuis was testing his mettle as a mathematician, the liveliest debate in the Paris Academy centered on dynamics, mainly the question of the proper measure for the force of moving bodies. Labeled the *vis viva* controversy, after Leibniz's term for "living force" (the mass of a body multiplied by the square of its velocity, or $mv^2$), it circled around philosophical, experimental, and theoretical matters that were not easily resolved.[10] These included conservation (whether there was something conserved in the universe, and if so what it might be), how to analyze the effects of collision or gravity, and whether impact or action at a distance could be intelligible explanations for phenomena. The nature of matter was implicated as well: was it made of discrete, hard, incompressible particles, or was it rather infinitely divisible and therefore inherently elastic? Then there was the further question of how to solve problems, and especially what kind of mathematical techniques to use. These contentious matters underlay all discussions of dynamics up to the middle of the century.

The roots of the *vis viva* dispute went back to Leibniz's critique of Descartes's conservation law for "quantity of motion," or the product of mass and speed ($mv$). Leibniz argued that *vis viva*, not motion, was the true measure of force and the quantity conserved in any dynamic system. The controversy flared up again following the publication of the Leibniz-Clarke correspondence in 1717, in which the

10. See Hankins, "Eighteenth-Century Attempts to Resolve the *Vis Viva* Controversy"; Iltis, "Madame du Châtelet's Metaphysics and Mechanics"; Hankins, *Jean d'Alembert*, 204 ff.

argument about matter and force was compounded with theological contentions. After this, the proper measure (or mathematical description) of force moved to the center of the controversy, with conservation (and theology) in the background. In the 1720s, the Leibniz-Descartes opposition was complicated by the participation of self-professed Newtonians from England and the Netherlands, and by new experiments to measure the effect of forces, notably by 'sGravesande.[11] The positions of the various protagonists, especially in Paris, do not correspond straightforwardly to the Newtonian, Cartesian, and Leibnizian labels, because the tenets of these three were adopted selectively and combined at will.[12] As we shall see, the labels were used by contemporaries to castigate their opponents or to solidify support for particular positions.

In 1724, the Paris Academy posed a prize question obliquely concerned with *vis viva:* "What are the laws according to which a perfectly hard body, put into motion, moves another of the same kind, which it encounters at rest or in motion, in the void or in the plenum?"[13] The renowned Swiss mathematician Johann Bernoulli submitted a response that started by denying the existence of hard bodies "in the vulgar sense" and went on to model the force of collision on the compression and extension of springs, nicely compatible with the Leibnizian measure of force. The analysis was predicated on the fundamental elasticity of matter, following Leibniz's teaching, in which hardness, or "rigidity," was equivalent to perfect elasticity. By looking at the motion imparted by springs to "rigid" bodies, Bernoulli was able to show that the force of the spring was proportional to the square of the velocity it gave to the body. He framed his systematic and comprehensive analysis of collisions not just as a defense of Leibniz, but as a *demonstration* of something that he claimed Leibniz could only prove "indirectly," namely the true measure of living force, or what Bernoulli called the force of motion.[14]

11. Willem 'sGravesande, Dutch author of a Newtonian textbook (1720) conducted experiments by dropping hard balls into soft clay and measuring the impressions; he concluded that the effects of impact must be measured by *vis viva,* reversing his own earlier position on the question. 'sGravesande published a series of papers in *Journal littéraire de la Haye,* starting in 1722. Samuel Clarke sent a letter to the *Philosophical Transactions* in 1728 (after Newton's death) arguing that a vote for *vis viva* was a vote against the probity of the great man, and 'sGravesande responded. See Hankins, "Eighteenth-Century Attempts to Resolve the Vis Viva Controversy."

12. This kind of classification leads either to oversimplified dichotomies or to hybrid categories. See Hine, "Dortous de Mairan, the 'Cartonian.'"

13. "Quelles sont les loix suivant lesquelles un corps parfaitement dur, mis en mouvement, en meut un autre de même nature, soit en repos, soit en mouvement, qu'il rencontre, soit dans le vuide, soit dans le plein?" (Maindron, *Les fondations des prix,* 18.)

14. Bernoulli represented bodies as elastic springs. Bernoulli, *Discours sur les loix de la communication du mouvement,* in JB, *Opera Omnia,* vol. 3. See also Shea, "Unfinished Revolution," and Harman, "Dynamics and Intelligibility."

Although this essay came to be regarded by partisans of *vis viva* as a foundational document, the judges preferred the effort of Colin Maclaurin, the Scottish mathematician, who rejected *vis viva.*[15]

Bernoulli was well known to the older generation of academicians, since he had been a frequent correspondent on mathematics and natural philosophy in the early years of the century. In fact, the Leibnizian calculus came to Paris largely through his efforts, since he had taught the authors of two French texts on the subject, the marquis de l'Hôpital and Pierre Varignon. His failures to convince the Parisians with his arguments about *vis viva* caused Bernoulli some bitterness, though academicians were by no means unanimous in their condemnation of *vis viva.* The depth of the division over this and related questions became apparent over the next few years and points up the inadequacy of any account of the Academy as exclusively promulgating one or another "system." Although the prize commission voted for the "Cartesian" approach to the laws of motion, the prizewinner (Maclaurin) was known as a Newtonian, and believed in empty space as well. Bernoulli, while failing to win the Academy over to his view of dynamics, had a number of allies within the institution who defended his model of elastic bodies and who promoted his distinctly Leibnizian mathematical methods for doing analytical mechanics. Bernoulli was also resolutely Cartesian (by his own admission) in his attacks on "unintelligible" Newtonian explanations that required forces to act across void space.

In 1728 and 1729, the *vis viva* question provoked particularly animated debate in the Academy. Bernoulli doggedly attempted to press his point privately with Jean-Jacques Dortous de Mairan, even after Mairan went public with a long and rather inept paper arguing against *vis viva* as a measure of motive force.[16] Joseph Saurin, one of Maupertuis's café associates, defended Bernoulli explicitly in March 1728; things came to a head soon thereafter when a new junior member, Charles-Etienne Camus, elaborated on Bernoulli's spring model of motive force, eliciting vocal opposition from some of his colleagues.[17] Bernoulli heard about it and reported to a correspondent on the scene in the Academy. "The Abbé Camus having begun to read a memoir in the meeting of academicians, a great whisper started to rise up

15. Bernoulli resubmitted his essay two years later for a new prize on the laws of motion of elastic bodies, but failed to win that as well. Greenberg has called this the low point of Bernoulli's influence in the Paris Academy (*The Problem of the Earth's Shape,* 245).

16. Dortous de Mairan, "Dissertation sur l'estimation & la mesure des forces motrices des corps," *MAS* 1728, 49. Bernoulli did not see this paper right away; the *Mémoires* for 1728 were not published until 1730. See JB to Mairan, 15 February 1729, BEB.

17. Camus gave two papers on *vis viva* in the winter of 1728, one on 4, 11, and 14 February and another on 10, 13, and 17 March (AS p-v). No manuscript survives. Published as Camus, "Du mouvement acceleré par des ressorts, et des forces qui resident dans les corps en mouvement," *MAS* 1728,

that he was going to argue for *vis viva,* as if this were an heretical opinion condemned by the Academy, but notwithstanding M. Camus, supported by several partisans of my doctrine, completed the reading of his paper."[18] This is an unusually explicit picture of dissension in the Academy, the kind of confrontation that was carefully masked in official publications.

## England

At the height of the *vis viva* controversy in Paris, Maupertuis took a trip to London. Nothing in the historical record suggests that he had any particular interest in Newtonian natural philosophy in this period. He may have read Keill's version of Newton or even Newton's works themselves, but he did not mention doing so. He would have seen the prizewinning essay by Maclaurin on the laws of motion, although later correspondence with Bernoulli indicates that he had not read Maclaurin carefully. But in May of 1728, armed with a letter of introduction for Hans Sloane of the Royal Society, Maupertuis traveled to London, where he spent three months socializing with Newtonian mathematicians and natural philosophers.[19] Eighteenth-century accounts of Maupertuis's life uniformly agree that this was a logical move for the man who was later hailed as France's first Newtonian, and several twentieth-century commentators follow this line, reading his subsequent career back into this early journey to Newton's home territory.[20] As David Beeson points out, however, Maupertuis did not betray any such partisanship before or even immediately after his trip, and there is no reason to believe either that contact with Newton's followers made him a convert or that his Newtonian tendencies endeared him to the English.[21] He undoubtedly recognized Newton's work as a force to be reckoned with, as many Frenchmen did, and he may have gone to England with the idea of learning more about English natural

---

159–96. Saurin's unpublished paper was recorded in the *proces-verbaux:* "Défence de la démonstration a priori de M. Bernoulli sur les forces vives," AS p-v, 6 March 1728, fol. 89r.

18. JB to Thiancourt, 29 May 1729, BEB. Bernoulli had heard the story from Cramer, who had heard it from someone in Paris: "Mr. l'Abbé Camus ayant commencé à lire un mémoire dans l'assemblée des Academiciens, il s'eleva dabord un grand murmure de ce qu'il alloit soutenir les forces vives, comme une opinion heretique et condamnée par l'Academie, mais que non obstant cela Mr. Camus, fortifié par quelques partisans de ma Doctrine, acheva la lecture de son Memoire" (JB to Cramer, 4 September 1728, BEB).

19. The older literature, based on La Beaumelle's biography, mistakenly made this a visit of six months; however, Maupertuis was writing letters from Paris in early September 1728 and attending meetings of the Academy. The chronology is straightened out in Beeson, *Maupertuis,* 64–65.

20. Brunet agrees, saying Maupertuis was a Newtonian before he left France, but he has no evidence for this. See also Brown, "From London to Lapland and Berlin," for a similar perspective.

21. Beeson, *Maupertuis,* 66. Beeson thinks it was the "empiricism" of the English that attracted Maupertuis.

philosophy. French hostility to Newtonian "attraction" theory, evident in Fontenelle's recent eulogy of Newton (1727), as well as in the Jesuit press, may also have piqued Maupertuis's curiosity.[22]

Whatever he learned in England, no traces of Newtonian inspiration appear in his work until 1731, and then only after he had worked closely with Johann Bernoulli, the foremost opponent of English mathematics and defender of Cartesian vortices. The journey to London is rather more intelligible as an adventure, intellectual and otherwise, that he thought might prove useful in his ongoing efforts to make a reputation. It was not yet clear that he would try to push the Newtonian angle for his own advantage. In fact, the letter of introduction he carried to Sloane from the botanist Bernard de Jussieu recommended him on the basis of his "taste for natural history, [experimental] physics, and mathematics," tastes as broad as those of the Royal Society.[23] For the time being it was enough to cross the Channel, an unusual step for a young academician, and to circulate in English scientific circles as a witty and sociable Frenchman.[24] The Royal Society was one stop on the tour, though it did not meet in July and August, the bulk of Maupertuis's sojourn. Expatriate Huguenots, as native French speakers, were undoubtedly his primary contacts and go-betweens. He was welcomed to the Society and elected as a fellow at the end of June on the recommendation of Abraham de Moivre, and he made the acquaintance of Pierre Desmaiseaux, at the center of the social circle in a well-known Huguenot coffeehouse. He also became friends with Martin Folkes, who later became secretary of the Society.[25] He met the astronomer James Bradley and the instrument maker George Graham, as well as de Moivre, James Stirling, Theophilus Desaguliers, and Brook Taylor.[26] These contacts, especially with Bradley and Graham, played a major role in his geodetical work some years later. Bradley had just completed a systematic series of obser-

22. Fontenelle, "Eloge de m. le chevalier Newton." *MAS* 1727, 151–72. On Castel's treatment of Newtonian mathematics in the Jesuit *Mémoires de Trévoux*, see Greenberg, *The Problem of the Earth's Shape*, 258–65. The journal was frequently hostile to the Academy as well.

23. Jussieu to Sloane, British Library, Sloane MS 4049, fol. 168.

24. Relations between French Academy and the Royal Society were not as strained as the simplified Cartesian-Newtonian bifurcation might suggest. Newton's optical experiments had been appreciated and replicated in France; French astronomers had traveled to London to observe a solar eclipse in 1715; books were regularly sent back and forth. Voltaire was also in England at this time (1726–1729), though Voltaire and Maupertuis were apparently not acquainted until later. See generally Guerlac, *Newton on the Continent.*

25. RS, Journal Book, 20 June 1728; the election took place on 27 June (Jussieu to Sloane, British Library, Sloane MS 4049, fol. 168). For Desmaiseaux, see Sgard, *Dictionnaire des journalistes*, 119–21.

26. See M to JB, 12 September 1731, BEB, for Maupertuis's personal relations with Taylor; for friendly relations with de Moivre, see M to Desmaiseaux, 29 April 1732, British Library, Additional MS 4285, fol. 212.

vations of apparent stellar motions with Graham's specially designed zenith sector, detecting the aberration of starlight.[27] This work represented the latest in precision astronomy and inspired Maupertuis's admiration for the English instruments he was later to use in Lapland. At the time of his London visit, however, he would have known little of the practice of astronomy and learned of these matters as a neophyte rather than an expert. He learned something of English mathematics as well, through de Moivre, Stirling, and others, although he did not find it to his taste. As he remarked upon receiving a gift of Stirling's latest work on differential methods using infinite series, "This business of series, the most disagreeable thing in mathematics, is no more than a game for the English; this book and that of M. de Moivre are the proof."[28]

## From Paris to Basel and Back

Upon his return from London, Maupertuis continued to investigate the mathematics of curves and their transformations. Looking for more generality and for efficient methods of calculation, several papers used analytical mathematics to reveal properties of curves not obvious at first glance. Fontenelle noted that this kind of work was "a spectacle pleasing to the geometric mind [*esprit géometrique*],"[29] analytically revealing "invisible" properties. The spectacle was not simply amusing, however; Maupertuis pointed out that it solved a "mystery" inherent in the theory of curves: "We know in general that a curve can be cut by a straight line in as many points as its equation has degrees: nevertheless it is not always possible to find all these points of intersection, and one might doubt the truth of the principle."[30] Here he pushed his analysis to say something novel about an accepted theory and to display his own insight, his *esprit géometrique.* These early papers make no grandiose claims, however, remaining narrowly focussed on particular classes of problems dating back to the literature of the turn of the century.

27. On Bradley and Graham, see Sorrenson, "George Graham." Aberration is a small apparent motion of the stars, resulting from the finite speed of light and the orbital motion of the earth. See chapter 5 below.

28."Cette affaire de suites qui est tout ce qu'il y a de plus désagréable dans les mathématiques n'est qu'un jeu pour les Anglois; ce livre et celui de M. de Moivre en sont une preuve" (M to Mairan, 20 October 1730, in "Lettres inédites de Montesquieu et de Maupertuis," *Revue d'histoire littéraire de la France* 37 (1908): 111–12). His tour also took him into female society; he apparently came back to France with a case of syphilis. He traveled to Montpellier for a cure from January through March 1729. Years later, La Condamine suggested that the disease may have been contracted in England (La Condamine to JBII, 3 August 1761, BEB). On dating the trip to Montpellier, see Beeson, *Maupertuis,* 74.

29. Fontenelle, "Sur les soudeveloppées," *HAS* 1728, 59.

30. Maupertuis, "Sur quelques affections des courbes," *MAS* 1729, 281.

Maupertuis could have continued in this direction, as Saurin and Nicole had done, but instead he took the unusual step of approaching the most famous living mathematician for instruction and inspiration, much as l'Hôpital had done in the previous generation. Johann Bernoulli was famous for his analytical brilliance, to be sure, but also for a persistent contentiousness concerning the ownership of mathematical methods and results. He had played a headlining role in Newton's priority dispute with Leibniz, as well as conducting a vicious polemic with his own brother Jacob over the calculus of variations.[31] Since the death of Jacob in 1705, he had been professor of mathematics at the University of Basel. Bernoulli had heard from his informants in Paris that Maupertuis was one of those "zealous partisans in favor of *vis viva*," and that the young partisan wanted to come to Basel to study with the great man in person.[32] Without mentioning the possibility of traveling to Switzerland, Maupertuis initially sounded Bernoulli out by asking for help with a problem he wanted to present to the Academy. He was looking for a more general understanding of the properties of curves at points of inflection or discontinuity than he had been able to find in l'Hôpital's *Analyse des infinimens petits* (1696). This was still the standard French text for learning the methods of the calculus, and Maupertuis knew that it had been written under Bernoulli's close supervision. Struggling to elaborate on l'Hôpital's examples, he recognized that he could do no better than to go directly to the source and see if Bernoulli could clear up his confusion.[33]

Maupertuis wrote his first letter as an homage to the acknowledged master, asking for his assistance and approval. As we have seen, this was a key moment in the *vis viva* controversy in Paris, and Maupertuis took advantage of it, without publicly declaring a position. (None of his own papers concerned *vis viva*.) Favorably impressed by the Frenchman's reputation and his flattering manner, Bernoulli answered graciously, in considerable detail.[34] Thus encouraged, Maupertuis made up

31. On the Newton-Leibniz priority dispute, Bertoloni Meli, *Equivalence and Priority*. The Bernoulli brothers' argument is documented in Goldstine, ed., *Die Streitschriften von Jakob und Johann Bernoulli*.

32. Johann Bernoulli heard about Maupertuis from his friend Thiancourt: "Vous dites, Monsieur, qu'il est un de mes zelés partisans pour le sentiment des forces vives; si cela est, il ne fera guere de plaisir aux Anglois qu'il estime et dont il est estimé, car les Anglois sont les mortels Ennemis et haïsseurs de la doctrine des forces vives" (JB to Thiancourt, 29 May 1729, BEB). Thiancourt's letter is lost. (Bernoulli noticed that Maupertuis was on the membership list of the Royal Society.) See also JB to Cramer, 4 September 1728, BEB.

33. M to JB, 28 January 1732, BEB, where he refers to a visit to Basel as "returning to draw from the wellspring of geometry [retourner puiser à la source de la géometrie]."

34. M to JB, 18 May 1729, BEB; JB to M 14 June 1729, BEB.

his mind to travel to Basel to immerse himself in the techniques of Leibnizian mathematics.[35] In so doing, he was taking his place in a long chain of illustrious disciples and students of Bernoulli, including l'Hôpital, Varignon, Leonhard Euler, the younger Bernoullis (Daniel and Nicolas), Jacob Hermann, and Christian Goldbach. He arrived there in September 1729, returning to Paris only in the following July. He charmed the aging professor and gradually was welcomed into his family circle, making a particular friend of the youngest son, Johann II. (Daniel, also a mathematician, held a position in the Saint Petersburg Academy during this period; Maupertuis made his acquaintance later.) Bernoulli found his student's attentions flattering, and in turn took Maupertuis's efforts seriously, supervising his work closely. They worked on many of the classic problems that had been instrumental in the development of the calculus, such as the brachistochrone, as well as more recent work published by Euler and others in Saint Petersburg, and by the English mathematicians. By his own account, Bernoulli, though getting on in years and suffering from painful gout, worked hard to come up with new results to share with his student.

The history of his often vitriolic engagement with competitors was never far from Bernoulli's mind as he initiated his French visitor into Leibnizian mathematics and physics. Here was a promising and receptive courier who might take the Swiss-Leibnizian perspective back to Paris, where it had once carried considerable weight. From Maupertuis's point of view, the same history meant that Bernoulli was the perfect guide to help him negotiate the fraught terrain of contemporary mathematics, since he had not only been party to the development of the Leibnizian calculus, but had also studied the English mathematicians carefully in order to refute them. Bernoulli was certainly one of the very few on the Continent who had read Newton's *Principia* with any degree of understanding, and he recognized its brilliance as well as its faults. He turned out to be a vital resource for Maupertuis's attempts at assessing and understanding Newton.

Bernoulli perhaps overestimated his new student's mathematical aptitude, but Maupertuis's tenacious questioning definitely stimulated the elder mathematician, who reported to Dortous de Mairan in Paris:

> As for M. de Maupertuis, I must tell you that I find him well-versed in all sorts of sciences, apart from the noble qualities of his soul, I am charmed by his conversation; I do not know if he profits from it as much as I. At least I can say sincerely that prompted by him, I have made some new mathematical discoveries that I might

35. The paper discussed in the first exchange of letters was read to the Academy 16 July 1729; published as "Sur quelques affections des courbes," *MAS* 1729, 277–82.

> never have made without him. He seems to have come here more to instruct me than to be instructed by me.[36]

Although this praise is undoubtedly an exaggeration, Bernoulli, who was easily roused to anger when contradicted, responded well to the younger man's deferential enthusiasm and found his company invigorating.

> I can sincerely say that I have done everything I could to give him the satisfaction he sought from me: all my studying during the nine months of his visit was nothing but revealing to him the deepest parts of my small stock of wisdom [*mon petit savoir*], without hiding anything from him. He noticed perfectly well that I am not at all mysterious, as people customarily are when they are persuaded that they are the only professors of certain small tricks. . . . I hope that in a short time he will become my master, and that he will in his turn return to me with interest, whatever he was able to learn from me.[37]

Evidently, Bernoulli welcomed the opportunity to mold a Parisian mathematician into his own image, revealing his secrets without the possessive jealousy he had expressed so vituperatively for many other mathematicians. His high hopes for a "return with interest" rested on Maupertuis's ability to make inroads into the Paris Academy; Bernoulli predicted that his pupil would one day be "the Phoenix of France." He was not above flattering the younger man: "Seeing in you a completely extraordinary talent to make noteworthy progress in the matter of new discoveries, you also have the gift of penetrating to the bottom of things where the other Geometers of France content themselves, as it seems, with skimming across the surface, a manifest example being M. de Mairan."[38]

36. "Quant à Mr. de Maupertuis, il faut que je vous avoue que je le trouve tres riche de toutes sortes de Sciences, outre les belles qualités d'ame qu'il possede, je suis charmé de sa Conversation, je ne sai s'il en profite autant que moi, au moins je puis dire sincérement qu'à son occassion j'ai fit de nouvelles decouvertes en matière de Mathematiques que je n'aurois peutétre jamais fait sans cela. Il semble étre venu ici plutôt pour m'instruire que d'étre instruit par moi" (JB to Mairan, 13 April 1730, BEB).

37. "Je puis dire avec sincerité que j'ai fait tout mon possible, pour lui donner la satisfaction qu'il cherchoit auprès de moi; Toute mon étude pendant les 9 mois, qu'il me frequentoit ne consistoit qu'à lui decouvrir le plus interieur de mon petit savoir, sans lui rien cacher, aussi s'est il bien aperçû que je ne suis rien moins que misterieux, comme le sont ordinairement ceux qui se persuadent étre les seuls Professeurs de certains petits artifices. . . . J'espere qu'en peu de temps il deviendra mon Maitre, et qu'il me rendra à son tour avec usure, ce qu'il peut avoir appris de moi" (JB to Mairan, 6 July 1730, BEB).

38. "Vous serés un jour le Phénix en France. . . . Voyant en Vous, . . . un talent tout à fait extraordinaire pour faire des progrés insignes en matiere de nouvelles decouvertes, aussi avés vous le don de penetrer dans le fond des choses au lieu que les autres Geométres de France se contentent

Bernoulli, nearing the end of his career, was committed equally to getting the French to adopt his approach to mechanics and to keeping the English at bay. He also sought to promote his techniques for solving differential equations, again at the expense of English methods.[39] He set out to teach not just the mathematical techniques he himself used to such powerful effect, like integration by parts and the substitution of variables, but also the application of these methods to physical problems through consideration of *vis viva.* Maupertuis arrived at just the right time to take on the role of disciple. Bernoulli interpreted his recent lack of success in the Paris competitions as the result of unreasonable prejudice against *vis viva,* particularly galling since he had earlier enjoyed a substantial influence in the Academy, at least in mathematical matters.[40] Maupertuis, at the beginning of his career, had not committed himself to any particular physics or metaphysics, at least publicly. He was looking to make a name for himself, and sharpening his mathematical skills was one route to his goal. In Paris, he had already learned all he could from Nicole and Saurin. As an associate member anxious to rise to one of the few pensioned positions, he needed to break new ground. He knew his own abilities well enough to realize that he would not be able to do this alone. In allying himself with Bernoulli, he was exploring possibilities for himself rather than adopting a cause, just as he had explored English mathematics and astronomy, without taking up the Newtonian banner.

Maupertuis's training in Basel equipped him to attack problems he would not have attempted otherwise and sparked a renaissance of what we might call Bernoullian mathematics in Paris.[41] As soon as he got back, he felt pressure to show the fruits of his prolonged absence. He wrote to Bernoulli: "The desire, or rather a kind of requirement, to read something to the Academy from time to time perhaps has made me hurry too much. Nevertheless, I will not risk reading anything on this until you have told me your opinion and whether the thing deserves to be written up."[42] Indeed, over the next few years, whenever he had a result, he sent it to Bernoulli for corrections and comments before presenting it to the Acad-

comme il semble d'en effleurer la superficie, exemple manifest en Mr. de Mairan" (JB to M, 29 August 1730, BEB).

39. For ongoing antagonism to the English, see JB to Cramer, 4 September 1728, BEB.

40. See Greenberg, *The Problem of the Earth's Shape,* esp. 232–46.

41. John Greenberg has analyzed this revival of analytical mathematics in the Paris Academy in compelling detail. Greenberg, "Mathematical Physics" and *The Problem of the Earth's Shape.*

42. "Lenvie ou plutost une espece de necessité de lire de tems en tems quelque chose a l'Academye ma peutetre fait trop haster. Je ne hazarderay cependant pas de rien lire sur cela que vous ne mayés dit votre sentiment et si la chose meritte que jy donne une forme" (M to JB, 31 July 1730, BEB).

emy. There was no one in Paris whose judgment he trusted in the same way. His first effort was closely related to many of the classic problems he had investigated under Bernoulli's tutelage, an analysis of the curve described by a falling body (accelerated by gravity) such that it covers equal increments of vertical distance in equal times. This problem, known as the *descensus aequabilis,* had been solved in the early days of the calculus, by Bernoulli among others, for motion in a void, but Maupertuis sought to make it more general by considering the effect of a resisting medium on the curve. He followed the method learned from his mentor, but barely trusted his own result. "Do me the favor, sir, to tell me if I am mistaken, since I would not dare to take a step without you."[43] This was a request he would make many times, relying on Bernoulli again and again to find his mistakes, improve his constructions, and approve his conclusions. Bernoulli seems not to have minded, and for the most part responded with good grace and generosity. He recognized that the Paris audience might not follow Maupertuis's calculation unaided, and he suggested modifications in notation to make it more intelligible. "Since you wish to write a piece for the public, you must try to avoid anything that could trip up [mathematically] weak readers."[44] He also gently suggested a way of simplifying the construction of the curve. When he wrote up his paper several months later, Maupertuis presented his solution as the heir to the historic work of Leibniz, Bernoulli, and Varignon, pointing to the novelty of his results but also to their place in the Leibnizian mathematical literature. He reminded his audience of his close personal ties to Bernoulli by including his teacher's own version of the curve's construction: "Having communicated this solution and this construction of the curve to M. Bernoulli, he sent me a way to perfect the construction, worthy of its illustrious author."[45] Soon thereafter, he read a paper by the master himself on a related problem. Bernoulli had suggested that Maupertuis read this as his own work, but the younger man recoiled from the suggestion. Instead he painstakingly translated Bernoulli's piece from the Latin, sent it back to Basel for approval, and arranged for its publication.[46]

43. "Faittes moy la grace, monsieur de me dire si je me trompe car je n'ozerois faire un pas sans vous" (M to JB, 31 July 1730, BEB). The resistance of the medium varies as $1/r^2$. He followed standard Bernoullian practice to solve this problem in the most general way possible: formulate a differential equation that describes the curve; when it turns out to be transcendental, hence not integrable, construct the curve by separating variables.

44. "Comme c'est pour le public que vous voulés faire une piece, il faut tacher de prevenir tout cequi pourroit arréter les faibles" (JB to M, 10 August 1730, BEB).

45. Maupertuis, "La courbe *descensus aequabilis* dans une milieu résistant," *MAS* 1730, 236–37.

46. J. Bernoulli, "Méthode pour trouver les tautochrones, dans des milieux résistants, comme le quarré de la vitesse," *MAS* 1730, 78–101. Johann Bernoulli told Maupertuis he could present it to

Maupertuis became Bernoulli's representative in Paris, soliciting his papers for the Academy, reading them publicly, correcting proofs, and sending news. He sent gifts of the best telescope lenses Paris had to offer, as well as eyeglasses for the older man and his wife. When Bernoulli had difficulty getting copies of his prize essay from the Academy's printer, Maupertuis had several dozen printed at his own expense and sent to Basel.[47] Bernoulli also hoped that Maupertuis might exert his influence to get the Academy to set a prize question on the physical cause of gravity so he could submit his alternative to Newton's attraction theory.[48]

Although this was a productive period for Maupertuis, he recognized his limitations when it came to the complex integrations that were Bernoulli's stock in trade. Several times, he thought he had achieved some general result, only to have his mentor point out its fatal flaws. After one such exchange, he reflected with a touch of bitterness, "I would have been much more surprised if I had found the isochrones, than I was to have failed in doing so. It was outrageous temerity to have undertaken it, and I will have to think about your lessons for several years before being able to attack such problems with any hope of success."[49] Thanking Bernoulli for showing him how to do a troublesome integration, he went on, "I admit that I could not have [found the integral] even for the simplest case, and indeed that would have required more skill than I have."[50] But he realized that

Academy "under your name" (JB to M, 9 November 1730, BEB). Translation and correction are discussed in M to JB, 8 December 1730, BEB; M to JB, 2 January 1731, BEB; JB to M, 18 January 1731, BEB; M to JB, 7 February 1731 BEB. The paper was presented at Academy as "un écrit de M. Bernoulli sur les Courbes Tautochrones" on 28 Feb. and 3 March 1731, AS p-v.

47. For difficulties with obtaining the prize essay volume, see M to JB, 31 July 1730, BEB; 18 August 1730, BEB; 2 October 1730 BEB. For corrections to the printed text of Bernoulli's essay, see JB to M, 12 October 1730, BEB. For telescope and spectacles, see M to JB, 18 August 1730, BEB; JB to M, 29 August 1730, BEB.

48. "I have my own theory on this subject, by which I explain the system of M. Newton [i.e. the inverse-square law of gravity] without the need for . . . attraction and the void, principles that are appreciated only in England, and that are regarded everywhere else and above all in France as chimeras" (JB to Thiancourt, 17 September 1730, BEB). Johann Bernoulli had raised this with Mairan as well (JB to Mairan, 6 July 1730, BEB). The prize question was never set in this form; in 1734 the prize (shared by Bernoulli with his son Daniel) asked for the cause of the inclination of planetary orbits (Maindron, *Les fondations des prix*, 18).

49. "Jeusse eté bien plus surpris si jeusse trouvé les isochrones que je ne lay eté de les avoir manquées. Cetoit une temerité outrée que de les avoir entreprises et il me faudra encor mediter vos lecons plusieurs années avant que de pouvoir attaquer de tels problemes avec quelque espoir de reussir" (M to JB, [October] 1730, BEB).

50. "Je vous avoue que je ne lavois peu faire du cas le plus simple et en effect il falloit pour cela plus d'art que je nen ay" (M to JB, 11 March 1731, BEB).

reading papers to the Academy would be a prerequisite to advancement. "You will no doubt find [the paper] too long but that is the style of our Academy, and although I do not think much of this piece I want to read it in one of our meetings, since there are many men in the Academy who think one is clever [*habile homme*] as long as one reads."[51] Above all, at this time Maupertuis wanted recognition for his cleverness, in mathematics as in conversation.

As he cranked out mathematical papers, Maupertuis continued to look for elegance and generality. By elegance he meant simplicity of expression combined with clarity. The algorithmic form of the Leibnizian calculus appealed to him, while the infinite series used by Newton and his followers seemed mysterious, and above all, too difficult. "This business of series has always seemed to me a terribly thorny [*epineuse*] subject, which my mind quite refuses," he admitted to Bernoulli.[52] They had discussed this "business of series" in the course of reading works by Stirling, Craige, and de Moivre, as a means for Bernoulli to point out the inadequacies and logical inconsistencies of the English mathematicians. From Bernoulli's Leibnizian perspective, series could never be more than ad hoc tools because they could not be reduced to algorithms, or "method."[53] Maupertuis explored the difficulties and limitations of current methods in a paper on the separation of finite variables in first-order differential equations. The technique for separating variables came directly out of his previous paper; Bernoulli had pronounced it "ingenious."[54] Maupertuis presented it to the Academy prefaced by reflections on the gulf that separated the Newtonian method of infinite series from the Leibnizian algorithmic ideal.

> As soon as one is versed in the new Methods one knows how much still remains to be desired in the integral Calculus. In vain will one have spent much work and trouble to arrive at the differential equation that contains the solution of the Problem; if this Equation is not integrable, by a lucky accident, or at least one of those in which one can separate the variables, one will be reduced to abandoning it; or at least one will not be able to get the roots except by infinite series. It is true that this Method of series that we owe to M. Newton is general, and the only absolutely general Method that the integral Calculus has; but it is also true that the solutions it

51. M to JB, [October] 1730, BEB.

52. "Cette matiere de suittes ma toujours paru une chose fort epineuse et a laquelle mon esprit le refuse assez" (M to JB, 30 October 1730, BEB).

53. On the comparison of Leibnizian and Newtonian perspectives on the calculus, see Greenberg, *The Problem of the Earth's Shape,* 254–55.

54. JB to M, 7 January 1731, BEB. Bernoulli elaborated on the compliment by saying, "Je ne m'en serois pas avisé quoique peutétre j'eusse pû trouver la meme chose par une autre methode; mais sans doute moins belle et plus longue" (ibid).

gives are very far from the elegance of solutions done by integration or quadratures; and that one should only consider it as the last resort in desperate cases.[55]

Here is the tension between the "absolutely general" but often impracticable method of series and the "elegant" but elusive method of quadratures, which required luck (or maybe inspiration) to bring to fruition. The mathematician in this paragraph is buffeted, powerless, between these two extremes, painfully and repeatedly drawn up short by "desperate cases" that resist his attacks. The separation of variables ("the primary object and the anguish of geometers") is a strategy to deal with this desperate situation, but it is hardly without its drawbacks. Even Bernoulli's admirable method for "all equations where the sum of the powers of the variables is the same for all terms" lacks generality. Maupertuis saw these shortcomings, but could not find his way around them, and recounted his struggle with recalcitrant problems in pursuit of generality. Having vented his frustration at the limitations of available techniques, he went on to give his own method for certain types of equations. The first portion of the paper was never published; it is preserved only in the manuscript record of the meeting. The printed version omitted all the agonizing about inadequacies and limitations, and simply presented the method as another technique in the mathematical arsenal.[56] Whether edited out by the author or by the secretary, who often shortened papers for publication, the published version masked the dismay and anxiety of the mathematician grappling with the obscurity of the integral calculus. Maupertuis recognized his own inability to find the philosopher's stone of a truly general method that would not rely on the kind of virtuosity characteristic of Bernoulli's calculations. This frustration ultimately drove him to look for other sorts of problems that would have physical and even metaphysical consequences.

As Maupertuis became one of the most active contributors to the Academy, Bernoulli persistently pushed him to challenge the "adversaries" of *vis viva.* With

55. "Pour peu qu'on soit versé dans les nouvelles Methodes l'on sait assez combien il nous reste encore à desirer dans le Calcul integral. En vain aura t'on employé beaucoup d'industrie et de peine pour parvenir à l'Equation differentielle qui contint la solution du Problème; si par un heureux hazard cette Equation n'est integrable, ou du moins de celles dans lesquelles on peut separer les indéterminées, on sera réduit à l'abandonner; ou du moins on n'en pourra avoir les racines que par les suites infinies. Il est vray que cette Methode des suites que nous devons à M. Newton est générale, et la seule Methode absolument generale qu'ait le Calcul integral; mais il est vray aussi que les solutions qu'elle donne sont fort eloignées de l'élegance des solutions qui se font par l'integration ou par les quadratures; et qu'on ne doit la regarder que comme la derniere ressource dans les cas désesperez" (Maupertuis, "Sur la séparation des indéterminées, "AS p-v, 28 April 1731, fols. 82v–83r).

56. Greenberg discusses this paper in detail (*The Problem of the Earth's Shape,* 249–55). The edited version was published as follows: Maupertuis, "Sur la séparation des indéterminées dans les équations différentielles," *MAS* 1731, 103–09.

his persuasive charm and sharp intellect, might not the protégé "convert" Mairan to the Leibnizian side of the *vis viva* controversy?[57] When he read Mairan's long *vis viva* paper in the *Mémoires* for 1728, Bernoulli just couldn't believe that the arguments he had showered on Mairan in the course of their correspondence could fail to settle the matter definitively: "It is almost impossible to misunderstand the true meaning we give to the term 'living force,' nevertheless he takes it in the opposite sense all the time, as if on purpose to smother the truth, out of fear that it might triumph over popular error."[58] The other vocal opponent of *vis viva* in the Academy was the chevalier de Louville, an astronomer. He had long made known his critical views, and in 1729 he presented a new refutation of Bernoulli's prize entry on elastic collisions.[59] This paper reached Bernoulli only in October 1731 and put him once again on the warpath. "These gentlemen sing out their victory at full voice, seeing that no one dares to respond to them in defense of *vis viva*."[60] He added that they could work together on such a defense when Maupertuis came back to Basel.

Although he was sympathetic, Maupertuis persistently kept a low profile on this question, at least in print. He had been known in the Academy as a defender of *vis viva* even before his trip to Basel, but he did not follow through to vindicate his mentor, as a true disciple would have done. This was not, however, a rejection of *vis viva*, which he continued to regard as a useful physical and mathematical concept. Mairan's obstinacy did not irk him as it did Bernoulli, because he had no personal stake in the question. "To convert [Mairan] is not possible; I have tried to engage him several times on this matter and he did not want to hear of it. He regards it, or pretends to, as settled by his paper [of 1728], and we cannot hope that he will see clearly on this until he gets to the other world."[61] Maupertuis agreed that Mairan's paper was misguided and obscure, as well as prolix. As a polemical piece, it failed miserably. But he did not think it worthwhile, or politic, to force a

57. JB to M, 10 August 1730, BEB, shortly after his return to Paris.

58. "[I]l est presque impossible d'ignorer le veritable sens que nous donnons au mot de *force vive*, cependant il le prend de travers à tout moment, comme si de propos deliberé il eut pris à tache d'étouffer la verité de peur qu'elle ne triomphât sur l'erreur populaire" (JB to M, 1 April 1731, BEB). Note the use of the first person plural. (The 1728 *Mémoires* were published in 1730.)

59. Louville, "Sur la théorie des mouvemens variés . . . ; avec la maniere d'estimer la force des corps en mouvement," *MAS* 1729, 154–84.

60. "Ces messieurs chantent victoire à pleine gorge voyant que personne n'ose leur repondre pour defendre les forces vives" (JB to M, 11 October 1731, BEB).

61. "Le convertir n'est pas une chose possible; je lay voulu mettre plusieurs fois sur cette affaire il n'a jamais voulu en entendre parler. Jl la regarde, ou feint de la regarder comme decidée dans son memoire, et jl ne faut plus esperer quil voit clair sur cela que dans lautre monde" (M to JB, 11 March 1731, BEB).

confrontation with Mairan and his supporters in the Academy. In any case, these men (and especially Louville) were not adept at the mathematical analysis that Maupertuis had been trained to see as essential to mechanics, so in a sense they were not worthy adversaries. "In the end," he wrote to Bernoulli, "I do not think this piece does great harm to our forces [*nos forces*]. And if you wanted to take the trouble to write something else on this, I am persuaded that in ten lines you would destroy volumes of objections."[62]

Although he did not sympathize with what he saw as Mairan's stubborn reliance on simplistic arguments, Maupertuis could not see a way out of the conceptual impasse at the root of the debate. In some problems, quantity of motion was conserved; in others, ignoring *vis viva* led to misconceptions and mistakes. Experimental evidence could be adduced for both types of case. But *vis viva* was Bernoulli's battle, and Maupertuis did not see what he could do to extricate the antagonists from their deadlocked positions. He encouraged his mentor to publish an analysis of the question, insisting that Bernoulli could do this perfectly well on his own. "Nevertheless if you judge whatever I could do worthy of publication, for the cause, I will be honored to be your commentator and to make your ideas understood."[63] His exchanges with Bernoulli always took for granted their shared ground, that analysis of *vis viva* should be applied to problems involving elastic collisions. As he continued to struggle with the mechanics in Newton's *Principia*, however, Maupertuis came to appreciate the importance of collisions between perfectly hard bodies, where motion (or momentum) would apparently be lost, since perfect hardness would imply an absence of elasticity, and hence no rebound after collisions. Bernoulli, as we have seen, followed Leibniz on this matter and would not admit the existence of such truly inelastic bodies. To do so would have meant giving up the Leibnizian principle of conservation of active force (*vis viva*). Maupertuis would come back to this question years later, but for the moment he avoided discussing it with his mentor so that it would not become a bone of contention between them. He had no interest in antagonizing the one person who could help him through the obscure byways of the integral calculus.

## The Shape of the Earth

While Bernoulli was encouraging his protégé to keep the *vis viva* controversy alive in Paris, Maupertuis was exploring other areas that might yield novel results. Because he consulted Bernoulli frequently about his mathematical projects in the

62. Ibid.

63. "Cependant se vous jugés ce que je pourroy faire digne de paroitre pour la cause je me feray honeur detre votre commendateur et de tascher de faire comprendre vos idées" (M to JB, 23 October 1731, BEB).

early 1730s, we can follow in some detail the development of Maupertuis's interest in the problem of the earth's shape. This would occupy him with increasing intensity over the next decade, eventually resulting in considerable tension with Bernoulli, but in the beginning it was just another problem that seemed to hold out the promise of yielding to the techniques of the integral calculus. In spite of Bernoulli's frequently expressed hostility toward Newton and his followers, Maupertuis did not shrink from consulting him about his own difficulties in reading the *Principia.* Bernoulli had worked through Newton's book years earlier, and while admitting a grudging admiration for many of its results, he resented his rival's insistence on proceeding geometrically and deductively. He insisted that Newton's work was obscure and that his attack on Descartes's vortices was based on a "paralogism."[64] Maupertuis imagined that he might be able to clarify Newton's rather opaque treatment of dynamics for a French audience. Reading Newton alongside mathematical work by his colleagues in the Paris Academy and by Bernoulli's talented disciples in Saint Petersburg provoked Maupertuis to this daunting task. He very likely would not have undertaken it without the consistent critical support of his mentor in Basel.

The shape of the earth was just one problem of many that Maupertuis could have fixed on for this exercise. It was an interesting choice for several reasons. First, it represented a departure from his purely mathematical investigations of the properties of curves. The earth's shape—the result of centrifugal force and gravity acting on a rotating fluid body—was a mathematical problem with physical and empirical ramifications. Although the problem subsequently came to be interpreted as a test of Newtonian theory, Newton's original calculation was based on measurements made by French astronomers, so that from the beginning French observational practices were implicated in the question. Furthermore, there was no necessary link between the flattened earth and Newton's theory of gravity. Christiaan Huygens, working from a Cartesian theory of gravity caused by the pressure exerted by material vortices, had also calculated the earth to be slightly flattened at the poles. Both of these calculations referred to pendulum measurements taken near the equator by the French academician Jean Richer in 1672, showing the effect of gravity to be weaker there than in Paris.[65] By the early eighteenth century, French astronomers had another set of data: measurements of

64. JB to Thiancourt, 17 September 1730, BEB. Bernoulli later wrote to Maupertuis, "I confess frankly that his customary obscurity repels me, and that when I must read something in [Newton], I am no less terrified than I would be if I had just been condemned to the galleys" (JB to M, 30 December 1731, BEB).

65. On Richer's expedition, see Olmsted, "The Scientific Expedition of Jean Richer to Cayenne." The definitive treatment of the theoretical problem is Greenberg, *The Problem of the Earth's Shape.*

celestial arcs and terrestrial distances made for a new map of the kingdom.[66] In 1718, Jacques Cassini announced to the Academy that the lengths of about eight consecutive degrees decreased slightly from south to north, a result that implied an elongated rather than a flattened earth.[67] Extrapolating from the length measures to the shape of the earth, Cassini used geometrical techniques familiar to astronomers, without invoking any theory of forces or other causal explanation and without using the integral calculus. Nevertheless, it was the discrepancy between Cassini's conclusions and Newton's theory-laden calculation that made geodesy—the measurement of the earth's shape—a controversial problem.

Maupertuis looked into Newton's treatment of the shape of the earth after reading an old paper by Dortous de Mairan, whose misunderstanding of *vis viva* had annoyed Bernoulli.[68] Shortly after the announcement of Cassini's results, Mairan had tried to reconcile the new measurements with an ad hoc "law of attraction" that gave the earth an elongated shape.[69] Whatever the shortcomings of his theory and his mathematics, Mairan was responding to the pressure of new empirical measurements and not to any abstract or preconceived "system." But Maupertuis took a jaundiced view of Mairan's cumbersome prose and misguided geometrical reasoning, and tried to figure out what more accomplished mathematicians had done with the problem. "I have exerted myself in recent days on this matter," he wrote to Bernoulli, "and since I am content neither with M. Hermann's treatment of it nor that of M. Newton, allow me to send you what I have done on this."[70]

66. Konvitz, *Cartography in France.*

67. Jacques Cassini, *De la grandeur et la figure de la terre.*

68. After Bernoulli asked him to refute Mairan's paper on *vis viva,* Maupertuis read the earlier paper on geodesy: Dortous de Mairan, "Recherches géometriques sur la diminution des degrés terrestres," *MAS* 1720. "Je ne suis guerres plus content d'un autre grand memoire qu'il donna en 1720 pour terminer croyoit il encor la querelle entre les Anglois et les Francois sur la figure de la Terre" (M to JB, 11 March 1731, BEB). Desaguliers rebutted Mairan in a three-part paper, "A Dissertation Concerning the Figure of the Earth." (See Greenberg, *The Problem of the Earth's Shape,* 51–78.) Desaguliers's critique apparently went unread in Paris in 1725, although Maupertuis cited it in "Sur la figure de la terre," *MAS* 1733, 154. He later drew attention to Desaguliers's arguments in *Examen des trois dissertations que M. Desaguliers a publié sur la figure de la terre* , published anonymously in 1740.

69. Greenberg points out that the paper is not Cartesian in any meaningful sense. He calls its approach "phenomenological" because Mairan took the empirical evidence as primary and indisputable, and reasoned backwards to a primordial shape for the earth that would have resulted in the measured elongation under pressure from the celestial vortex. Greenberg analyzes Mairan's faulty mathematical inferences and his commitment to empirically based (rather than rationally deduced) principles and theories. See Greenberg, *The Problem of the Earth's Shape,* chapter 2.

70."[J]e me suis exercé ces jours passés sur cette matiere, et comme je nay eté content ny de la maniere dont la traitée m. Herman ny mesme de celle de m. Newton, permettés moy de vous

Newton addressed the shape of the earth in the context of his theory of mutual attraction where force varies inversely with the square of the distance between bodies. He had also demonstrated that for all point masses inside a sphere, the force of attraction varies directly as the distance from the center. But he did not always make his analysis or his assumptions explicit, and Maupertuis had trouble following the argument—he thought Newton had assumed first one law of attraction and then another.[71] He explained to Bernoulli that he was motivated by his dissatisfaction with Newton as well as with Mairan: "I do not understand anything of the method M. Newton followed; it seems to me that he starts by supposing the spheroid ellipsoidal, and that he then takes up, and rejects, the hypothesis of gravity proportional to the distance from the center. In this hypothesis one does find an ellipse for the shape of the terrestrial meridian, but with a different proportion between the two axes than [Newton] finds."[72] Maupertuis was not alone in his confusion; Bernoulli admitted that he also had read and reread the relevant sections of the *Principia* without understanding them, finding only "obscurity and impenetrability."[73]

Annoyed by Mairan's inadequate treatment of the problem, Maupertuis thought he might be able to do something better. Though complex, the problem of the shape of the earth was more specific and more concrete than the *vis viva* question; it was also potentially just as controversial, since it would be a commentary on Newton, whose theory of gravity was viewed with distrust in Paris. In choosing this direction, Maupertuis was running the risk of alienating senior members of the academy, starting with Mairan, with whom he had maintained cordial relations. Because of the geodetical measurements by Cassini's team, the Newtonian problem risked antagonizing the local astronomers as well. When he first looked into the problem, Maupertuis was not defending Newton's theory; he simply approached the problem slightly differently and wrote up his analysis as trans-

envoyer ce que jay fait sur cela" (M to JB, 11 March 1731, BEB). (Referring to Jacob Hermann's analysis in *Phoronomia* (1716).

71. Newton analysed the problem for ellipsoids of revolution that deviate only infinitesimally from the spherical.

72. Part of his trouble may have been caused by reading this section of book 3 of the *Principia* without having carefully followed the development of the theory of gravitation in book 1. "Je nentends rien à la methode que m. Newton a suivie; il me semble quil commence par supposer le spheroide, ellypsoidique, quil prend ensuitte et quitte lhypothese dune pesanteur proportionelle à la distance au centre. Dans cette hypothese on trouve effectivement une ellypse pour la figure du meridien terrestre mais une proportion entre les deux axes differente de la sienne" (M to JB, 11 March 1731, BEB).

73. JB to M, 1 April 1731, BEB.

parently as possible, using the analytic techniques he had learned from Bernoulli. The draft he sent to Basel concerned rotating fluid bodies subject to attractive forces varying as any power of distance and concluded that flattening at the poles results in all cases. That is, regardless of the theory of gravity, rotating bodies will deform by bulging slightly at the equator.

Maupertuis did not address the question of whether such an effect would be measurable for the earth; the problem was strictly algebraic and not yet geodetical. Although he set up the problem as Newton had done, by imagining the spheroid as a sum of columns radiating from the center to the surface of the fluid body, so that the weight of each was balanced by that of its opposite number, Maupertuis did not limit his analysis to an inverse-square law of gravity. Bernoulli returned the manuscript with very few corrections, declaring it "very good, clear and worthy of publication." As usual, he was more interested in the method than the results: "Do not worry about the fact that under the hypothesis of weight proportional to the distance from the center your solution gives you a different proportion between the axes than what M. Newton found; your method is clear, but that of Newton is obscure: why then hesitate between the clear and the obscure?"[74]

The first part of Maupertuis's paper was thus an abstract and general treatment of the relationship between weight and centrifugal force. Although he had, by his own admission, "tortured" himself trying to understand Newton on the shape of the earth, Maupertuis at this point was already leaning toward the operational view of gravity articulated by Newton: "until we know what attraction is, I think it is evident enough from the facts that we can make use of it."[75] He knew this was more than Bernoulli would cede to Newton—he jocularly addressed his mentor as an "enemy of attraction"—but he was banking on their shared antagonism to Mairan's simplistic (and nonanalytic) reasoning. Whatever their differences, Maupertuis and Bernoulli still needed each other at this point.

74. "Je doute si Mr. Newton lui meme a bien entendu tout ce qu'il a dit dans cet endroit, ayant souvent remarqué que quand il n'a pas bien pu se dépetrir d'une recherche qui l'embarassoit, il a cru qu'il lui étoit permis de se jetter dans un galimatias, et que les grands Mathematiciens tels que lui pouvoient s'arroger des licenses en fait de raisonnement comme les grands Poetes de profession tels que Virgile et Ovide se donnoient souvent des licences poetiques; . . . Ne vous inquiétés donc pas de ce que dans l'hypothese d'une pesanteur proportionelle à la distance au centre v.e solution vous donne une proportion entre les deux axes differente de celle que Mr. Newton a trouvé; votre methode est claire, mais celle de Mr. Newton est obscure, pourquoi donc balancer longtems entre le clair et l'obscur" (ibid.).

75. "Je m'etois bien donné la torture pour entendre ce qu'il dit sur la figure de la terre" (M to JB, 23 April 1731). "En attendant quon scache ce que cest [que] lattraction je crois quelle est assez donnée par les faits pour quon puisse s'en servir" (M to JB, 11 June 1731, BEB).

Although he treated the rotating-body problem purely in mathematical terms, he also ventured to speculate about possible physical consequences of his results. In an addendum to his solution, he argued that the rings of Saturn might have originated as captured cometary debris. Could a large planet attract such matter, which would then condense and flatten into ring-shaped bodies? Bernoulli objected that a comet's tail could not supply the necessary density for the rings (which reflect sunlight), remarking archly that this hypothesis "will be more appreciated in London than in Paris."[76] Maupertuis admitted to a fondness for this section, however, and it appears in the published version of the paper, with explicit responses to Bernoulli's objections. This kind of speculation, interpolated from mathematical arguments and empirical evidence, would become a hallmark of Maupertuis's scientific style.

He continued to work on the general problem of "the flood [*torrent*] of fluid matter circulating around a center outside the flood," sending it back again to Basel for corrections a few weeks later.[77] Although he had intended to read this work to the Paris Academy, he did not do so immediately. He was in line for a pensioned position, and he took care not to antagonize Mairan and Fontenelle at this crucial juncture.[78] His paper on rotating bodies could only be interpreted as a rejection of Mairan's approach and especially of the slipshod mathematics and ad hoc reasoning behind it. Instead of pushing forward with this challenge, Maupertuis presented an uncontroversial paper on the anatomy and toxicity of scorpions, based on observations he had made in Montpellier several years before. This paper contained some novel conclusions, debunking popular wisdom about scorpions, and affirmed Maupertuis's sympathies with René-Antoine Ferchault de Réaumur, the foremost naturalist in the Academy.[79] As for the work on rotating bodies, he translated it into Latin, and sent it to the Royal Society in London—after asking Bernoulli to read it one last time. "I do not at all wish to read this piece in our meetings where there are people who are shocked simply by the word 'attraction.'"[80]

76. Ibid.

77. M to JB, 18 May 1731, BEB.

78. Maupertuis mentioned his anticipation of election to the pension becoming available with the retirement of Saurin (M to JB, 18 May 1731, BEB).

79. Maupertuis, "Expériences sur les scorpions," *MAS* 1731. Maupertuis published five papers in the *MAS* for 1731.

80. "Je ne veux point lire cette piece dans nos assemblées ou il y des gens qui le mot seul dattraction epouvante" (M to JB, 11 June 1731, BEB). Maupertuis sent the manuscript to London with a covering letter to Desmaiseaux, 9 July 1731 (British Library, Additional MS 4285, fol. 211). The piece was not published until spring 1732.

## Newtonian Attraction

The cause of gravity, like *vis viva,* was an unresolved question that provoked debate in the Academy, as this comment about attraction—the word associated with Newton's theory—attests.[81] Fontenelle had reflected critically on "occult" forces in his eulogy of Newton several years before. Prize questions for 1728 (the cause of gravity) and 1730 (the cause of the elliptical shape of planetary orbits) had raised the question publicly as well. In his winning submission on the latter, Johann Bernoulli self-consciously pandered to his judges' imagined predilections, using the dynamics of vortices to derive elliptical orbits and to "respond to the strongest objections that have been raised in England as invincible weapons against vortices."[82] In 1732, the question of gravity came up again, as the Academy solicited essays on the cause of the inclination of orbits in the solar system. When they received no satisfactory submissions, the commissioners rolled the competition over to 1734, with a doubled prize.[83] This time, Bernoulli split the prize with his son Daniel.[84] Bernoulli's grumbling about conspiracy notwithstanding, prize essays did not represent the Academy's position in any obvious sense. Necessarily the work of outsiders, opposing views might even share a given prize. Nor did consensus rule the Academy on contentious issues, despite its equable public face on view in the *Histoire.* In fact, academicians discussed gravity more in private than they did in print, just because it was contentious. In May 1731, for example, Mairan and Bragelogne reported on a Latin work on the cause of gravity that had been submitted for the Academy's review by one Abbé Obrien.[85]

81. Fontenelle endorsed Privat de Molières's latest attempt to account for gravity by the motion of planetary vortices: "Il ne paroît donc pas nécessaire de supposer pour le système de l'univers des attractions qu'on ne peut concevoir, puisque des forces centrifuges bien constantes & bien avérées donnent tout ce que donneroient les attractions. . . . Le système général de Descartes mérite que non-seulement la nation Françoise, mais toute la nation des philosophes, soit disposée favorablement à le conserver. Les principes en sont plus clairs, et portent avec eux plus de lumiere" (Fontenelle, "Sur les mouvemens en tourbillon," *HAS* 1728, 103).

82. Bernoulli, *Nouvelles pensées sur le systême de Descartes,* 136. Bernoulli did not consider his essay a completely accurate account of his theory of planetary motion; it was crafted to satisfy the judges in Paris. Bernoulli wanted the prize as a vindication after his two consecutive defeats in 1724 and 1726, but also because he needed the money. For his financial troubles, see JB to M, 5 May 1731, BEB.

83. Maupertuis was part of the 1732 prize committee, along with Cassini, Réaumur, Mairan, and Nicole; the 1734 committee was the same (M to JB, 20 October 1732, BEB; and 21 September 1733, BEB).

84. Maindron, *Fondations de prix.* On the prize competition, see Aiton, *Vortex Theory.* Winning essays and runners-up were published by the Academy in Académie des Sciences, *Receuil des pièces.*

85. "Il y a dans tout l'ouvrage des idées assez ingénieuses, des vues nouvelles, qui pourront faire naître dans l'esprit des lecteurs des pensées propres à éclairer cette matière" (AS p-v, 6 May 1731).

Although they did not exactly endorse it, they approved of its mechanical approach, based on vortices of ether in the interstices of matter. Shortly after this report, which must have occasioned some discussion, Maupertuis reported contemptuously to Bernoulli that the vortex theory published by Villemot in 1707 was much admired by some of his colleagues. "Between us, they think themselves more knowledgeable about this than they actually are and [they] believe in general that all that is required to resolve everything is to imagine more vortices."[86]

Maupertuis realized that any work predicated on gravitational forces acting in a void would provoke outrage among some of his senior colleagues, especially if couched in sophisticated mathematical terms. And so he sent his paper surreptitiously to the Royal Society in London, "the doctrine accepted therein being a bit odious in this country where I had initially thought to give it." He had actually written a "justification [*apologie*]" for attraction for his Paris audience, but decided to suppress it, at least for the time being. "I did not have the courage to give it in a country where it seems that they don't think deeply enough and where they do not do justice to M. Newton's system."[87] He was undoubtedly being particularly cautious in anticipation of the imminent election for the vacant pension. He carefully considered how to present the separate parts of his work—the mathematical solution to the rotating-body problem, the speculation about Saturn's rings, and the justification for the reasonableness of attraction—to different audiences. Maneuvering to secure his own place, as an academician, as an author, and as a mathematician, he was developing a sensitivity to matters of genre, style, and reception. And in July, the Academy endorsed him for the position of *pensionnaire géomètre.*[88]

Bernoulli never objected to Maupertuis's plan to send his paper to the Royal Society. His remarks about the likely negative reaction in Paris may even have prompted the decision.[89] He carefully corrected and commented on the paper, in

86. M to JB, 18 May 1731, BEB. On Villemot, see Aiton, *Vortex Theory.* Bernoulli was just waiting for an opportunity to present his own model for celestial mechanics reconciling Newton's law of gravitation with the dynamics of a fluid space-filling aether. See JB to M, 5 May 1731, BEB.

87. "[L]a doctrine qui y est repandue etant un peu odieuse dans ce pais cy ou javois d'abord pensé à la donner. . . . je n'ay pas eu le courage de la doner dans un pais ou il me semble qu'on ne medite pas avec assez dattention et ou lon ne rend pas assez de justice au systeme de m. Newton" (M to JB, 30 July 1731, BEB).

88. AS p-v, 20 July 1731.

89. "Sur ce que vous m'avés dit et que me crois aussy quelle seroit mieux receue en Angleterre qu'icy jay envie de lenvoyer aux transactions philos. et pour cela je l'ay mise en latin" (M to JB, 11 June 1731, BEB).

all its versions, just as he had done for Maupertuis's previous work.[90] But he did view it with some consternation, perhaps even as a betrayal, given his own long history of hostile relations with the English mathematicians, and he wondered how it would play in Paris. As always, Bernoulli saw the Paris Academy as inquisitorial and autocratic, and drew a parallel between his own defense of *vis viva* and Maupertuis's newfound interest attraction.

> I don't know if you will court your compatriots more effectively by publishing your justificatory piece on attraction in England, than if you had published it in Paris, for do you not fear being treated as a deserter, in going to defend elsewhere an opinion that passes in your hometown as heretical physics? My experience with *vis viva* in the Academy, so poorly received that it caused me to lose the prize, should terrify you; at least I would fear for your advancement to the rank of *pensionnaire* if you had not already achieved it.[91]

Bernoulli linked himself to Maupertuis as a martyr to Parisian stubbornness, implying that they were equally vulnerable to the whims of the powerful "Cartesian" contingent in the Academy, and he quite astutely noticed the relevance of Maupertuis's timing to the question of his advancement. But Maupertuis, for all his openness to mathematics and physics emanating from England, Switzerland, and even Russia, was maneuvering for a place at the Parisian center. Now secure in the Academy, he could afford to be more audacious about championing an unpopular position, and he resented Bernoulli's insinuation that he was constrained by the theoretical commitments of his French colleagues. "I will never adopt this or that position [*sentiment*] for political reasons, and the Academy does not require it either."[92]

## *Vis Viva* Redux

Although he had not yet had the "courage" to go public with his defense of attraction, Maupertuis insisted in the face of Bernoulli's pressure that he would

90. He even suggested changes in the Latin vocabulary of the translation (JB to M, 26 June 1731, BEB).

91. "Je ne sai si vous ne ferés mieux votre cour à Mrs. vos Compatriotes en publiant en Angleterre votre piéce apologetiq sur l'attraction, que si vous l'aviés publiée à Paris, car ne craignés vous pas d'etre traité de transfuge, en allant defendre ailleurs une opinion qui passe chez vous pour heretiq en fait de physiq. Mon exemple sur les forces vives si mal reçus dans l'academie, qu'elles m'ont fait perdre le prix, vous devroit faire trembler; au moins je craindrois votre avancement à la dignité de Pensionaire, si vous ne l'aviés deja emportée" (JB to M, 12 August 1731, BEB).

92. "[J]e ne seray jamais de tel ou tel sentiment par politique et Lacad.ie ne lexige pas non plus" (M to JB, September 1731, BEB).

defend the cause of *vis viva.* "[I]f I thought, after the excellent things you have said about living forces, that they still needed to be defended, and that I could add something to illuminate this matter, I would not hesitate to do it openly in the Academy, but fortunately the cause needs nothing further to be decided."[93] From Bernoulli's point of view, given the recent publication of Louville's attack on *vis viva,* the matter was far from settled, and he pushed his protegé to make good on his boast and take a public stance. Bernoulli provided a detailed litany of Louville's mistakes, asking Maupertuis to write up a response "either under your name or under mine" in the form of a *mémoire* for the Academy. "The rebuttal, witten by your pen, will have a thousand times more grace than it would have if it came only from my pen." He also remarked that "no one dares to respond to [Mairan and Louville]."[94] Maupertuis took this as a challenge and responded testily, "In spite of your doubts about my courage with regard to received opinions in the Academy, I can assure you that I would be capable of doing more daring things for the truth."[95] Throughout this exchange, as in their discussion of attraction, Maupertuis conceived his actions and potential actions in terms of bravery and danger. Only a few months before, he had admitted lacking the "courage" to present his elaboration on Newton in Paris; now he claimed to be ready to do anything, no matter how risky, for the truth.

It took him several months, but eventually Maupertuis did write up a critique of Louville, incorporating Bernoulli's objections, and sent it off to Basel. "I do not know if what I have done deserves to appear in a matter where your name alone can do more than all my reasons; at least it will demonstrate to you my zeal."[96] His reluctance was tempered by a strong sense of obligation to Bernoulli; he knew full well his dependence on the older man's mathematical acumen. He also knew that open confrontation was frowned on in the Academy, and certainly would not get

93. "Si je croyois qu'apres les excellentes choses que vous avés dittes sur les forces vives elles eussent encor besoin detre soutenues et que je peusse ajouter quelque chose à leclaircissement de cette matiere. Je n'hesiterois pas à le faire en pleine Acad.ie mais la cause heureusement n'a plus besoin de rien pour etre decidée" (M to JB, 12 September 1731, BEB).

94. "La reponse, faitte de votre plume aura mille fois plus de grace qu'elle n'auroit si elle partoit seulement de ma plume" (JB to M, 11 October 1731, BEB). Fontenelle's summaries in the *Histoire* had also been sharply critical of Bernoulli, accusing him of "specious reasoning" (Fontenelle, "Sur la force des corps en mouvement," *HAS* 1728, 81).

95. "Malgré le soubcon que vous aviés sur mon courage à legard des opinions receues dans lAcad. je peux vous asseurer que je serois capable de faire des choses plus hardies pour la verité" (M to JB, 28 January 1732, BEB).

96. "Je ne scay pas si ce que jay fait meritte de paroitre dans une affaire ou votre nom seul peut plus que toutes mes raisons; Du moins il vous marquera mon zele" (M to JB, 28 January 1732, BEB).

into print. He had already had his work censored for critical remarks about other authors on several occasions.[97] "In our Academy, they may find the style a bit harsh, we shall see; it is difficult to praise people while telling them they are mistaken."[98] As they exchanged a few more letters about how to argue most forcefully for *vis viva,* Maupertuis continued to fuss about what form to give the paper, and how to present it to the Academy. If he weighed in on the *vis viva* controversy, he wanted it to be clear to his colleagues that he was acting as a mouthpiece for the Swiss mathematician, but he was also loath to play this subsidiary role just when he was coming into his own. He asked whether he should incorporate Bernoulli's comments into the text, or leave them distinct as answers to his own questions. "In this way you would not seem to have deigned to respond to M. de L. but only to have kindly strengthened the work of one of your disciples. I assure you that it pains me a bit to presume to put my name to a *mémoire* in which I had as little part as this one."[99] Throughout this correspondence, Maupertuis tried to keep a distinctive voice for himself, while Bernoulli pushed him to take on the polemic wholeheartedly.

Finally, the *mémoire* was finished, but he put off presenting it. He did circulate it privately to some of his friends, with favorable results. Saurin, in particular, signed on as a supporter of the critique.[100] Why was Maupertuis so reluctant to go public with this paper? Surely, the answer involves more than the philosophical, or mechanical, grounds for contention. Maupertuis's letters clearly indicate that he considered himself a proponent of *vis viva.*[101] But he had to decide what battles to fight in the Academy, and especially whether he wanted to revive an old battle that seemed to have no end. In addition, Louville was a marginal figure, and his contributions to the Academy made his inadequacy as a mathematical

97. See Greenberg, *The Problem of the Earth's Shape,* 114–15.

98. "On trouvera peutetre dans notre Acad.ie le stile un peu dure, nous verrons; il est difficile de louer les gens en leur disant quils se trompent"(M to JB, 28 January 1732). As usual, Bernoulli returned the manuscript with copious commentary (JB to M, 2 March 1732, BEB).

99. "De cette derniere maniere vous n'auriés pas lair pour cela davoir daigné repondre à M. de L mais seulement d'avoir bien voulu fortifier lecrit d'un de vos disciples. [J]e vous avoue que je souffre un peu d'oser mettre mon nom à un memoire auquel j'ay aussy peu de part qu'a celui ci" (M to JB, 10 March 1732, BEB). JB responded point by point, saying that M should do as he wishes with the form of the paper (JB to M, 13 April 1732, BEB).

100. "[C]eux de nos m[essieu]rs a qui je l'ay fait voir m'en ont paru tres contents, et pensent comme nous" (M to JB, 14 April 1732, BEB). Saurin's opinion is mentioned in the same letter.

101. He did have reservations about the metaphysical aspect of the *vis viva* question, but this was not the basis of his criticism of Louville. Maupertuis did not question the conservation of *vis viva,* but wondered if there might be other quantities conserved in different situations (M to JB, 12 May 1732, BEB).

physicist only too apparent.[102] By August, months after the initial promise to read the refutation, his adversary's health had become sufficiently precarious to make an attack inappropriate. Sure enough, in October, Maupertuis reported that "M. Louville has gone to learn Dynamics in heaven," conveniently letting his earthly opponent off the hook.[103] The refutation was never read to the Academy, and the manuscript does not survive.

## Return to the Problem of Attraction

In 1731, Maupertuis was actively searching for problems where his efforts would allow him to enhance his reputation as an original thinker. The list of topics he addressed in the Academy attests to this energetic, if somewhat restless, search: papers on the separation of variables, ballistics, the anatomy and physiology of scorpions, "curves of pursuit," the aurora borealis. During a long sojourn from August to November 1731 in Saint-Malo, where he had retreated to attend to family matters following his mother's death, he continued to slave over the *Principia* and the dense mathematical papers collected in the *Commentarii* of the Saint Petersburg Academy, looking for inspiration.[104] The Saint Petersburg volume, with papers by Daniel Bernoulli, Bilfinger, Hermann, and others, made him think further about *vis viva* and left him with a sense of the futility of moving in that direction himself.[105]

Maupertuis's initial engagement with Newton's analysis of central forces had been somewhat tentative, especially on the question of the ontological status of gravitational attraction. But gradually he came to suspect that he could contribute novel results (with Bernoulli's help on the mathematics) that might be noticed even beyond the walls of the Academy. Though he had little patience for English mathematics, he was increasingly aggravated by the simplistic and uncompre-

102. "[I]l nous leut lautre jour un vilain petit memoire dans le quel il pretendoit comparer la force de la pression avec la percussion, et ou il concluoit qu'on setoit trompé jusqu'icy de croire que les corps qui circulent tendent à sechaper par la tangente" (M to JB, 12 May 1732, BEB). Louville was respected as an observational astronomer. As a retired army officer, he had special permission to live outside of Paris while maintaining his academic position.

103. M to JB, 4 August 1732. Bernoulli replied, "[S]'il ne revient pas [de son Apoplexie] on pourra laisser tomber la dispute ou donner un autre tour à votre piece, afin qu'il ne paroisse pas, qu'elle soit directement contre lui" (JB to M, 17 August 1732, BEB). Two months later, Maupertuis reported, "M. Louville est allé apprendre la Dynamique au Ciel" (M. to JB, 20 October 1732, BEB). Bernoulli complained that all his opponents (and there were many) died before their conflicts could be resolved.

104. He had planned a trip to England in this period, hoping to reestablish connections made three years earlier, but changed his plans and never went.

105. M to JB, 23 October 1731, BEB.

hending response of his colleagues to Newton's physics, and he took it upon himself to look for a topic in Newton that he might translate into the Leibnizian calculus.[106] At least one of his efforts brought him up short, after he had tried to work through Newton's analysis of cometary orbits. He was prompted to this effort by Jacques Cassini's reflections on comets in a long paper arguing, contrary to Newton, that all comets orbit the sun in the same direction as the planets, from west to east. Cassini set out to show, based on a review of all available cometary observations, that the motion of comets did not contradict the "system of vortices," implicitly defending Cartesian cosmology from Newton's challenge. According to Maupertuis, Cassini's presentation included an explicit challenge to Newton's claim that three observations would determine an orbit; on the contrary, Cassini insisted, the problem must remain inherently indeterminate. Maupertuis went on to tell Bernoulli, "Such an error did not seem possible in a man like M. Newton, and in effect I was soon able to show M. Cassini and the Academy that M. Newton had not been mistaken and that the problem was in fact determinate (all this between us, please)."[107] The secretary did not record this exchange; it can hardly have endeared Maupertuis to Cassini.

But if Maupertuis could find the holes in Cassini's mathematical reasoning, he had more trouble following Newton on comets, and he implored Bernoulli to help him.

> [I]f you had not yet thought about this and wanted to give it a few hours, you could give us a solution more beautiful than that of M. N[ewton], . . . because with all the confusion that N. and his commentator Gregory leave in it, one can hardly call what we have a solution at all. I have been brash enough to work on it and to try to reduce the problem to algebra. I prefer such solutions to those held up by the scaffolding of geometry; but I saw that the thing is too far beyond my powers.[108]

106. Varignon had done something similar at the turn of the century, when he developed analytic algorithms, based on the Leibnizian calculus, to reconfigure Newton's treatment of central forces in the *Principia.* See Blay, *La naissance de la mécanique analytique.*

107. M to JB, 23 October 1731, BEB. This discussion about calculating orbits from a limited number of observations does not appear in the published version of Cassini's paper. J.Cassini, "Du mouvement véritable des Cometes à l'égard du soleil et de la terre, " *MAS* 1731, 299–346. For Fontenelle's refusal to publish anything that could be construed as direct criticism of another author, see Greenberg, *The Problem of the Earth's Shape,* 114–19.

108. "Et si vous ny avés point pensé encor et que vous y voulussiés mettre quelques heures vous nous pourriés donner une solution plus belle que celle de m. N., . . . car avec tout lembarras que m. N. et son comentateur m. Gregori y laissent on ne peut presq pas apeller ce quon a une solution. jay eté assez temeraire pour y travailler et tacher de reduire le probleme à lalgebre dont je prefere les solutions a celles qui se font par des echaffaudages de geometrie; mais jay veu que la chose est trop audessus de mes forces" (M to JB, 23 October 1731, BEB).

Bernoulli had no wish to embroil himself in the analysis of cometary orbits, recognizing the inherent difficulty of the problem as well as the "impenetrable obscurity" of Newton's treatment.[109] Evidently no breakthroughs could be expected from this direction.

Meanwhile, Maupertuis's piece on rotating fluid bodies had been read to the Royal Society in July, but it was not immediately prepared for publication. He wrote in some consternation to Hans Sloane in September, asking what had happened to it, since he had heard nothing and was worried that he had offended someone.[110] Apparently, the paper had been sent to the mathematician John Machin at Gresham College, where the manuscript languished until the following spring. Finally, Machin wrote to Desmaiseaux, who relayed the response to Maupertuis.[111] Machin wanted to reassure Maupertuis that "it was so far from being thought slight or trivial or unintelligible that on the contrary every member who has any knowledge in these matters was extremely well satisfied with it as a performance which discovered a great skill and address in the author."[112] However, he did object to the implication that the examples represented physical phenomena rather than mathematical abstractions. Machin here put his finger on a confusion in Maupertuis's argument, where the law of gravity is assumed as a given in order to find the shape of the rotating body (such as the earth). But, Machin pointed out, the expression for effective weight (which he calls the "law of gravity") actually depends on the shape of the body, and will vary as the shape changes. "And had he confined himself to the solution of the Problem as it is merely mathematical, I can't see any objection to afford to it."[113] When he got word of this critique, Maupertuis agreed that the paper gave "mathematical solutions rather than true and exact physical explanations."[114] He asked to have a slight revision made before publication, adding "a little scholium" to reassure the reader and to allay

109. JB to M, 13 November 1731, BEB.

110. M to Sloane, 17 September 1731, British Library, Sloane MS 4052, fol. 14.

111. Desmaiseaux, a Huguenot journalist and man of letters, acted as go-between in this correspondence. Maupertuis wrote to Desmaiseaux on 29 April 1732, after hearing about Machin's letter: "[J]'ay eu de la peine de voir qu'elle n'eust point été inserée dans les Transactions Philosophiques parce qu'ayant voulu temoigner mon respect et mon zele à la Société Royale par quelque chose qui luy pust etre agreable, j'ay eu lieu de craindre au contraire d'avoir fait quelque chose qui lui avoit déplu"(British Library, Additional MS 4285, fol. 212). Only some of this correspondence survives.

112. Machin to Desmaiseaux, 29 March 1732, British Library, Additional MS 4285, fol. 214.

113. Ibid. Beeson describes Maupertuis's scholium as an advance in his understanding of Newton, not realizing that the revision is a direct response to Machin's reading (Beeson, *Maupertuis,* 81).

114. M to Desmaiseaux, 29 April 1732, British Library, Additional MS 4285, fol. 212.

any suspicion that he meant to contradict Newton. This addition was a direct response to Machin's critical reading of the paper. "Those not sufficiently on top of [*au fait*] this subject might hold it against me when they see that I give other numbers for the axes of the earth than those given by M. Newton."[115] Maupertuis had hoped to impress his English colleagues, and was relieved that his paper finally appeared in the *Philosophical Transactions*, although the response from mathematicians was modest at best. Machin was certainly polite, but did not see anything pathbreaking in Maupertuis's work.[116]

By the time the paper was printed in London in the spring of 1732, Maupertuis had decided that he would no longer hide his new interests from his compatriots. One year after sending his manuscript to England, he told Bernoulli that "the approbation you gave it made me brash enough to appear in public."[117] He decided to do so by taking the unusual step of publishing his work on rotating fluid bodies as a book, rather than presenting it to the Academy, and adding to it "a preliminary section on gravity [*pesanteur*], exposing the different ideas of the Cartesians and Newtonians."[118] This was the "apologia" for attraction he had mentioned the summer before; Bernoulli had not yet seen it, nor would he do so for some months. The book embedded the mathematical problems in a discussion of Cartesian and Newtonian physics, designed to interest a broader spectrum of readers. This discussion drew on the longstanding exchange with Bernoulli, but it betrayed the extent to which Maupertuis had been thinking about the metaphysical status of the forces that his teacher considered anathema to good physics. Although it never said so explicitly, the *Discourse on the Various Shapes of the Celestial Bodies, with an Exposition of the Systems of Mssrs. Descartes and Newton*, published in Paris in 1732, was a reponse to Bernoulli, as well as to Fontenelle, Mairan, and others in the Paris Academy.[119]

115. Ibid. He adds, "Si d'apres les memes hypotheses que luy mes solutions differoient des siennes, je n'hésiteroient pas à les jetter au feu" (ibid.).

116. Machin's letter is substantial; he had read the paper carefully. He also passed on Edmond Halley's apologies for not having time to send comments. The paper appeared as Maupertuis, "De figuris quas fluida rotata induere possunt problemata duo," *Philosophical Transactions*, no. 422 (1732), 240–56.

117. "[L]'approbation que vous y donates m'a rendu assez hardy pour paroitre en public" (M to JB, 4 August 1732, BEB).

118. "[J]'y ay joint un preliminaire sur la pesanteur dont jexpose les differentes idées qu'ont les Cartesiens et les Newtoniens" (ibid.).

119. *Discours sur les différentes figures des astres* (Paris, 1732). Reprinted in slightly revised form in *Oeuvres*, 1:81–170. He gave the book to the Academy for its approbation in November or December 1732; on 19 November, he had not yet sent the book to Bernoulli, though it was already printed (M to JB, 19 November 1732, BEB).

In his winning 1730 prize essay, Bernoulli had attacked Newton's mathematical critique of vortices, claiming to show just what Newton had declared impossible: that the periods of the layers of the vortices follow Kepler's law of areas exactly.[120] At this very moment, Bernoulli was working on a new prize essay for the Paris Academy, again on a question that involved gravity, namely the inclination of the planetary orbits to the plane of the ecliptic. As he told Maupertuis, he planned to construct a "system of the world" that would combine the best of Descartes and Newton.[121] This synthesis would of course rely on impulse mechanics rather than long-range forces acting across empty space and would challenge Newton's objections to vortices on mathematical grounds. Familiar as he was with Bernoulli's categorical opposition to Newtonian gravity and empty space, Maupertuis well knew that his own book might give offense in Basel.

Bernoulli had tried, in his way, to reconcile Descartes and Newton by arguing that the inverse-square law could be accommodated to fluid mechanics and hence to vortices. Maupertuis instead dispassionately compared the Cartesian model of celestial vortices with Newton's universal gravitation, to the detriment of vortices. He accepted Newton's critique of Descartes, based on the incompatibility of Kepler's planetary laws with fluid dynamics, explaining it in relatively simple terms, and slid almost imperceptibly into a defense of the Newtonian position on void space and action at a distance. In doing so, he rejected Bernoulli's compromise, which had impressed him originally. Maupertuis warned Bernoulli, before sending him a copy of the book, that his "extremely succinct" comparison of Descartes and Newton gave the advantage to the latter.[122] But in the event, Bernoulli chose to read *The Shapes of Celestial Bodies* as a strictly "geometrical," or hypothetical, treatment of gravity, and accepted at face value the author's professions of agnosticism about the physical cause of gravity. He pointedly took no offense at Maupertuis's book, apparently reluctant to cede his disciple to his old rival Newton, even though that rival was now dead. "If you appear English to me, Monsieur, you appear to be a reasonable Englishman, who does not push the principle of attraction beyond geometry into physics, as the rigid followers [*sectateurs*] of Newton do, more than Mr. Newton himself did."[123]

120. Bernoulli, *Opera Omnia*, 3:145 ff.

121. JB to M, September 1732, BEB. Bernoulli was actually writing for the 1734 competition, which had been held over from 1732 for lack of a successful entry in that year. In his essay on the inclination of planetary orbits, Bernoulli used the nonspherical shape of planets to account for the inclination of orbits. He shared the prize with his son Daniel (Maindron, *Les fondations des prix*, 18).

122. M to JB, 10 November 1732, BEB.

123. JB to M, 9 August 1733, BEB. John Machin had pointed to exactly this distinction in his reading of Maupertuis's *Philosophical Transactions* paper, when he had cautioned against extending the

Bernoulli had his reasons for wishing to maintain cordial ties. (Among other things, he was competing for the 1734 Paris prize and hoped that Maupertuis, as a prize commissioner, would cast a vote in his favor.)[124] For Maupertuis, his first book marked a gradual lessening of his reliance on his Swiss mentor and a burnishing of his Parisian image. Intent on enhancing his stature in literary and social circles, he did not plan to limit himself to operating as Bernoulli's Paris representative. Although he had devoted considerable effort to establishing his reputation in the Academy, substantially assisted by Bernoulli's generosity in coaxing along his mathematical abilities, he had also been building his reputation as a man of wit and brilliance outside the Academy. With the final illness and death of Houdar de la Motte, he became the center of the lively society of the Café Gradot.[125] His friendships with La Condamine and Alexis Clairaut date from this period; Voltaire moved in the same circles.[126] Evidently, Maupertuis was known as an authority on Newton even before his book appeared, since Voltaire asked for help with the theory of universal gravitation in October of 1732. At this point, Voltaire was completing his account of Newtonian natural philosophy for the *Lettres philosophiques.* "To whom can I better address myself than to you, sir, who understands it so well, who even works on [Newton's] philosophy, and who is so capable of confirming the truth or of demonstrating the error of it?"[127]

By writing a clear and elegant exposition of the rival accounts of cosmology, Maupertuis acknowledged that he might benefit from presenting the fraught question of gravity to a wider public. This dramatic turn to the public was buttressed by the technical virtuosity of his academic work. Although it incorporated mathematics, the book expanded on the specific problems, both philosophically and rhetorically. Publishing *before* presenting his analysis of rotating bodies to the

---

analysis to phenomena (Machin to Desmaiseaux, 29 March 1732, British Library Add. MS 4285, fol. 214).

124. JB to M, 9 October 1732, BEB.

125. La Beaumelle, *Vie de Maupertuis;* Charles Collé recalled seeing Maupertuis in the retinue of Houdar de la Motte many years earlier (Collé, *Journal historique,* August 1759, in *Journal et mémoires,* 189).

126. Voltaire recalled a dinner at Charles Dufay's with Maupertuis and La Condamine, who was dressed as a Turk, having just returned from a journey to the Orient (Voltaire to M, [c. 15 December, 1732], Volt., *Corr.,* 86:261).

127. "A qui puis-je mieux m'addresser qu'à vous monsieur, qui l'entendez si bien, qui travaillez même sur sa philosophie, et qui êtes si capable ou d'en confirmer la vérité ou d'en démontrer le faux?" (Voltaire to M, 30 October 1732, Volt., *Corr.,* 86:243–45). This was Voltaire's first letter to Maupertuis. Working from Pemberton's book on Newtonian natural philosophy, Voltaire was confused about the analogy between the moon in its orbit and a body falling on earth. Maupertuis was able to straighten him out, over the course of several further letters.

Academy was a strategic move informed by ambition. The apparatus supporting his philosophical arguments remained technical, but he could say things about the contrast between Cartesian and Newtonian views in a book that he could not say in the Academy. Making his case in this way, he also avoided the long delay attendant on publication of the Academy's *Mémoires.* This made his book topical, like that of Voltaire—if not nearly so dangerous. Though Maupertuis later claimed that it did not enjoy the kind of success he had hoped for, the book did make its author into something of a public figure. He published it with some trepidation, "barely daring to compare [attraction] to impulse," for fear of the attacks it would provoke.[128] In retrospect, he may have exaggerated the audacity of this move, but he was effectively circumventing Fontenelle's control of dispute in the Academy by challenging the secretary's own assessment of the relative merits of action at a distance and impulse mechanics. It was done with a finesse that even Fontenelle could appreciate. Nowhere in the book did Maupertuis openly attack the proponents of vortices, which would have meant breaching the bonds of civility. "It is not for me to pronounce on a question that divides the greatest Philosophers, but I am permitted to compare their ideas."[129]

The book demystified Newton, and gravity, for French readers, by insisting that the force of impulse was no more intelligible than the force of attraction.[130] In order to develop this argument, he rejected the stylistic constraints of the academic *mémoire.* After recapitulating the history of attempts to derive the shape of the earth, either from measurements or from theory, he embarked on what he called a "metaphysical discussion" of gravity. (Metaphysics in general was discouraged in academic discourse.) In spite of the fact that Newton treated gravity as a lawlike effect rather than a cause, Maupertuis noted that many people recoil from "attraction," calling it an occult cause or a "metaphysical monster." So he devoted some attention to just this question of whether attraction, "even when considered as a property of matter," entails any absurdity or logical contradiction.[131] In a sense, he overstated the Newtonian case by looking at the implications of attraction as a property of matter, analogous to extension or impenetrability. "It would be ridiculous to wish to assign to bodies other properties than those that

128. Maupertuis, "Sur l'attraction," *Lettres de M. de Maupertuis,* in *Oeuvres,* 2:285. D'Alembert also portrayed Maupertuis as the courageous defender of Newton (*Preliminary Discourse,* 88–89). According to Voltaire (not always a reliable source), the first edition of the *Discourse* sold fewer than 200 copies (Voltaire to J. B. N. Formont, [c. 1 June 1733], Volt., *Corr.,* 86:343). No information about the number of copies printed is available.

129. *Discours sur les différentes figures des astres,* in *Oeuvres,* 1:90.

130. Attraction and impulse, as causes of motion, are equally mysterious, "attraction being no less possible in the nature of things than impulse" (ibid., 1:132).

131. Ibid., 1:93–94.

experience has taught us are found there; but it would be even more so, after barely knowing a small number of properties, to wish to decree dogmatically the exclusion of all others; as if we could measure the capacity of subjects, when we know them only by this small number of properties."[132] The argument rests on the profoundly enigmatic connection between attributes and the subjects to which they belong, whether those attributes be controversial (attraction) or not (impenetrability). Since attraction does not contradict any other accepted property, there is no need to reject it out of hand. This claim does not entail a position on the actual existence of the property, but it does insist that attraction cannot be ruled out of court on logical grounds. It is not "metaphysically impossible." And if it is not, experience must be consulted in deciding on its validity. "Attraction is no more, so to speak, than a question of fact."[133] Here we are back on clear Newtonian ground.

The apologia, then, accomplished two parallel objectives: to deny the absurdity or logical impossibility of attraction, and to call into question the transparent intelligibility of impulse as the cause of motion. The challenge to Cartesian epistemology rested on a willingness to countenance a lack of intelligibility in nature, or a limit to human understanding: "I do not believe that it is permitted to us to ascend to first causes, nor to understand how bodies act on each other."[134] Having gotten this far, Maupertuis went on to compare the two rival "systems." The brunt of his analysis of vortex physics rested on the failure of all attempts to match the model of swirling fluid with the observed phenomena of planetary motion, especially with Kepler's laws of planetary motion. Newton's well-known challenge to Descartes and Huygens certainly inspired the analysis, though Maupertuis did not wholeheartedly endorse Newton's fluid mechanics; instead he followed Bernoulli, referring favorably to the 1730 prize essay.[135] By the end of the chapter on vortices, he concluded that although no one had yet managed to reconcile vortices with the phenomena, he could not declare this impossible in principle. "Nothing is more beautiful than the idea of Descartes, who wanted to explain everything in Physics by matter and motion: but if we wish to conserve the beauty of this idea, we must not allow ourselves to suppose matters and motions with no

132. "On seroit ridicule de vouloir assigner aux corps d'autres propriétés que celles que l'expérience nous a appris qui s'y trouvent; mais on le seroit peut-etre davantage de vouloir, après un petit nombre de propriétés à peine connues, prononcer dogmatiquement d'exclusion de toute autre; comme si nous avions la mesure de la capacité des sujets, lorsque nous ne les connoissons que par ce petit nombre de propriétés" (ibid., I:96–97).

133. Ibid., I:103.

134. "[J]e ne crois pas qu'il nous soit permis de remonter aux premières causes, ni de comprendre comment les corps agissent les uns sur les autres" (ibid., I:93).

135. Ibid., I:110.

other rationale than our need for them."[136] Newton fares somewhat better, since he has shown "with reasonings of the most certain Geometry" that a central attractive force, acting across a void, can be perfectly reconciled with all three of Kepler's laws. Without recourse to equations, Maupertuis explained the synthetic power of Newton's insight about the law of gravity, which describes the orbits of the planets, the moon, and the fall of bodies on earth. Throughout, Maupertuis preserved a rhetoric of evenhandedness, describing both systems in some detail and refusing to tell the reader which should prevail. "For myself, I confess that I do not know what this gravity [*pesanteur*] of matter is; I know no better what impulsive force is. If one could show that the one depends on the other, that would definitely simplify the systems; but in the meantime, I believe that without pronouncing on the claims that one may have over the other, we can use both of them."[137]

The weight of the argument, notwithstanding this disclaimer, rested with Newton. Fontenelle saw through Maupertuis's rhetoric of impartiality, as he was no doubt intended to do. The Academy's secretary, in his 1727 eulogy of Newton, had openly objected to the reification of gravity as a mutual attraction between bodies, articulating a reading of the *Principia* that resonated with the perspective of many of his academic brethren.[138] Setting the tone for subsequent interpretations, including that of Maupertuis, Fontenelle framed his analysis as a contrast between Descartes and Newton. Descartes, Fontenelle reminded his audience, had built his physics on the "clear and distinct" notion of impact as the cause of all motion. The only way one body can affect another is through collision; objects fall to earth as a result of the pressure of swirling particles of matter impinging on them. Newton, cavalierly turning his back on conceptual clarity, used "attraction" to denote "the active force of bodies."

> [This] force is actually unknown, and he does not pretend to define it. But if it could act as well by impulse, why would this clearer term not be preferred? Everyone will agree that it was hardly possible to employ the two terms indiscriminately; they are

136. Ibid., 118.

137. "Pour moi je confesse que je ne sçais ce que c'est que cette pesanteur de la matière; je ne sçais pas mieux ce que c'est que la force impulsive. Si l'on pouvoit faire voir que l'une dépend de l'autre, cela simplifieroit assurément les systèmes; mais en attendant, je crois que sans prononcer sur les droits que l'une peut avoir sur l'autre, on peut se servir des deux" (Maupertuis, *Discours sur les différentes figures des astres,* 1st ed. (1732), 45). This passage was deleted from later editions, starting in 1752, by which time it would have been absurd to maintain that both systems held equal weight.

138. Fontenelle, "Eloge de Newton," read at public assembly in November 1727, and printed in pamphlet form shortly thereafter, so it circulated well in advance of the appearance of the *Histoire* for that year. There are also four English editions from 1728. See Delorme, "Tableau chronologique."

> too opposed to each other. The continual usage of the word "attraction,"... at least familiarizes the reader with an idea proscribed by the Cartesians, and which all other philosophers have agreed in condemning; one must be on one's guard to avoid imagining some reality in it, for one is susceptible to the danger [*péril*] of believing what one hears.[139]

From Fontenelle's perspective, Newton had needlessly given up a "clear" notion for a dangerously equivocal one. The inability to define attraction in appropriately intelligible terms implied to Fontenelle its limitations as the basis of a viable theory.

Fontenelle ignored the details of Newton's attack on Cartesian celestial mechanics. He noted several signal successes of the law of gravity in matching up with empirical evidence, as in the moon's orbit and the perturbation in the orbits of Jupiter and Saturn. As long as Newton stuck to modelling phenomena mathematically, Fontenelle had only praise for his work. But, as we have seen, he discreetly accused Newton of leading readers down the primrose path toward accepting attraction—"the active force of bodies"—as a physical reality.[140] Although he recognized the magnitude of Newton's accomplishment, he managed to describe the physical theory in terms that made it seem too cumbersome to be true. As a mathematical description, he was willing to let the law of gravitation stand, but as physics, he could not endorse it.

> All bodies, according to M. Newton, weigh [*pèsent*] on each other, or attract each other in proportion to their masses; and when they orbit around a common center, by which in consequence they are attracted, and which they attract, their attracting forces vary in proportion to the inverse square of their distances from this center; and if all, together with their common center, orbit around another center common to them and to others, there are then new relations that make a bizarre complication. Thus each of the five satellites of Saturn weighs on the four others, and the others on it; all five weigh on Saturn, and Saturn on them: the whole weighs on the Sun and the Sun on this whole. What Geometry was needed to untangle this chaos of relations![141]

Fontenelle's artful prose reflects the repetitive tangle of mutual attractions, overwhelming the reader with the confusing complexity of a theory that seemed to abandon clarity and distinctness. Although Fontenelle applauded the "sublimity"

139. Fontenelle, "Eloge de M. Newton," *MAS* 1727, 151–72, reprinted in *Oeuvres complètes*, 7:118.

140. Ibid. The association of attraction with "active force" made it theologically questionable, according to the Jesuit writers in the *Mémoires de Trévoux*. This theological danger was not Fontenelle's concern.

141. Ibid.

of Newton's mathematics, and the novelty of the theory, he presented it as arcane rather than conceptually simple. He could not countenance the rejection of Cartesian first principles, although he acknowledged Descartes's failure to supply a causal picture that could stand up to mathematical scrutiny. "The one [Descartes] sets out from what he understands clearly to find the cause of what he sees; the other [Newton] sets out from what he sees to find its cause, whether it be clear or obscure. The clear [*évidens*] principles of the one do not always guide him to phenomena as they are; phenomena do not always guide the other to clear enough principles."[142] This is hardly an unreserved endorsement of Descartes's physics, but it does spell out Fontenelle's reluctance to stray from Cartesian philosophical ideals.

Five years after Fontenelle delivered his eulogy of Newton, Maupertuis took his first public steps into the debate by endorsing Newton's mathematical attack on vortices *and* by refusing to be disturbed by the physical meaning of attraction as an inherent force residing in matter. Though he did not mention Fontenelle by name, his readers could hardly have ignored the subversive attitude of his enterprise. For one thing, he implicitly questioned Fontenelle's ability to speak for the Academy; for another, he was pushing a concept that had been labelled dangerous by theologians. (If matter can attract other matter across empty space, then materialism cannot be far behind.) For the moment, he sidestepped the question of matter's inherent activity, stressing the congruence of Newton's mathematical physics with terrestrial and celestial phenomena. The viability of gravity as the key to a mathematical account of the universe overrode questions about its ontological status. Although he encouraged readers to decide for themselves on the merits of attraction, Maupertuis did attribute a realist definition of gravity to Newton, who "discovered *that gravity was nothing other than a phenomenon resulting from a force distributed [répandue] throughout matter, by which all its parts attract each other in proportion to the inverse square of their distance.*"[143]

The "metaphysical" discussion yielded to the mathematical in comparisons of different possible theoretical formulations of gravity, and their implications for the shapes of rotating bodies. Equations finally made an appearance in problems about rotating fluid bodies, the same problems recently published in the *Philosophical Transactions.* The equations did not, of course, represent Newton's mathemat-

142. "L'un part de ce qu'il entend nettement pour trouver la cause de ce qu'il voit; l'autre part de ce qu'il voit pour trouver la cause, soit claire, soit obscure. Les principes évidens de l'un ne le conduisent pas toujours aux phénomènes tels qu'ils sont; les phénomènes ne conduisent pas toujours l'autre à des principes assez évidens" (ibid., 7:121).

143. Maupertuis, *Oeuvres,* 1:136, italics in original; Roger Cotes gave a similar reading in his preface to Newton.

ics as such, but rather a Leibnizian approach to problems inspired by Maupertuis's struggles with the *Principia*. By the end, then, the book came full circle, back to problems related to the shape of the earth, which had introduced the rival cosmologies. The problem solutions display hard-won technical credentials to legitimate the larger argument, while the chapters on Cartesian and Newtonian physics gave the calculations a broader polemical context. In explicating that context he developed the smooth rhetorical style with which he had begun his academic career some years before, when he reflected on the shapes of musical instruments. The book is a hybrid of genres, part mathematics, part astronomy, part metaphysics, part polemical tract, and part "curiosity." Although it includes mathematics appropriate to the academic identity of the author, the chapter on metaphysics reads as conversational philosophy: "I flatter myself that no one will stop me here to tell me that this property of bodies to gravitate toward each other is less conceivable than a property [like impenetrability] recognized by everyone."[144] The author appealed to an audience of his peers—those who agree, without necessarily sharing his mathematical perspicacity, on standards of argument and style. He distinguished this elite readership from "the people," who confuse true understanding and simple familiarity. Such simple minds have never stopped to think about how attributes inhere in subjects. "Ordinary people [*le peuple*] are not surprised when they see a body in motion communicate this motion to others; their habit of seeing this phenomenon prevents them from perceiving what is marvelous in it: but Philosophers will take special care not to believe that the force of impulse is more conceivable than the attractive force. What is this impulsive force? How does it reside in bodies?"[145]

By adopting a conversational tone, and using it to undermine assumptions Fontenelle had championed, Maupertuis adapted Fontenelle's stylistic moves to his own ends. He took issue with Fontenelle's evaluation of Newton, not just by demonstrating the explanatory power of the inverse-square law of gravity, but also by denying that the notion of action at a distance was either unintelligible or dangerous. He wavered between the purely "geometrical" use of attraction and the "physical" reality of force as a property of matter. Nominally, he stuck to the former, while strongly implying that he was leaning toward the latter. This equivocal

144. "Je me flatte qu'on ne m'arretera pas ici, pour me dire que cette propriété dans les corps, de peser les uns vers les autres, est moins concevable que celle que tout le monde y reconnoit" (ibid., I:98).

145. "Le peuple n'est point étonné lorsqu'il voit un corps en mouvement communiquer ce mouvement à d'autres; l'habitude qu'il a de voir ce phénomene l'empeche d'en appercevoir le merveilleux: mais des Philosophes n'auront garde de croire que la force impulsive soit plus concevable que l'attractive. Qu'est-ce que cette force impulsive? comment réside-t-elle dans les corps?" (ibid., I:98).

posture not only avoided accusations of reviving occult causes, and preserved the civil comportment required of an academician, but also gave readers a variety of options in interpretation. "It seems that one could, without risking very much, decide in favor of [universal gravitation as a property of matter]. However, since all I have to say works equally well with all three [theories], I leave it to each reader to think of it what he will: he will be able equally well to adapt his ideas to the explanation of phenomena that I am going to propose."[146] However carefully he disguised it, the reader could not help but see that Maupertuis had been swayed by Newton's arguments, and especially by his mathematical demonstrations. Although the conclusions were different, the style of polite argument was not so far from Fontenelle.

The book ends with a chapter on astronomical phenomena.[147] This section appealed to readers' appetites for "curious" natural phenomena and for speculation about the varieties of objects in the cosmos. Here too, contemporaries would have recognized the stylistic inspiration from Fontenelle, arguably the most admired of French writers on scientific (and especially astronomical) matters at this time. Some passages of the *Shapes of Celestial Bodies* might well have reminded readers of the elegantly formulated speculations of Fontenelle's fantastically successful *Conversations on the Plurality of Worlds.*[148] For example, Maupertuis asks, "And why would the kind of uniformity that we see in a small number of planets keep us from suspecting at least some variety in the others that the immensity of the heavens hides from our view? Relegated to a corner of the universe with feeble organs, why should we limit things to the little that we perceive of them?"[149] This glib and provocative kind of questioning recalled Fontenelle's reflections on the inhabitants of other worlds. Certain subjects, like comets or the rings of Saturn, also evoked the colloquial ruminations of Fontenelle's *Conversations* about causes or consequences of cosmological phenomena.[150] Following a discussion of the deleterious effects of cometary encounters with planets, Maupertuis remarked casu-

146. Ibid., I:138.

147. In later editions, this was broken into several chapters. By 1752, in the first edition of his collected works, Maupertuis appended a chapter on "Conjectures about attraction," so that the *Shapes of Celestial Bodies* ended with metaphysical reflection rather than astronomy. See below.

148. Originally published in 1686, the *Conversations* had gone through many editions by 1730, many of them with substantial revisions and additions incorporating new astronomical discoveries.

149. "Et pourquoi l'espece d'uniformité que nous voyons dans un petit nombre de planetes nous empecheroit-elle de soupçonner du moins la variété des autres que nous cache l'immensité des Cieux? Relégués dans un coin de l'Univers avec de foibles organes, pourquoi bornerions-nous les choses au peu que nous en appercevons?" (Maupertuis, *Oeuvres,* I:141).

150. Even in his academic writing, Maupertuis occasionally lapsed into this style, e.g. à propos of Prop. 70 of Newton's *Principia:* "Dans une Planete creuse, les animaux pourroient aller librement

ally, "But instead of these sinister catastrophes, an encounter with comets might add new marvels, and useful things to our earth."[151] In explaining the origin of Saturn's rings, based on the Newtonian theory of gravity, he continued the conversation with the latest astronomical developments, just as Fontenelle had done. Apparently, Maupertuis was making a play for Fontenelle's constituency, even as he updated the older man's libertine Cartesianism by challenging readers to discard what had become canonical assumptions about the transparent clarity of matter and motion. In this sense, the *Shapes of Celestial Bodies* asserted its claim to the forward-looking, "modern" status defended by none other than Fontenelle himself in the literary battle of ancients and moderns at the end of the seventeenth century.

The reviewer in the *Journal des sçavans* appreciated Maupertuis's hybrid style, remarking on the "manner both curious and solid" that characterized his explanations. The book "spread light and pleasure on the most abstract truths" and would appeal to even those readers who lacked knowledge of "sublime Geometry."[152] Louis-Bertrand Castel, on the other hand, in a more hostile review in the Jesuit *Mémoires de Trévoux*, found the argument unconvincing. "A little less calculation and a little more discussion would give a more distinct idea of all these spheroids." Maupertuis's facility for calculating hypothetical consequences led him beyond the limits of observations, Castel objected. "We have absolutely no observations about the shape of the fixed stars. But M. de Maupertuis conjectures nevertheless that they are completely flattened."[153]

Fontenelle himself reviewed the book in the Academy's *Histoire* for 1732 (published only in 1735). The bulk of this notice, which extends over some eight quarto pages, concentrated on the mathematical treatment of gravity with respect to the shape of the earth. "M. de Maupertuis sought the shape of the supposed spheroid in any possible hypothesis of the action of gravity . . . and he gives a general formula for it that came to him after a fairly difficult and fairly adroit calculation."[154] The review went on to mention Maupertuis's treatment of various astronomical phenomena, some observed and some imagined, but remained extremely

---

de tous côtés, sans recevoir aucune impression de la pesanteur." (Maupertuis, "Sur les loix de l'attraction," *MAS* 1732, 353).

151. Maupertuis, *Oeuvres,* 1:160.

152. Review of "*Discours sur les différentes figures des astres* . . . , par M. de Maupertuis . . . ," *Journal des sçavans,* April 1733, 206–07.

153. Castel, Review of *Histoire de l'Académie Royale des Sciences de Berlin,* vol. 2 (1746), *Mémoires de Trévoux,* April 1733, citations on 713, 705–06. On Castel's opposition to Newtonian natural philosophy and his defense of English mathematics, see Greenberg, *The Problem of the Earth's Shape,* 258–65.

154. Fontenelle, *HAS* 1732, 87.

circumspect about the controversial features of the book. Fontenelle did point out that, for the earth, Newton's flattened spheroid had already been contradicted by measurements made by the Academy's own astronomers in the course of charting the meridian in France. "It is clear that actual measurements ought to be preferred to the results of geometrical theories founded on a very small number of very simple assumptions, from which one has purposively separated off all the complexity of physics and reality."[155] The comparison of vortex physics to action at a distance was almost entirely absent from Fontenelle's summary. Only at the very end did he remark,

> Here attraction unveils itself, since all the tendencies of bodies toward a central point can always be brought back to mechanical ideas, or at least it seems possible that they could be: but as soon as one body acts by its mass on another distant body, we can no longer pretend that it is not attraction as such. And M. de Maupertuis does not dissemble on this point. . . . He draws a parallel between the sentiments of Descartes and of M. Newton, and all the advantage is with the English philosopher.[156]

Fontenelle's discreet treatment of the contentious aspects of the *Shapes of Celestial Bodies* betrayed his disapproval of its agenda, as well as his acknowledgment that gravity had found a way into academic discourse, if only by the back door.

## Newton in the Paris Academy

In the fall of 1732, with his book in press, Maupertuis continued to struggle with Newton. His letters to Bernoulli recount his frustrations, with the *Principia* and with several of his colleagues who persisted in attacking Newton in the Academy. "Our Abbé de Gamaches has undertaken to reverse all that [Newton] says on the loss of motion in resisting media. This matter seems to me of such great delicacy that although I have tried hard to study it in M. Newton, I understand nothing of it: I do not understand the force of his reasoning well enough to judge it." He was afraid that without Bernoulli's help he would never be able to get it, nor would his colleagues who "do not understand it any better than I."[157] Instead of pursuing this recalcitrant problem, he went back to some of Newton's theorems

155. "Il est evident que les mesures actuelles doivent etre préférées a ce qui resulte de Théories géometriques fondées sur un très-petit nombre de suppositions tres-simples, d'oû l'on a ecarté à dessein toute la complication du Phisique et du réel" (Fontenelle, *HAS* 1732, 92).

156. Ibid., 93.

157. M to JB, 24 September 1732, BEB. Gamaches was in Fontenelle's "modern" camp during the literary debate earlier in the century.

about the gravitational force of spherical masses and read a long commentary on three sections of the *Principia* to the Academy, stretching over seven meetings.[158] This was the first formal presentation to his colleagues of the fruits of his encounter with Newton, and it was also the first substantial mathematical work that he did not submit to Bernoulli for corrections. In fact, he only mentioned it to Bernoulli a month after he finished his reading, describing it as "quite a daring enterprise."[159] He chose, for his "explication, or commentary," several of Newton's propositions about the attractive forces of spheres and spherical surfaces. These propositions did not deal with the shape of rotating bodies, but simply with the attraction of masses of various shapes on point particles. The paper is an exercise in making the notoriously opaque geometrical demonstrations of the *Principia* accessible to an audience familiar with the conventions of Leibnizian notation—without openly proselytizing for Newton's theory of gravity. "I do not examine whether attraction repudiates or supports sound philosophy [*la saine philosophie*]. I treat it here only as a mathematician; that is to say, as a quality, whatever it is, whose effects are calculable, if we consider it to be distributed uniformly throughout all matter, and acting in proportion to its quantity."[160] The subsequent discussion explored some of the "remarkable" consequences of forces that decrease with the square of the distance between attracting masses. The paper's four "problems" derive general equations (for gravity proportional to any power of the distance) for the force of various solids and surfaces of revolution on an external point mass. Several of Newton's key propositions follow from the general equations, as for example the null force of a spherical shell on a mass inside the shell. The paper was carefully crafted to show the intelligibility of Newton's results, when reformulated by means of Leibnizian integral calculus, and in particular to show some consequences of the inverse-square law. Instead of Newton's demonstrative geometrical proofs, Maupertuis derived equations as solutions to general problems of central forces which could then be applied to particular cases. So instead of proving consequences of inverse-square forces, he showed how Newton's

158. He read the installments on 23 December; 21 and 31 January; 11, 18, 25 and 28 February. AS p-v 1733; text on fols. 40–51. The shortened version was published as Maupertuis, "Sur les loix de l'attraction," *MAS* 1732, 343–62. Newton's propositions are *Principia*, book I, sections 12 ("Attractive forces of spherical bodies"); 13 ("Attractive forces of non-spherical bodies"), and 14 ("The motion of minimally small bodies acted on by centripetal forces"). On this paper, see Greenberg, "Geodesy in Paris," 250.

159. "[J]'ay executé ces jours passés à l'Acad une entreprise assez hardi, c'était de lire l'analyse et un commentaire que j'ay fait sur les trois sect[ions] de m. Newton qui regardent lAttraction par rapport aux differentes figures des corps . . ." (M to JB, 31 March 1733, BEB).

160. Maupertuis, "Sur les loix de l'attraction," *MAS* 1732, 343.

propositions about such forces followed as special cases of equations for forces acting according to any power of the distance between attracting bodies.

The published version of the paper included a remarkable section exploring, in decidedly un-Newtonian fashion, the metaphysical ramifications of the inverse-square law.[161] Although it is missing from the manuscript copied into the minutes, Maupertuis later referred to it as part of the paper he had read to the Academy. When he reorganized and revised the *Shapes of Heavenly Bodies* for his collected works in 1756, he added the metaphysical reflections from the earlier *mémoire* under the heading "Conjectures on Attraction," indicating that he did not regret or disavow these speculations.[162] Leaving aside the question of the physical mechanism of gravity, this passage reflects on God's possible reasons for choosing a particular form for the mathematical law. "If God had wished to establish a law of attraction in nature, why would this law follow the proportion that it seems to follow? . . . In that infinity of different relations that seem to have an equal right to being employed in nature, was there some reason to prefer one over another?"[163] He approached these questions by looking for such grounds of discrimination in the geometrical properties of the law, starting with what he calls "uniformity," a notion that incorporates symmetry and homogeneity. Suppose that God had wanted symmetrical macroscopic bodies like spheres to exhibit the same property of attraction that characterized their smallest particles: "to attract in the same proportion on all sides." If this were the criterion, then all possible systems would not be equivalent. "Once the metaphysical reason for preference was posited, mathematical necessity excluded an infinite number of systems, in which there could not be agreement of the same law in the parts and in the whole." Of course, there are many hypothetical force laws that would fit this criterion. What was special about the inverse-square law? Here he brought in just the kind of result that he was about to explicate in the mathematical portion of his paper.

> It is true that [according to the inverse-square law] when a body is placed inside a solid sphere, gravity [*attraction*] no longer follows the same law, it works in that case in direct proportion to the distance to the center: but the attraction of spheres on bodies inside them ought not to be analogous to the attraction of the smallest parts

161. The manuscript, copied into the *procès-verbaux*, includes comments and problems based on propositions from sections 12, 13, and 14 of the *Principia*. The published version contains only the material on sections 12 and 13, along with a section on God's reasons for establishing an inverse-square law of gravity that is missing from the manuscript records.

162. Maupertuis, *Oeuvres*, 1:160–70. He mentions reading the text to the Academy on page 162.

163. Maupertuis, "Loix de l'attraction," *MAS* 1732, 346–47.

> of matter, of which the attraction could never affect any but bodies outside of them, since they are the ultimate particles of matter.[164]

So what looked like greater consistency for an attractive force operating simply as a multiple of distance (which would be uniform in all directions, and was the same both inside and outside the sphere) was actually a disadvantage when considered according to the parts-whole argument. In other words, the peculiar mathematical consequence of the inverse-square law that the force on the same small body differs depending on whether it is inside or outside a large mass, made it appropriate for revealing the rational consistency of nature. Presumably, God would have recognized this subtlety and would have made his choice accordingly.

This argument only makes sense (if it does) for a universe made up of discrete particles. The notion that physics reflects God's rationally optimized choice sounds Leibnizian, but the particular form of the argument denied Leibniz's (and Descartes's) vision of continuous matter filling space. The appearance of this argument in a paper ostensibly agnostic about the actual mathematical form of the law of gravity certainly lent weight to Newton's inverse-square law; if God had chosen it, rational men should do the same. Maupertuis concluded his metaphysical digression by apologizing for the "temerity" of attempting to unravel such mysteries. "But anything may be proposed, as long as one does not give it more weight than it has."[165] This willingness to speculate, even on abstruse metaphysical questions with theological implications, would resurface many times in his subsequent work. Bringing God into the pages of the *Mémoires* was a radical move at this point, especially in the context of speculating about divine choices.

Fontenelle, in his summary of this paper for the *Histoire*, did not so much as mention the metaphysical discussion. He did explain just what was new about Maupertuis's approach:

> As M. Newton gave only the particular propositions he needed on this subject, M. de Maupertuis wanted to ascend to a general theory that would be the source of everything. To simplify the thing, he considers only an infinitely small body . . . with respect to another body that consequently will attract it without being attracted by it; the large body is a spheroid described by the revolution of any curve around an axis, and the corpuscle is placed somewhere on this axis extended to whatever distance from the spheroid. The action of the attractive force will vary according to

164. Ibid., 347. This is Newton's Prop. 73. Maupertuis discussed this proposition (the force on a corpuscle inside a solid sphere), ibid., 354.

165. Ibid., 348

any power whatever of the distance, this is principally what comprises the generality [of the analysis].[166]

As far as Fontenelle could see, the question was still open as to which law was the correct one. The strange consequences of the Newtonian analysis of forces on bodies inside spheres led him to conclude that "the original [*primitive*] attraction is much changed, not only is the corpuscle attracted according to the simple distance, but even in direct proportion to this distance, which is to say, more strongly attracted by the sphere the farther it is from its center, which is completely contrary to the first idea we had established [of the inverse-square relation]."[167]

Where he had resisted getting embroiled in the *vis viva* controversy as Bernoulli's spokesman, Maupertuis now carefully positioned himself not exactly as Newton's spokesman, but as a French mathematician who could see beyond the prejudices of his compatriots to the rationality in Newton's work. To take this position meant overcoming the obfuscation of Newton's own style and mathematics, and giving it a form intelligible to French readers. This was what appealed to Voltaire about Maupertuis's project and what disturbed Mairan and Fontenelle. His promotion of Newtonian natural philosophy was as much about style as about theory or the ontology of gravity. The style in question was mathematical, as well as rhetorical. Fontenelle remarked that Maupertuis had undertaken his analysis "undoubtedly beckoned by the opportunity to use the most subtle Geometry."[168] The split in the Academy developed over the uses of this "subtle geometry" as much as over the occult forces themselves. Fontenelle thought the notion of mutual attraction actually obscured understanding, because so many factors (size, shape, direction, the forces of other bodies) had to enter into calculations, and he distrusted this complexity. Privat de Molières, arguably the most dyed-in-the-wool Cartesian in the Academy, and not much of a mathematician, articulated this position in a paper delivered just a month after Maupertuis had finished his long explication of Newton's propositions on gravity. "The mechanical forces of the vortex will furnish us more precisely with the astronomical laws as they actually are, than will the purely metaphysical forces of M. Newton, which give them with too much geometrical precision [*une trop grande précision géometrique*]."[169] Those academicians who adopted and refined the techniques pioneered by the Bernoullis were the ones who were ready to give up vortex mechanics in favor of action at a

166. Fontenelle, "Sur l'attraction Newtonienne," *HAS* 1732, 113.

167. Ibid., 116.

168. Ibid., 112.

169. Molières, "Les loix astronomiques des vîtesses des planetes dans leurs orbes, expliquées méchaniquement dans le Systême du Plein," *MAS* 1733, 311; presented to Academy 24 March 1733 (AS p-v).

distance. "Cartesian" became a shorthand label, used disparagingly by Maupertuis in private, for the mathematically inept and philosophically unadventurous among his colleagues, mostly of Fontenelle's generation. The issue at stake was not whether Cartesian rationalism should be replaced by British-style empiricism.[170] There was a longstanding focus in the Academy on collecting empirical evidence, as there was in England; the *Mémoires* are full of detailed records of peculiar phenomena, as well as astronomical events and measurements. The contention over the uses and types of mathematics, and over the meaning of gravity, cannot be cast as a battle between abstraction and observation, or theory and experiment.

## Mathematics outside the Academy

In the early 1730s, the practice of mathematics in the Academy was invigorated by a growing cohort of young men, most of whom became friends and allies of Maupertuis, among them Alexis-Claude Clairaut, Etienne Camus, and, slightly later, Alexis Fontaine.[171] Maupertuis's alliances with the new generation of mathematicians in the Academy extended into a wider social circle as well. He had been well connected socially and politically through his father, and he had made his way in café society as a young man; now he was being cultivated by Voltaire and the marquise du Châtelet, each of whom moved flamboyantly through aristocratic circles.[172] Voltaire described Newtonianism with outspoken enthusiasm as a "sect" and painted the conflict with Cartesianism in ironically religious terms that made Maupertuis into a crusader or a priest. "Your first letter baptized me in the Newtonian religion, your second gave me my confirmation. I thank you for your sacraments."[173] Voltaire did as much as anyone to make Newtonian philosophy fashionable and to associate it with a contrarian, slightly subversive intellectual stance. He relied on Maupertuis for explanations and corrections to the sections on gravity and optics in the *Philosophical Letters.* But while Maupertuis enjoyed the prerogatives of his flourishing reputation as man of letters and bon vivant, he did not seize on Voltaire's suggestion that he become "head of the sect" when Voltaire himself had to leave town in 1734 under threat of imprisonment. Because of his clash with the censors, Voltaire represented himself as the victim of perse-

170. Beeson claims that Maupertuis went through a transformation from "a priori rationalist" to Newtonian empiricist as he shifted from mathematics to geodesy (Beeson, *Maupertuis,* 87).

171. Camus joined in 1727, Clairaut in 1731, and Fontaine in 1733. The only new member who did not fall into line behind Maupertuis was Bouguer. See Greenberg, *The Problem of the Earth's Shape,* 107–14 for details of their rivalry.

172. Reliable evidence about social contacts in this period is scarce. It seems that Maupertuis was a regular at the salon of the Brancas family, which Voltaire and Châtelet also frequented.

173. Voltaire to M, [15 November 1732], Volt. *Corr.,* 86:251. On the meaning of Newtonianism for Voltaire, and its connection to the Enlightenment, see Shank, *Before Voltaire.*

cution, equating the philosophical, political, and religious forces of evil: "I am beginning to suspect that it is the partisans of vortices and innate ideas who are instigating the persecution against me. Cartesians, Malebranchists, Jansenists all are turning themselves loose on me."[174] Maupertuis, however, was not in exile, and he would hardly have sought to alienate his colleagues in the Academy any more than necessary. His defense of Newtonian gravity had been decidedly polite by comparison with the critical tone that shocked Voltaire's readers.[175] Maupertuis seems to have remained on cordial terms with Fontenelle, for example, in spite of their philosophical differences.[176] Neither was Maupertuis ostracized; he was thriving in the Academy, chosen by his peers for key tasks such as judging prize competitions and serving as deputy director of the Academy in 1735.[177]

Reliable sources on interpersonal dynamics inside the institution are lacking. La Beaumelle's biography describes "Newtonian dinners" given by Maupertuis for his protégés. "On meeting days, he gave a dinner for several young Newtonians, whom he brought to the Louvre full of gaiety, of presumption and of good arguments. He turned them loose on the old academicians [*la vielle academie*], who from then on could not open their mouths without being assailed by these *enfants perdus,* ardent defenders of attraction. One of them overwhelmed the Cartesians with epigrams, another with demonstrations. . . . This little troupe was animated with the liveliness, sometimes caustic, of their leader."[178] La Beaumelle's colorful retrospective account made the young Newtonians sound like an organized gang of provocateurs, but it seems likely that he exaggerated it in the retelling.

Voltaire used similarly polarized and hyperbolic language in his fawning admiration of Maupertuis, who interpreted Newton for him. "Who would have

174. Voltaire to M, 29 April 1734, Volt. *Corr.,* 86:468. Only Voltaire's side of this correspondence survives.

175. For one reader's scandalized reaction to Voltaire's *Lettres philosophiques,* see Le Blanc to Bouhier, 15 April 1734, Volt. *Corr.,* 86:460–62.

176. La Beaumelle, *Vie de Maupertuis,* mentions Maupertuis's correspondence with Fontenelle on philosophical topics at this time; the letters do not survive.

177. This was an elected post that carried honorific as well as remunerative value; the choice had to be approved by the minister. The deputy director normally served as director the following year, deferred for Maupertuis because of his journey to Lapland in 1736.

178."Les jours d'assemblée il donnait à diner à quelques jeunes newtoniens, qu'il menait au Louvre pleins de gaieté, de présomption et de bons arguments. Il les lâchait contre la vieille academie, qui désormais ne pouvait ouvrir la bouche sans être assaillie par ces enfants perdus, ardents défenseurs de l'attraction. L'un accablait d'épigrammes les cartésiens, l'autre de démonstrations. . . . Cette petite troupe était animée de l'enjouement quelquefois caustique de son chef" (La Beaumelle, *Vie de Maupertuis,* 33). The account was based on La Beaumelle's conversations with Maupertuis, and on the recollections of La Condamine, who was part of the group (La Condamine to JBII, 25 June 1769, BEB).

thought fifty years ago that the same power caused the motion of the planets and gravity? Who would have suspected the refrangibility and those other properties of light discovered by Newton? He is our Christopher Columbus. He has brought us to a new world, and I would very much like to travel there, following you. What a lot of questions (perhaps poorly posed) I would like to ask you!"[179] Voltaire might also have asked who would have thought, fifty years earlier, that a poet and playwright would have been so passionate about such things. Maupertuis's association with Voltaire no doubt enhanced his image as a gadfly, especially among the social elite. Along with other academicians, like La Condamine and Buffon, he came to be identified as an adventurous defender of Newtonianism.[180]

In the 1730s, mathematics became fashionable in Paris. Ladies of quality took up "geometry" as the latest fad, with Maupertuis at the center of the trend. Though hardly a mass phenomenon, the unexpected connection between women and mathematics did not go unnoticed by contemporaries. Voltaire, arriving back in the capital in 1735, complained, "Verses are hardly in style any longer in Paris. Everyone [*tout le monde*] is playing the geometer and the physicist. They undertake to reason. Sentiment, imagination, and the graces are banished."[181] Abbé Le Blanc, who had spent several years in England in the early 1730s, came back around the same time to observe the Parisian scene with evident distress. Something had happened while he was away:

> How many *femmes savantes* there are who, having taken a course of experimental philosophy with the Abbé Nollet, reason ceaselessly about the shape of the earth, or the rings of Saturn, and think themselves up on all the most hidden mysteries of nature. . . . Today the mania of women in France is to believe themselves fit for the abstract sciences. They have developed a taste for calculation, just as they once did for novels. Newton has replaced *Le grand Cyrus* on their dressing tables. A lady of high style cannot create a sensation [*faire de bruit*] unless she has a geometer in her retinue, and the geometer who does not circulate in *le monde*, looks a sorry sight among his fellows.[182]

179. "Qui auroit pensé il y a cinquante ans que le même pouvoir faisoit le mouvement des astres et la pesanteur? Qui auroit soupçonné la réfrangibilité et ces autres propriétés de la lumière découvertes par Newton? Il est notre Christophe Colomb. Il nous a menez dans un nouvau monde, et je voudrois bien y voiager à votre suitte. Que de questions (peutêtres mal fondées) je vous ferois!" (Voltaire to M., 3 November 1732, Volt., *Corr.,* 86:246).

180. Buffon had come to Paris in 1732, though he was only elected to the Academy in January 1734. The following year, he published a translations of Stephen Hales's *Vegetable Staticks,* with a Newtonian introduction. See Roger, *Buffon.*

181. Voltaire to Cideville, 16 April 1735, Volt., *Corr.,* 87:132.

182. Le Blanc, *Lettres d'un français,* 3:369–70.

Le Blanc made calculation, transplanted from the study to the boudoir, has the analogue of novel reading, with Newton taking the place of Mlle. de Scudéry, the fashionable seventeenth-century novelist. Le Blanc must have had Maupertuis in mind as the prototype of the geometer performing for well-born ladies, since the shape of the earth and the rings of Saturn were his trademark problems, familiar to the public from *The Shapes of Celestial Bodies.* Le Blanc's hostile but perceptive eye identified the reciprocal aggrandizement of the relation between savant and lady. The feminine "mania" promoted the worldly reputation of the mathematician, whose place in *le monde* put his less sociable fellows to shame.

After Voltaire, Maupertuis's next conquest was Emilie du Châtelet, who turned out to be quite serious about learning mathematics. Just re-entering society after the birth of her third child, the marquise was eager to integrate intellectual substance into her frenetic social life. Voltaire probably introduced her to Maupertuis, shortly after the publication of *The Shapes of Celestial Bodies.*[183] Le Blanc may well have based his lampoon of the *femme savante* on her, as her "lessons" with Maupertuis sparked considerable gossip. Indeed, their tutorials led to a brief romantic liaison.[184] By some chance, her letters to him survive, giving a rare glimpse into the many-layered relationship between them. A series of hurried notes from January 1734 combine demands for emotional and erotic attention with pleas for more help with algebra.[185] "I sent for you at the Academy and at your home, Monsieur, to tell you that I will spend the evening at home today. I have spent it with binomials and trinomials. I can study no more if you do not assign me a task, and I desire one intensely."[186] "I was very happy with your two manuscripts. I spent all yesterday evening in profiting from your lessons. I would very much like to be worthy of them. I fear losing the good opinion that people have given you of me, I feel that it would be paying dearly for the pleasure that I take in learning the truth embellished with all the graces that you have lent it."[187] "You do not wish to

183. For the network of aristocratic homes visited by Maupertuis in this period, see Badinter *Les passions intellectuels,* 57.

184. Brunet does not mention the sexual relationship. Brunet, *Maupertuis,* 1:28–30. See generally Besterman, ed., *Les lettres de la marquise du Châtelet.*

185. We have only Châtelet's letters (and occasional third-party comments) for evidence of her relations with Maupertuis; original autograph letters in BN, Fr. 12269. The notes from early 1734 are short, frequent, and undated. On Châtelet, see Terrall, "Emilie du Châtelet and the Gendering of Science"; Ehrman, *Mme. du Châtelet;* Badinter, *Emilie, Emilie.*

186. Châtelet to M, [Jan 1734], *Lettres,* 1:36.

187. Châtelet to M, [January 1734], *Lettres,* 1:30. The manuscripts would have been papers presented to the Academy about gravity, very likely including the reflections on God's reasons for choosing the inverse-square law. Voltaire read them as well; see Voltaire to M, January 1734, Volt. *Corr.,* 86: 443–44).

encourage your pupil, since I do not know if you have found my lesson well done. Tomorrow we are planning to go, Mme. de St-Pierre and myself, to the Jardin du Roi and we very much hope to see you there."[188] "I have led a disordered life these past days, I am dying, my soul needs to see you as much as my body needs rest. Come soon, alone or in company."[189] "I have studied a lot and I hope that you will be a bit less unhappy with me than the last time. If you would like to come and judge [my work] tomorrow I will be infinitely obliged to you. . . . I leave it until tomorrow to demonstrate to you all *my gratitude*."[190] Although her frustration is evident in these letters, she was also tenacious enough to keep working at her equations, perhaps to her tutor's surprise. "It is not for myself that I wish to become a geometer, it is out of vanity [*amourpropre*] for you. I know that someone who has you for a teacher is not permitted to make such mediocre progress, and I cannot tell you how ashamed I am about this."[191] Their tutorial relations continued by letter when she was away from Paris, as she became more serious about mathematics and physics. Once again, it seems that charm and flair, or "imagination" in Châtelet's words, made mathematics palatable. "I swear I understand nothing of M. Guisnée [an algebra text] and I think that it is only with you that I can learn *a* minus 4*a* with pleasure. You strew flowers on a path where others cause one to find only brambles; your imagination knows how to embellish the driest subjects without taking away their correctness and their precision."[192]

Maupertuis's appeal as a lover, as a tutor, and as a conversationalist were all tangled up together in his relations with the marquise. These talents served him well in the coming years, as he came to personify the man of science as man of the world. His development as a mathematician paralleled his social successes, and he cultivated his reputation as a defender of Newton in both settings. His public—his readers and admirers—saw his interpretations of Newton as a mark of his daring and his wit.

188. Châtelet to M, [January 1734], *Lettres,* 1:35.
189. Ibid., 1:31.
190. Ibid., 1:33 (Châtelet's emphasis).
191. Châtelet to M, 28 April [1734], ibid., 1:37.
192. Châtelet to M, 7 June 1734, ibid., 1:44.

# 4

# The Expedition to Lapland

SOON AFTER THE PUBLICATION of Maupertuis's *Discourse on the Shapes of the Heavenly Bodies* in 1732, geodesy became a major topic of discussion in the Academy.[1] This later metamorphosed into a new version of the debate about gravity, but initially contention centered on measurement and calculation techniques, rather than on theory. Interpretations of cosmology were overshadowed by arguments about how to perform and evaluate observations and calculations, about who should be entrusted to do this work, and about how they should present it to the public. To demonstrate its ability to enhance the glory of the crown while producing useful knowledge, the Academy sent two scientific expeditions out to the ends of the earth—one to the equator and one to the Arctic circle.[2] These expeditions generated intense public interest in what might seem an arcane scientific question: how much did the shape of the earth deviate from the spherical? The northern voyage, under the direction of Maupertuis, served as the keystone of his life, a defining moment for his persona as man of science. As such, it fed a story line, originating with Maupertuis himself and perpetuated in many subsequent accounts, about his achievement and its place in the history of science. He referred back to his Arctic expedition frequently, in many different ways for the rest of his life, and he quite explicitly made it an integral part of his identity. In taking on this dramatic and potentially dangerous task, he was gambling on a favorable outcome; if it had failed, he would have needed another framework on which to hang his reputation.

## How to Determine the Earth's Shape

Latitude is defined as position on a sphere along a north-south line known as a meridian, where the equator is at 0° and the pole is at 90°. True latitude would be

1. Greenberg, "Geodesy in Paris"; idem, "Degrees of Longitude"; idem, *Problem of the Earth's Shape;* Beeson, "Lettre d'un horloger anglois"; idem, *Maupertuis;* Lafuente and Peset, "La question de la figure de la terre."

2. On the expeditions, see J.-P. Martin, *Figure de la terre;* idem., "P.L.M. de Maupertuis: Maître d'oeuvre de l'expédition de Laponie (1736–1737)," in Hecht, ed., *Pierre Louis Moreau de Maupertuis;* Trystam, *Procès des étoiles;* Lafuente and Delgado, *La geometrización de la tierra.*

determined by the angle between a radius of the earth at the equator and another where the observer is located on the meridian. Since this number could only be observed directly from the center of the earth, in practice it is approximated by the elevations of fixed stars. To avoid the distorting effects of refraction near the horizon, astronomers take these measurements relative to the zenith, the point directly overhead. The same star observed from two different points on the meridian will yield a difference in angular position corresponding to the difference in latitude between the two locations. That arc corresponds to a linear distance on the ground; the length along the meridian corresponding to one degree of latitude is the "length of the degree" at that distance from the equator. On the surface of a perfect sphere, all degrees of latitude are equal, regardless of how far north or south of the equator. But if the sphere is flattened at the poles, the meridians are no longer perfect circles, and the length of a degree (measured by the stars) decreases from north to south, and conversely for an elongated spheroid. In the eighteenth century, there were two ways to measure such deviations from the spherical: by comparing the speed of pendulum clocks at different latitudes (to measure variations in the magnitude of gravity) and by measuring lengths on the ground to correlate to stellar latitude observations. The latter depended on triangulation, the same method used by land surveyors and mapmakers, to turn the angles of a chain of connected triangles into distances. If all three angles and one side of a triangle are known, the other two sides can be calculated—this is simple Euclidean geometry. Since the triangles share sides, the length of a single side can give all the other lengths through a series of calculations, and the whole distance from one endpoint of the chain to the other can be extrapolated. The virtue of the method is that angles are much easier to measure precisely than linear distances over varied terrain. The triangles must be laid out roughly along the meridian, with their vertices at prominent points clearly visible to observers, who move from point to point collecting angle measurements.

Jacques Cassini, sometimes called Cassini II, was the son of the Italian émigré astronomer Jean-Dominique Cassini, brought to France by Louis XIV to found the Paris Observatory in the 1670s. In 1718, Cassini II had announced that triangulation along the Paris meridian implied an elongated (prolate) rather than a flattened (oblate) earth—these were the results that Mairan had tried to accommodate in his 1720 paper.[3] Virtually uncontested for fifteen years, the model of the elongated earth attracted renewed attention in the early 1730s, when several

3. Cassini, *De la grandeur et figure de la terre.* For Maupertuis's critique of Mairan's paper on geodesy, see chap. 3 above. Measurements of the Paris meridian had been started by Jean Picard in 1669, and continued off and on over the decades by Jean-Dominique Cassini (Cassini I) and his associates at the Paris Observatory, including his son, Jacques (Cassini II).

mathematicians questioned the adequacy of local measures for settling the matter. As we have seen, Newton and Huygens had both derived a flattened earth from different theories of gravity. There was no necessary link between Cartesian theory or vortex mechanics and an elongated earth; however, because of the coincidence of Cassini's empirical data and his personal commitment to vortex physics, the prolate model became associated with the kind of Cartesianism espoused by Fontenelle and others in the Academy.

## Geodesy in the Paris Academy

Renewed attention to geodesy in 1733 was actually sparked by a conjunction of circumstances that had nothing to do with theories of gravity. For one thing, in 1730 the new controller-general of France, Philibert Orry, decided to revive the project to produce a state-of-the-art map of the kingdom. Colbert had initiated this project in the 1680s, and work had proceeded fitfully since then. To move the mapping forward, Orry sent Cassini out to measure the east-west perpendicular to the previously measured Paris meridian.[4] Cassini's mission was to determine the coordinates of towns and landmarks; geodesy was only a secondary concern. Since his earlier work had resulted in the unexpected prolate spheroid, however, and since he was the heir to the prestigious tradition of observational astronomy his father had established at the Observatory, he had a substantial stake in the question of the earth's shape.[5] From June through November of 1733, Cassini was in the field with a team of Observatory astronomers measuring a string of triangles westward from Paris to Saint-Malo.[6] As it happened, this new surveying effort coincided with the publication of a critique of the earlier geodetic conclusions. Some years before, the Italian mathematician Giovanni Poleni had written a pamphlet proposing a new way of determining the earth's shape from longitude, rather than latitude, measurements. (Longitude is measured along east-west lines, parallel to the equator; latitude is measured in a north-south direction.) This did not cause a stir in France when it was first published in 1724, but a second edition was reviewed at length in early 1733 in the Dutch *Journal historique de la république des lettres.* The anonymous reviewer not only described the virtues of geodetical calculations based on measuring along a parallel, but also questioned Cassini's method of

4. On the mapping project, see Gallois, "L'Académie des sciences et les origines de la carte de Cassini"; Konvitz, *Cartography.*

5. Ironically, Cassini I (Jean-Dominique), renowned for his telescopic discoveries, had observed the oblate shape of Jupiter, a bit of evidence cited by Newton in his discussion of the flattening caused by centrifugal force in rotating planets. (*Principia,* book 3, prop. 19).

6. Cassini II outlined the project in May: "Dessein de la parallele de Paris," AS p-v 20 May 1733.

determining the prolate shape.[7] The review openly challenged Cassini's calculation, rather than simply proposing a new approach to geodesy, as Poleni himself had done.

As Cassini was preparing to go off on his surveying expedition, Maupertuis was reading Poleni's book, in all likelihood prompted by the *Journal historique* review. In June, a few days after the astronomers left town, he gave a paper in which he explored the proposal to measure degrees of longitude as a means to geodetical calculations.[8] Poleni's basic point was that the difference in length of degrees of latitude along the Paris meridian was so small as to be within the margin of error of the measuring instruments. He argued that the difference between oblate and elongated spheroids would be more readily detectable by comparing the length of a degree of *longitude* and comparing this to the expected length for a perfect sphere.[9] Maupertuis took this claim seriously and showed how to calculate the ratio of the earth's axes from longitude data. With the new east-west measuring effort underway, suggestions about using longitude to determine the earth's shape seemed particularly relevant, and a number of academicians who were not themselves involved in the mapping project became excited about the possibilities. All participants recognized that the difficulty in practice would be the accurate determination of the longitude of the endpoints of the arc being measured, but they assumed that this could be overcome.[10]

Although he did not challenge Cassini directly, judiciously waiting until his senior colleague left Paris, Maupertuis's shift in this paper from a priori reasoning about gravity to what he called "the facts of the matter" (*la question de fait*) marked a tentative move into the astronomer's territory. His approach remained mathematical, since he derived equations for relating the axes of the spheroid, degrees along

7. *Journal historique de la république des lettres* (January–February 1733): 105–18; Greenberg thinks the reviewer was probably Elie de Joncourt, the editor of the journal (*Problem of the Earth's Shape,* 79). For a detailed discussion of Poleni and the review, see Greenberg, "Geodesy in Paris"; idem, "Degrees of Longitude." The *Journal historique* was published in Leiden in 1732 and 1733. See Sgard, *Dictionnaire des journaux.*

8. Maupertuis, "Sur la figure de la terre, et sur les moyens que l'astronomie et la géographie fournissent pour la déterminer," AS p-v, read on 3, 6, and 10 June, 1733; published under same title, *MAS* 1733, 153–64 (quotation from Poleni's Latin text on 153). Poleni had won the Academy's prize in 1733 for an essay about determining the speed of ships at sea.

9. Poleni's reviewer had also pointed out that Cassini had extrapolated the decrease in length of degrees from south to north by averaging over the measurements, because the difference from one degree to the next was too small to be observed directly. Greenberg, "Geodesy in Paris," 247; idem, "Degrees of Longitude," 155–56.

10. The standard method for determining differences in longitude in the eighteenth century used observations of the motions of the satellites of Jupiter.

a parallel (longitude), and degrees along a meridian (latitude), but he was proposing "different astronomical and geographical means" for deciding the question. By showing that the shape of an ellipsoid could be extrapolated equally well from two arcs along a meridian or two arcs along a parallel, he detached the geodetical question from any particular theory of gravity.[11] He was also introducing his colleagues to Poleni's proposal, with its implicit rejection of Cassini's earlier results. Immediately after the presentation, several of his colleagues took these suggestions in a more practical direction. Louis Godin, one of the few astronomers not involved in Cassini's expedition, presented a method for tracing a parallel on the ground, and La Condamine designed an instrument for making such measures.[12] After the mapping expedition returned to Paris for the winter, there was another flurry of activity by astronomers and mathematicians. Cassini, who had not yet seen the review of Poleni's book, maintained that his numbers were unequivocal and his methods above reproach. He reported that the length of a degree of longitude on the perpendicular to the Paris meridian was shorter that would be expected for a sphere, confirming his earlier conclusions about the earth's elongation.[13] Cassini's son, Cassini de Thury, not yet a member of the Academy but already working as part of the Observatory team, presented his first paper on the shape of the earth.[14] Eustachio Manfredi, a highly-regarded Italian astronomer and a corresponding member of the Academy, sent a paper on a method for using lunar parallax for geodetical calculations. Clairaut also got in on the act, with a sophisticated analysis of methods for calculating the earth's shape from the perpendicular to the meridian. Although Clairaut kept his analysis abstract and did not say Cassini was wrong, his paper implied that Cassini's calculations were too simple, because they were based

11. Maupertuis, "Sur la figure de la terre, et sur les moyens que l'astronomie et la géographie fournissent pour la déterminer," *MAS* 1733, 156. Greenberg shows that Maupertuis followed the reasoning of Poleni's anonymous reviewer quite closely. Greenberg, "Geodesy in Paris," 250–52. Beeson insists on Maupertuis's "evolution from a theoretician to an experimentalist" (*Maupertuis*, 104), but overlooks the complex interaction of theory and experiment, and the use of methods of calculation foreign to Cassini's calculational practice. Maupertuis was developing an approach to geodesy that blended theory, mathematics (analysis rather than traditional spherical geometry), and observation.

12. Godin, "Méthode pratique de tracer sur terre un parallele par un degré de latitude donné," *MAS* 1733, 223–232; La Condamine, "Description d'un instrument qui peut servir à déterminer . . . tous les points d'un Cercle parallele à l'Equateur," *MAS* 1733, 294–301.

13. Jacques Cassini, "De la carte de France, et de la perpendiculaire à la méridienne de Paris," *MAS* 1733, 389–405; discussion of elongated earth on p. 402. This paper was presented to the Academy of Sciences public assembly on 14 November 1733.

14. No text survives; the procès-verbaux, however, record that "M. Cassini de Thury a presenté une mémoire sur la figure de la terre" (AS p-v, 12 December 1733).

on spherical geometry, whereas the geometry of the surface of a spheroid is more complicated.[15]

Cassini insisted that his longitude measurements had confirmed the same elongated spheroid he had found earlier, but others continued to argue that definitive comparisons would require going farther afield. A greater difference in latitude would mean a greater disparity in the length of a degree, widening the allowable margin of error.[16] In December 1733, Godin argued for making measurements at the equator and volunteered to make the journey himself.[17] By January 1734, Maupertuis reported to Bernoulli, "A propos of M. Godin, they are talking in the Academy about sending him with a few others to Peru to measure several degrees of the equator. . . . For myself, I do not believe that anything certain can be concluded about the shape of the earth from all [Cassini's] measurements."[18] Although it was not formally approved until early the next year, planning for the equatorial expedition kept geodesy a top priority in the Academy. While Cassini returned to the field to continue his measurements in the easterly direction, Maupertuis went back to explicating and expanding on Newton. In the summer—again after Cassini had left Paris for the measuring season—he returned to geodesy, especially as treated in the *Principia.* This paper continued his longstanding project to translate Newton's geometrical presentation into the Continental infinitesimal calculus. He worked out the ratio of the earth's axis to its diameter assuming various hypothetical theories of gravity, including that of Newton. "I will try to clarify what M. Newton said about this, which is neither one of the least splendid parts of his book, nor one of the easiest to understand."[19] A rash of papers by Pierre Bouguer, Clairaut, and Maupertuis over the next few years explored alternative methods of observation and calculation, using a kind of mathematical analysis foreign to the astronomers. Cassini's book on the size and shape of the earth contained some geometrical diagrams and demonstrations, and many tables of observations, but no equations.

15. Clairaut, "Détermination géometrique de la perpendiculaire à la méridienne tracée par M. Cassini; avec plusieurs méthodes d'en tirer la grandeur et la figure de la terre, " *MAS* 1733, 406–16. Clairaut presented this paper on 5 December 1733; Cassini was present (AS p-v, 5 December 1733).

16. Desaguliers had made this argument in his 1725 critique of the precision of Cassini's measures. Desaguliers, "Dissertation Concerning the Figure of the Earth."

17. Louis Godin, "Sur l'utilité d'un voyage sous l'équateur," AS p-v 23, December, 1733; no text for this paper survives. This is the first mention of a possible expedition in the Academy archives.

18. M to JB, 2 January 1734. BEB.

19. Maupertuis, "Sur la figure des astres," AS p-v, 28 Aug. 1734; published as "Sur les figures des corps celestes," *MAS* 1734, 55–100. Greenberg calls this paper "the first exposition in the French language of the theory of the earth's shape appearing in the *Principia,* third edition" (*Problem of the Earth's Shape,* 119).

Maupertuis and Clairaut, on the other hand, directed much of their work toward showcasing mathematical methods, often by considering various observational programs. When Cassini finally saw the review of Poleni in 1734, he responded with an outraged defense of his methods to the Academy. By this time, his audience was well aware of the issues at stake.[20]

This overview of the many different kinds of work on the theory and practice of geodesy shows how the interchange of results and ideas focused a substantial group of academicians on interlocking mathematical, theoretical, and empirical problems. One paper or solution sparked another, whether as a challenge or a clarification. This kind of scientific sociability was not always collaborative or even friendly, but it gave the subject a momentum that continued to grow with the departure of the first expedition.

## Long-Distance Expeditions

The South American expedition left in May 1735 for the Spanish colony of Peru, with the approval of the king of Spain. Led by Godin, the other key men on this expedition were Maupertuis's friend La Condamine, always ready for an adventure, and Bouguer.[21] Fontenelle announced their impending departure at the public meeting in April, playing up the dangers of working in a "wild" and inaccessible land. "How many hardships, and fearful hardships, accompany such an enterprise? How many unforeseen perils? And what glory must not redound to the new Argonauts?"[22] In Basel, Bernoulli was less excited at the prospect. He doubted whether this adventure to the equator would produce reliable data. "Tell me, do the observers have a predilection for one or the other of the two sentiments? Because if they favor the flattened earth, they will find it flattened; if on the contrary, they are imbued with the idea of the elongated earth, their observations will not fail to confirm its elongation: the difference between the compressed spheroid and the elongated is so slight, that it is easy to be mistaken if one wants to be mistaken

20. J. Cassini, "Réponse aux remarques qui ont été faites dans le *Journal historique de la République des Lettres* sur le traité *De la grandeur et de la figure de la terre,*" read to the Academy in December 1734 and published in *MAS* 1732, 497–513. Normally, the paper would have been held until the volume for 1734 went to press, several years later. For the substance of Cassini's response, see chap. 5 below.

21. The primary accounts for the South American expedition are Bouguer, *La figure de la terre;* La Condamine, *Mesure des trois premiers degrés du méridien;* idem, *Journal du voyage fait par ordre du roi à l'équateur.* Louis XV stressed the utility of the academicians' efforts "not only for the progress of the sciences, but also for commerce, in making navigation more exact and easier." Copy of Louis XV's order concerning the Peru expedition, 13 February 1735, in BN, Fr. 9674.

22. Fontenelle, "Discours prononcé à l'Académie des sciences . . . sur le voyage de quelques académiciens au Pérou," in *Oeuvres complètes,* 1:29–30.

in favor of one or the other opinion."[23] This problem was no less severe for the Observatory astronomers than for the long-distance travelers, once Cassini had invested so heavily in the elongated earth. But Bernoulli had put his finger on a problem that would plague geodesists for years to come.

After the departure of the South American expedition in the spring of 1735, the mathematical and observational aspects of the problem continued to occupy the Academy's attention. Maupertuis brought up a possible expedition to the north as soon as his colleagues had sailed. In May, he read a paper deriving an analytic expression for the shape of the earth and summarizing the recent history of geodesy; a week later, he reiterated the importance of traveling as far as possible from France to get measurements for comparison. Although no one in Paris was ready to openly question the accuracy of the Cassinis' numbers, he noted dryly that "observations seem to give the earth the shape of an elongated spheroid, and solid reasoning seems to give it the opposite shape. . . . We owe to France the measure of the degrees of [latitude] over an extent of 8° 31′. It seems that the resolution of the great question on the elongation or the flattening is reserved for her as well."[24] Professing ignorance about both the cause and the amount of the earth's deformation, he used the geometry of ellipses and some fairly simple algebra to derive an equation relating the axial and equatorial diameters and the lengths of two degrees of latitude along a meridian (fig. 2).[25] This equation, he claimed, was "much easier and more practical" than any presented earlier, including his own of the previous year, because it avoided infinite series. "Since [the equation] contains the first degree of latitude, another degree at any latitude (with its sine), and the ratio of the diameter of the equator to the axis, it is easy to conclude from it the point to which these voyages are useful or necessary, and the distance to which they should go, once one has agreed on the greatest error that could be committed by

23. "Mais dites moi, Monsieur; les Observateurs, ont ils quelq[ue] predilection pour l'un et l'autre des deux sentiments? Car s'ils sont portés pour la terre applatie, ils la trouveront seurement applatie; si au contraire, ils sont imbus de l'idée pour la terre allongée, leurs observations ne manqueront pas de confirmer son allongement: le pas du spheroide comprimé pour devenir allongé est si insensible, qu'il est aisé de s'y tromper, si on veut etre trompé en faveur de l'une ou l'autre opinion" (JB to M, 8 May 1735, BEB). Bernoulli had built Cassini's elongated earth into his essay on the inclination of planetary orbits, so he had some vested interest in preserving the result.

24. Maupertuis, "Sur la figure de la terre," 25 and 27 May, 1735, AS p-v 1735, fol. 125. The text was revised and shortened for publication, omitting the history of the debate. On 8 June 1735, the minutes record a "second paper" by Maupertuis, "Sur la figure de la terre et de quelle utilité peuvent être les voyages pour la déterminer."

25. Maupertuis, "Sur la figure de la terre," *MAS* 1735, 98–105. For detailed summary and commentary, see Greenberg, *Problem of the Earth's Shape*, 80–83.

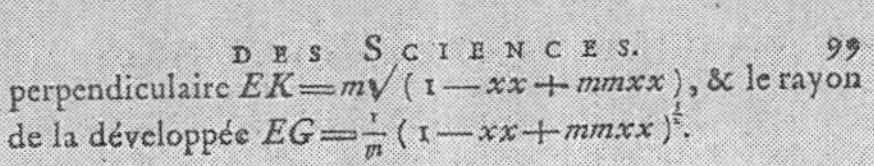

DES SCIENCES. 99

perpendiculaire $EK = m\sqrt{(1 - xx + mmxx)}$, & le rayon de la développée $EG = \frac{1}{m}(1 - xx + mmxx)^{\frac{3}{2}}$.

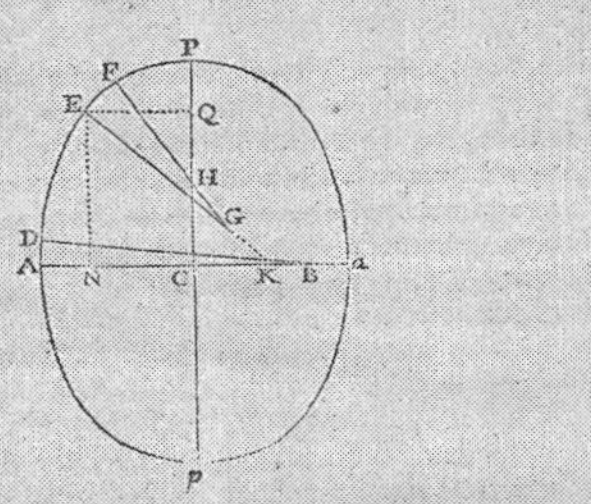


Soit maintenant le ſinus de l'angle $AKE$ qui eſt l'angle de la latitude $= s$ pour le rayon $= 1$, on a $1 . s :: m\sqrt{(1 - xx + mmxx)} . m\sqrt{(1 - xx)}$, ou $xx = \frac{1 - ss}{1 - ss + mmss}$; & mettant cette valeur dans l'expreſſion du rayon de la développée, on a $EG = \frac{mm}{(1 - ss + mmss)^{\frac{3}{2}}}$.
Comparant maintenant deux arcs du méridien, l'un $AD$ qui commence à l'équateur avec un autre $EF$ quelconque, mais dont les amplitudes ſoient les mêmes, & ne ſoient pas grandes comme deux arcs chacun d'un degré, on a $AD . EF :: AB . EG$, ou (nommant $A$ le 1$^{er}$ degré de latitude, & $E$ le degré en $E$) on a $A . E :: mm . \frac{mm}{(1 - ss + mmss)^{\frac{3}{2}}}$, ou $A . E :: (1 - ss + mmss)^{\frac{3}{2}}\ 1$, ou $E = A \times (1 - ss + mmss)^{-\frac{3}{2}}$.
Cette équation exprimant la relation entre le premier degré de latitude, un degré quelconque, ſa latitude, le diametre de

N ij

Fig. 2. Diagram from derivation of equation relating degrees of latitude shape of the earth. Note the elongated shape of the generic diagram, emphasizing the generality (or at least the absence of prejudice) of the mathematical result. Maupertuis, "Sur la figure de la terre," *MAS* 1735, 99.

skilled observers."[26] One consequence of this equation was a result stated by Newton, but not demonstrated or formulated analytically in the *Principia,* concerning the increase in the length of degrees from equator to pole on a flattened spheroid.[27] Maupertuis showed that his equation holds for either a flattened or an elongated spheroid, as long as the deviation from the spherical is very small. Furthermore, although he discussed the flattened earth as one of several possibilities, he explicitly disregarded changes in weight at different latitudes because this issue could not be separated from a theory of gravity. His result, he claimed, was "independent of any system." It could then be used to illuminate observational problems without the taint of theoretical prejudice.[28]

This position provided safe ground from which to discuss observational error, a discussion relevant to evaluating Cassini's measurements, as well as to planning

26. Maupertuis, "Sur la figure de la terre," *MAS* 1735, 101.

27. Newton, *Principia,* book 3, prop. 20, problem 4. Maupertuis quotes Newton ("Sur la figure de la terre," *MAS* 1735, 100).

28. Maupertuis, "Sur la figure de la terre," *MAS* 1735, 100–11.

new ones. The treatment of the earth's shape in 1735 is thus markedly different in emphasis from Maupertuis's work of a few years before, when he had shown how all revolving spheres subject to central forces would become flattened, regardless of the particular theory of gravity. That argument had been primarily an exercise in hydrostatics, working on the hypothesis of a homogeneous earth with a fluid core. When he first ventured into this subject, he had been more interested in penetrating Newton's abstruse geometrical arguments than in actual measurements, and he had certainly had no intention of producing his own observations to compete with those of Cassini. Once the expeditions became a reality, however, the travelers had to downplay their theoretical predilections while simultaneously foregrounding the technical rationale for their measurements.

Maupertuis also found a mathematical justification for sending expeditions to points differing as much as possible in latitude. By going over (not for the first time) the several contending ratios defining the earth's ellipticity—from Newton, Huygens, and Cassini—he underscored the uncertainty of existing knowledge. All of these shapes deviated enough from the spherical for differences in length to be detectable, within the range of error permissible to "skilled observers." Furthermore, if the earth were really as prolate as Cassini claimed, the effect should be detectable without going as far as the equator. But then he destabilized the ground further by suggesting that the earth might differ even *less* from the spherical than either Newton or Cassini had claimed. "What would make the expedition to the north absolutely indispensable would be if the earth (as there is reason to believe) is less elongated or less flattened than Messieurs Cassini and Newton suppose it to be. This elongation or flattening could be such that one of the voyages alone would not suffice to decide the question."[29] Without supporting this claim numerically, his demonstration of Newton's result drew attention to the smallness of the earth's bulge, necessarily difficult to detect. In essence, Maupertuis was elaborating on the theme that the shape of the earth was a "question of fact" that could be settled only by direct measurements with unequivocal instruments and techniques, the sooner the better. Still, it would be a mistake to read this as a simple faith in the transparency of empirical data. The claim lying just beneath the surface is that he and his colleagues, with the benefit of their mathematical insight, but with no experience in the field, were perfectly capable of making the delicate observations necessary for calculating the earth's shape to an unprecedented level of accuracy. They planned to do so with the aid of new instruments—some of them imported from England—and new equations. They

29. AS p-v 25 May, 1735, fol. 124. The explicit justification for going to distant latitudes was omitted from the published version of this memoir, which was not printed until 1738, after the expedition had returned.

were bringing their expertise at calculation and "sound reasoning" to bear on an observational problem that had previously been the domain of astronomers who saw no use for the analytic techniques of the calculus. Thus, Maupertuis held out the prospect of a definitive reconciliation of observation, analysis and theory.

The plea for gathering new data barely disguised an implicit denigration of the Observatory measurements. He ended one of his presentations with another consequence of his new equation, computing the point on the earth's surface where contiguous degrees along a meridian should differ maximally, and hence where the difference should be easiest to measure—this point occurs at 45 degrees, not so far from the latitude of France.[30] This barbed comment was made especially pointed by the application of the differential calculus, the very mathematics that Cassini never mastered. Given the discrepancies among the numbers Cassini published in 1713 and 1720, neither set of measurements could be considered definitive, in spite of the nearly optimal location for measuring the difference of contiguous degrees.

Unlike his previous contributions on geodesy, he delivered the papers to the Academy in May and June 1735 with Cassini in the room. Although the published version contains only carefully disguised criticism of the astronomers, the oral presentations were apparently more confrontational, pointing to shortcomings of the recent geodetical expeditions, with such comments as: "Therefore, if one had performed towards Bourges the same operations that had been done in Paris, to compare the differences in latitude with the distances measured on the meridian, the operations would have been more advantageous."[31] Anyone who had been following the dispute knew that the substantial investment in the equatorial expedition tacitly acknowledged the inadequacy of existing measurements to decide the issue. Even Fontenelle, finessing the question of the accuracy of the Observatory's work, noted, "It is certain that a comparison of the two voyages to the equator and to the north would produce more than either could produce alone, and that in taking the earth's measurements thus in all directions, we would finally force its true shape to reveal itself."[32] Fontenelle recognized that whatever the resolution of the question, the expeditions gave the Academy enhanced credibility and visibility, in the eyes of the crown and of the public. Whether the earth turned out to be flattened or elongated, the efforts to settle the question would be a dramatic contribution to the glory of king and nation.

The timing of Maupertuis's first public mention of the possibility of sending observers north as well as south—just after the equatorial team had left Paris—suggests that he saw an opportunity to capitalize on the crown's interest in the

30. *MAS* 1735, 104.
31. Maupertuis, "Sur la figure de la terre", AS p-v 8, June 1735, fol. 131 v.
32. Fontenelle, "La figure de la terre," *HAS* 1735, 51.

matter, as well as on heightened public attention to the dispute. He may also have been playing on concerns about how long it would take for results to come back from Peru, for he argued that the "polar" expedition would be shorter and potentially more efficient. Even though initially it would lag a year behind, "we will perhaps know as soon by this voyage as by that of Peru what the shape of the earth is; since if it is as elongated as M. Cassini thinks, the difference between a degree of latitude in [the Arctic] and a degree near Paris should be observable."[33] Maupertuis also knew that his counterparts in South America were planning to take measurements not only along the meridian, but parallel to the equator as well, following up on Poleni's suggestion. The longitude determinations promised to be difficult and were likely to extend their absence, especially since the travelers disagreed about how to proceed. However, a second costly expedition was by no means a forgone conclusion.

Once the question of the earth's shape became a primary issue in its own right, rather than a by-product of cartography, Maupertuis and his colleagues changed the rules of the geodetical game. When they argued for a northbound expedition to mirror the original proposal for measurements at the equator, the Paris mathematicians hoped to make Cassini's local measures largely irrelevant. They assumed that if both groups returned to France armed with new numbers from distant latitudes, the difference between the equatorial and Arctic results would be enough to decide the question with only passing reference to the Observatory work. The Lapland expedition was sold as a kind of insurance against the possibility that the effect was too slight to be detected by one set of measurements alone.

## Expedition to the Arctic

Technical, political, and personal factors came together to make the Lapland expedition possible.[34] Maupertuis had approached the problem of the earth's shape from many angles. He recognized the difficulties of Newton's *Principia* for a French audience, even a learned one, and he had already made several attempts to adapt parts of Newtonian physics for French mathematical and philosophical tastes, with substantial help from Johann Bernoulli. Although he has often been portrayed as struggling against the prejudices of a "Cartesian" academy, Maupertuis had actually earned the approbation of his fellow academicians even before mounting his expedition. The Arctic expedition gave him the chance to expand and strengthen his reputation, inside and outside the Academy. From the time

33. M to JB, 12 September 1735, BEB.

34. On the Lapland expedition, see Maupertuis, *La figure de la terre;* Outhier, *Journal d'un voyage au nord;* Nordenmark, *Anders Celsius;* Nordmann, "L'Expédition de Maupertuis et Celsius"; Martin, *La figure de la terre;* Lacombe and Costabel, *La figure de la terre.*

the king approved the venture, Maupertuis had a new kind of public visibility, exploited in public sessions of the Academy and reported in the press.[35]

The records of the Academy are silent about exactly when or how the decision to seek government approval for a second expedition was taken, or how the scientific team was put together. Talk of a new expedition was circulating publicly as early as June 1735, following Maupertuis's speech to the public session, and well before it had been officially sanctioned.[36] Although he had no experience at astronomical observation, Maupertuis put himself forward as the leader of such an expedition, and Clairaut agreed to go with him. There were in fact no senior academicians with astronomical training available to undertake this task: Cassini and Maraldi were busy with geographical and astronomical work in France; Godin was on his way to South America. Maupertuis and Clairaut had probably been talking about such an expedition for some time. In the fall of 1734, they had spent two months together in the Bernoulli household in Basel, and on their return they had set up house at an out-of-town retreat in Mont-Valérien, a peaceful spot just west of Paris. They were both extraordinarily productive in the next eighteen months; their retreat was a place to work, but also to entertain a select group of friends, including La Condamine, the marquise du Châtelet, the Italian poet Francesco Algarotti, and the Swedish astronomer Anders Celsius. Celsius had come to Paris on a grand tour of European observatories; he had met Algarotti, recently enamored of Newtonian natural philosophy, in Italy. When they arrived in Paris together at the end of the summer of 1734, Algarotti introduced Celsius to the group of young men around Maupertuis. Both of the foreigners were intrigued by the idea of traveling to the Arctic as part of the proposed expedition, although Algarotti had little to offer other than conviviality, and in the end changed his mind.[37] Celsius, however, turned out to be an invaluable asset for Maupertuis and provided the solid astronomical experience needed for the credibility of the team. The academic debate about geodesy was the formal context for the expedition; informally, this circle of like-minded friends, where amusement and science came together in sociable and witty terms, pushed each other to take more dramatic steps than they might have taken otherwise.

35. Maupertuis spoke at the public meetings in April 1736, just before embarking; in November 1737, just after returning to Paris; and in April 1738, just as his book about the expedition came off the press.

36. "M. de Maupertuis va au pôle mesurer la Terre" (Châtelet to Richelieu, [15 June 1735], in *Lettres* I:75; see also Caumont to Bouhier, 11 July 1735, in H. Duranton, ed., *Correspondance littéraire du président Bouhier*, fasc. 6, p. 94.

37. In October 1735, Algarotti was still planning to join the travelers (Châtelet to M, 3 October [1735], *Lettres* I:84).

Along with the technical issues and the potential for enhancing his scientific reputation, Maupertuis had personal reasons for wanting to make a dramatic move. His correspondence hints at restlessness and unhappiness, possibly linked to a love affair, but he did not reveal any details. "This voyage, which would hardly suit me if I was content and happy, seemed to me the best thing I could do in my present situation and I do it with all the favor that a powerful minister can procure for private citizens."[38] The minister's favor, a mark of prestige, almost offset whatever personal trouble he suffered. Châtelet too noticed that "he has a restlessness [*inquiétude*] of spirit which makes him very unhappy and which proves that it is more necessary to occupy his heart than his mind; but unfortunately it is easier to do algebra than to be in love."[39] However we evaluate these expressions of *ennui,* the lure of adventure certainly played its part as well, and he threw himself into the preparations for the expedition. "I have a real passion for traveling," he wrote to Bernoulli, announcing the impending journey. His old mentor was somewhat taken aback, not least because he suspected that the travelers had already made up their minds about what they would find. "You and I are made rather for the study," he replied, and went on to complain that Clairaut's myopia hardly suited him to astronomical observations. But Maupertuis was ready to get out of his study, and he wanted to have Clairaut's prodigious mathematical skills near at hand. They both knew that successful completion of their mission would require as much calculating as observing. In fact, by including a mathematician who lacked the visual acuity normally considered the *sine qua non* of the accomplished astronomer, the expedition planners brashly flaunted their differences with traditional astronomical practice.

Once he had laid the groundwork in the academy, Maupertuis used his personal connections to lobby for the necessary support from the crown for an expedition to the pole (as it became known). He called on the comte de Maurepas, minister of the navy and of the Maison du Roi, as well as honorary member of the Academy of Sciences, and well known to Maupertuis. In his history of mathematics, Montucla gives a fanciful account of how Maupertuis convinced the minister, who was at home recuperating from an illness: "Maupertuis was pleasant, he sang songs, he played the guitar, and that helped him to obtain the commission he asked for."[40] Whatever the appeal of his musical diversions, Maupertuis undoubtedly

38. M to JBII, 30 Dec.1735, BEB.

39. Châtelet to duc de Richelieu, [15 June 1735], *Lettres,* 1:75.

40. "Maupertuis étoit agréable, il faisoit des chansons, il jouoit de la guitare, et cela lui aida à obtenir la commission qu'il demandoit." Montucla, *Histoire des mathématiques* 4:149. Lalande (who knew Maupertuis in the 1750s) completed this volume after Montucla's death; he may have added this detail, which appears in a note.

also raised one of Maurepas's favorite subjects—the improvement of navigation techniques—and persuaded him to go to the finance minister, Cardinal Fleury, for approval of the necessary expenditure. The expedition was also discussed at Versailles on the occasion of the annual presentation of the volume of the Academy's *Mémoires* to the king.[41] In September, five months after the departure of the equatorial expedition, Maurepas transmitted the official royal orders, authorizing a second expedition and he went to work finding and provisioning a ship to take the travelers to Stockholm.[42]

As an accomplished observer on his way to England to collaborate with the best Newtonian astronomers, Celsius was an ideal ally for Maupertuis at just this moment. Celsius agreed to acquire several instruments made by George Graham, the most accomplished maker of astronomical instruments in London.[43] "Please have the sector made as soon as possible, and profit from the good will of Mr. Graham. We will also need one of those astronomical clocks [*pendules*], and the machine for measuring the pendulum. We have other pendula by Mr. Le Roy and the rest of our instruments will be made by Langlois."[44] Celsius also acquired a transit telescope, "a telescope perpendicular to, and moveable about an horizontal axis."[45] Graham had not only designed and made the portable zenith sector with which James Bradley had discovered the aberration of starlight, he had also done meticulous experiments on the effect of temperature on the rate of pendulum clocks.[46] Since pendulum measurements would be a crucial part of the Lapland program, as they were for the equatorial expedition, Graham's expertise could not have been more apt. Although Maupertuis had met Graham on his trip to London, Celsius's presence in London facilitated the expeditious construction of the custom-built sector. This instrument, more than any other, materially distinguished the Lapland measurements from those of the Cassinis; priced at

41. For an account of the ceremonial occasion, see M to JB, 12 September 1735, BEB.

42. On outfitting the expedition, see Nordmann, "L'expédition de Maupertuis et Celsius en Laponie."

43. Nordenmark, *Anders Celsius,* 240–43; Nordmann, "L'expédition de Maupertuis et Celsius," 80–81.

44. M to Celsius, 22 November 1735, RS archives (extract in Nordenmark, *Celsius,* 58). For a list of the instruments taken to Lapland and their makers, see Wolf, *Histoire de l'Observatoire de Paris,* 187–88. Celsius had the English instruments transported directly to Sweden (Nordmann, "L'expédition de Maupertuis," 82). On Graham, see Sorrenson, "Scientific Instrument Makers"; idem, "George Graham." Detailed description of sector in Maupertuis, *Figure de la terre,* 94–99.

45. Maupertuis, *Figure de la terre,* 39. An engraving of the instrument appears in Le Monnier, *Histoire celeste* (Paris, 1741).

46. On these experiments, see Sorrenson, "George Graham."

130 pounds sterling, it also represented a substantial monetary investment (fig. 3).[47] In turning to English instruments, Maupertuis once again posed a carefully veiled challenge to Cassini and his Parisian instruments. Although he also took several French quadrants made by Langlois and clocks by Le Roy, Maupertuis repeatedly called attention to the innovative design and consequent precision of the nine-foot English zenith sector, to be used for the crucial stellar observations that would enable the latitude calculations. Celsius made the most of his time in London, learning how to use the instruments, and even observed a lunar eclipse from Graham's house.[48] The Swedish astronomer thus provided a crucial, empowering link to England, the prime alternative source for precision astronomical instruments, and to Sweden, where state approval of the proposed operations would be required. He was also familiar with observing conditions in the north, and sug-

Fig. 3. George Graham zenith sector, made expressly for Lapland expedition. The nine-foot telescope hangs from the vertex of the wooden frame; observations are made in a reclining position. Maupertuis, *Degré du méridien entre Paris et Amiens* (1740).

47. For the instrument's price, see Celsius to Delisle, 12 March – 15 May 1737 (from Torneå), MS Obs. 1029a. The transit telescope cost 17 pounds sterling, according to Celsius.

48. Sorrenson, "George Graham," 216–17. Celsius was elected a fellow of the Royal Society during his visit to London; RS, *Journal Book*, 1 April 1736.

gested the Gulf of Bothnia as a possible site for the triangulations. The plan was to measure the base on the ice, assuring a smooth flat surface, and avoiding the inevitable visual obstructions encountered on wooded land.[49]

While Celsius was in England, Maupertuis and Clairaut spent the academic vacation period (September and October), enjoying the hospitality of the Cassinis, father and son, at their country estate in Thury, and "practicing astronomy."[50] Clairaut tried to get Algarotti to join them, since at that time he was still planning to go along to the north and knew nothing about astronomy. "M. Cassini charged me with asking you, and we would have the pleasure of sharing with you the charming life that we lead in this place."[51] All parties remained on cordial terms, collaborating on observations of the same eclipse viewed by Celsius in London.[52] The other academicians joining the journey to Lapland were the young astronomers Pierre Le Monnier and Charles-Etienne Camus, a specialist in the mechanisms of clocks and other instruments. Another essential participant was the Abbé Réginald Outhier, a corresponding member of the Academy with considerable experience with surveying and mapmaking. Outhier had worked with Cassini on measurements for the perpendicular in 1733 and was highly regarded for his surveying skill. He came to Paris and assisted with observations on simple pendula, analogous to those planned for the Arctic.[53]

It was well known that the period of a given pendulum varied depending on latitude; the first such comparative measurements had been done in the 1660s by Richer, who found that a pendulum beat faster at the equator than in Paris.[54] Apparently, the effect of gravity (whatever its cause) decreased toward the equator, where the increasing centrifugal force of the rotating earth partially counteracted the central force of gravity. Newton had used these data to derive an expression linking variation in weight to the sine of the latitude, and connected this expression to his calculation of the proportion of the earth's axis to its equatorial di-

49. Maupertuis discussed these plans in M. to Celsius, 22 Nov. 1735, RS Archives.

50. "Cleraut [Clairaut] sera du voyage et nous allons lun et lautre passer les vacances chez m. Cassini pour nous excercer à lAstronomie" (M to JB 12 September, 1735, BEB).

51. Clairaut to Algarotti, 6 September 1735, quoted in Badinter, *Passions intellectuels,* 75. See also Maupertuis to La Condamine, 8 September 1735, AS Fonds Maupertuis. This letter was also written from the Cassini estate in Thury.

52. J. Cassini, "Observation de l'éclipse de lune du 2 octobre de cette année 1735, faite à Thury," *MAS* 1735, 473–76. He observed the eclipse "at Thury with Mssrs. de Maupertuis and Clairaut."

53. Outhier, *Journal d'un voyage au nord,* 40. Outhier was the secretary of the bishop of Bayeux, who encouraged his astronomical interests. Outhier "worked with us on the description of the perpendicular from Caen to St. Malo" (Cassini, "De la carte de France," *MAS* 1733, 405).

54. Clock time was compared to the sidereal day, determined by observing the transit of a fixed star. On the Cayenne expedition, see Olmsted, "Scientific Expedition of Jean Richer."

ameter.[55] Eighteenth-century investigators, both English and French, used new clocks, paying careful attention to variables affecting their performance, in order to refine measurements of effective gravity. George Graham, the London instrument maker, had raised the standard of precision with his measurements of the effect of temperature on the length of the pendulum.[56] The French expeditions included pendulum experiments as well. La Condamine and Godin sent their first pendulum data back to Europe from Guyana in 1735, long before doing any surveying or astronomical work. The Lapland expedition would also include this type of experiment, and Maupertuis hoped for the greatest possible precision from the instruments obtained from Graham. Although calibration of such devices posed challenges, especially with respect to temperature, the measurements were relatively straightforward.[57] However, deriving the shape of the earth from the pendulum data required incorporating a theory of gravity into the calculations. Such measures could only answer the geodetical problem indirectly. For the purposes of the debate in France, where Newtonian theory itself was contested, they could be used only to corroborate the geographical and astronomical measurements.

In both Peru and Lapland, the academicians planned to do the same kind of surveying used by the Cassinis in France. This involved astronomical observations to determine latitudes of the endpoint of the arc being measured, and triangulation on the earth's surface to measure distance. Maupertuis intended to adapt the standard surveying methods of cartography to the conditions of the Arctic region. In early 1736, he gave a paper on minimizing error in determining the length (on the ground) of an arc. He derived an expression for the error accruing in a series of triangulations used in surveying distances. To find the optimum number of triangles for such a series, he differentiated to minimize observational error analytically, and showed his audience that he brought new ways of thinking to old practices. Using a very simple application of the calculus, he once again deflected discussion away from the geometrical terrain familiar to Cassini. He argued that a short arc could yield even more precision than a long one, like the eight degrees that had been measured in France, so long as all angles were observed with consistently low error rates. This sort of argument served a specific purpose in the context of the Arctic expedition, because he needed a chain of triangles that his team could measure in a relatively brief period. He also proposed novel methods

55. Newton, *Principia,* book 3, prop. 20.

56. On Graham's pendulum experiments and his central role in measurements taken for comparison in Jamaica, see Sorrenson, "George Graham," 210–11.

57. To avoid possible effects of temperature variations, Maupertuis maintained Paris temperature in the Lapland observing station by heating the room. Maupertuis, *Figure de la terre,* 174–75.

to avoid some sources of error. So, for example, he hoped to reduce distortion from atmospheric refraction by observing only stars near the zenith, when determining latitudes for the northern and southern endpoints of the arc. (This had not been common practice by the French cartographers.) These observations would have to be precise to several seconds: "Where are the instruments, and what is the degree of application that could assure us of those several seconds?"[58] he asked rhetorically. He was thinking of the Graham sector then under construction in London, though he did not say so. This was shortly after Cassini de Thury had spoken on techniques for measuring distance on the ground "as exactly as possible." He stressed the practical problems encountered in the field, problems most of the travelers had never encountered. "This practice, so simple in speculation, must demand great precautions in its execution." The young astronomer went on at great length about how to place measuring sticks with the greatest accuracy, and so on.[59]

Although Maupertuis's expedition would make use of the same basic combination of surveying and stellar observation used by three generations of Paris astronomers, the plans for the Lapland measurements contradicted the standard cartographic procedure of measuring as many triangles as possible and then choosing the best observations.[60] Having shown analytically that measuring an arc of only 1° would actually reduce the potential for overall error, Maupertuis proposed to measure a single series of triangles, with the stipulation that all three angles of every triangle be observed directly. This transformed the difficulties of working in the Arctic terrain, where it would be impractical to measure an arc longer than 1°, into an advantage and drew attention to the shortcomings of the accepted French practice of extrapolating some angles when direct observations proved impossible. On the eve of his departure, Maupertuis returned once again to the podium at the public meeting, with a speech summarizing the reasons for going to Lapland and the history to date of the measurements. The *Mercure de France* extracted the "succinct, but clear" presentation in considerable detail, noting that "the portion of the Public interested in the progress of the Sciences will very likely have the satisfaction of seeing the famous question of the shape of the earth decided." Maupertuis primed his public to be able to assess his future results, by explaining

58. Maupertuis, "Sur la figure de la terre," *MAS* 1736, 302–12; similar methods appear in his paper of the previous year.

59. Cassini de Thury, "Des opérations géometriques que l'on employe pour déterminer les distances sur terre et des précautions qu'il faut prendre, pour les faire le plus exactement qu'il soit possible," AS p-v, 25 Jan. 1736.

60. Cassini defended this practice in his response to the review of Poleni, presented to the Academy in December 1734 (see note 20 above).

how the measurements—by triangulation and with the seconds pendulum—related to the theoretical issues at stake.[61]

Once it was decided that a significant number of academicians would actually leave France to make their measurements, the problem of the earth's shape took on new dimensions. It became a complex practical problem with ramifications beyond the domains of mathematics or astronomy. Maupertuis and Clairaut were now applying the abstract algorithmic calculus, on which they had made their reputations, to actual measurements. The numbers, equations, and diagrams filling the pages of the *Mémoires* underlay an enterprise that required international diplomacy, the requisition and provisioning of ships, the purchase and shipping of instruments, and assistance from local officials and residents. The expedition also took the travelers temporarily outside the sphere of the Academy and freed them from certain constraints. They had to address the considerable problems of working in a strange environment on unknown terrain, of making their instruments work under these conditions, and of living in the extreme climate of the Arctic region.

## In Search of the Earth's Shape

A variety of observational methods for determining the earth's shape had been discussed in the Academy over the years. In spite of his examination of longitude measurements, Maupertuis and his companions decided early on to concentrate their efforts on measuring an arc of approximately one degree at the most northerly latitude possible. Numerous decisions would define the actual procedures adopted by the expedition, but there were three basic components to the overall task. These were conceptually simple, but delicate to achieve with the necessary accuracy. As a first step, the travelers would need to establish markers at elevated points along a meridian line to define a series of triangles. The two most distant markers would determine the endpoints of the terrestrial length and the arc of latitude. Standard surveying operations, to measure angles by sighting from point to point, would then determine a series of triangles constructed by connecting the observation points. Next, they would directly measure one side of one of the triangles by laying out measuring sticks end to end on the ground. From this baseline, the total length of the meridian segment could be calculated. Finally, they would take observations of fixed stars at each end of the chain of triangles to determine the exact difference in latitude between the endpoints. This would give the arc length corresponding to the measured linear distance on the curved surface of the earth, which could then be compared to the length of the degree in France, and eventually, in Peru. In addition, all measures would have to be checked, instruments

61. *Mercure de France* (April 1736): 756–64. The speech was read again to the closed session of the Academy after the expedition's departure; it was not published.

verified and calibrated, and numbers reduced to the plane of the horizon to improve accuracy. Each of these three operations—observing angles for the chain of triangles, measuring the baseline, and determining the position of certain fixed stars near the zenith—required different instruments and different calculations.

The lack of accurate maps and reliable information about local conditions complicated the technical problem of designing a series of operations that would produce acceptable measurements. Many crucial decisions had to wait until the travelers had reached the northern latitudes. In this sense, they really were going off into the unknown, and they had to adjust plans continually as they went along. In Stockholm, King Frederick's cartographers supplied the expedition with the latest maps of both coasts of the Gulf of Bothnia, where the Frenchmen assumed they would be able to conduct their operations using the many small islands for their observing stations.[62] The islands would be accessible by boat, and the gulf was conveniently oriented along a north-south axis. Only after journeying from Stockholm to Torneå, at the very north of the gulf, did they see that the islands along the western shore, although numerous, were too low to be visible at a distance, and hence useless as triangulation points.[63] After exploring the islands near the eastern shore and finding things no better there, they considered the alternatives. Celsius, the only Scandinavian among them, favored waiting for winter and measuring the whole length of the degree directly on the frozen surface of the gulf, without bothering with a chain of triangles.[64] (One degree of latitude corresponds to about sixty miles.) This would have meant a much longer and more physically demanding baseline measurement, but would have avoided the inevitable complications of measuring and reducing a whole series of angles. It would also have meant waiting for winter for the dark night sky necessary for the stellar observations that would define the endpoints. The French did not want to put off their operations, however; nor did they trust the ice in the gulf. LeMonnier suggested clear-cutting a path straight northward through the forests along the coast and measuring that distance directly, but this was also deemed impractical.[65]

Meanwhile, Maupertuis had made a preliminary foray north from Torneå and had discovered a landscape full of mountains high enough to be visible from neighboring peaks, but low enough to be accessible. They did have the drawback of being heavily forested and rather too close to each other, but by this time the men realized that the process of measurement would entail more physical effort

62. Outhier, *Journal d'un voyage au nord,* 52.

63. Camus to Geoffroy, 7 April 1736, BN, Fr. 9674.

64. Outhier, *Journal du voyage au nord,* 82.

65. Ibid., 82–83. See also "Extrait d'une lettre écrite de Pello . . . par M. Le Monnier" *Mercure de France* (December 1736): 2730–34.

than they had imagined before confronting the Arctic landscape. They took note of one promising feature of that landscape—the river that emptied into the gulf at Torneå followed the meridian even more closely than their maps had led them to believe, providing an unexpectedly convenient backbone for the chain of triangles. With considerable assistance from local residents and their boats, the expedition turned northward into relatively uncharted territory (fig. 4).[66]

When he arrived in Torneå, Maupertuis met a Swedish lieutenant colonel, one M. Duriez, who spoke French and who was eager to aid the expedition with logistical support and local knowledge. He commanded a contingent of soldiers, "the regiment of Westrobothnia," a reserve force of local "peasants" who spoke Finnish. Outhier reported that they were "men of courage, who were not afraid of fatigue," which was fortunate because they were called on to perform much grinding labor in the months to come.[67] Moreover, they all owned small boats suited to navigating the river, which proved essential to the work of the expedition.[68] Throughout the summer season, they ferried the men of science and their instruments up and down the river, portaging around occasional stretches of white water and cataracts. They bent their backs to the work of carrying instruments and supplies up mountains, clearing trees, and constructing markers to be visible from distant peaks. If they grumbled about the absurdity of their tasks, they complained in Finnish, and the Frenchmen did not understand. Torneå, a small town of seventy households, located at the mouth of the river, became the expedition's base of operations. The local notables, in addition to supplying soldiers for labor, facilitated the scientific work in other ways. The Swedish governor of the province welcomed Maupertuis upon his arrival in Torneå, and traveled up the river with him to reconnoiter. The governor introduced him to a young man who spoke Finnish and Latin, as well as Swedish and French, and who agreed to accompany them as translator.[69] Most of the travelers were housed by a prosperous merchant, whose son-in-law they had met in Stockholm. The pastor and other notables also lent their premises when needed. The arrival of the Frenchmen attracted the curiosity

66. On the outfitting of the expedition, see Nordmann, "L'expédition de Celsius et Maupertuis." For the decision to go north from Torneå, see Maupertuis, *Figure de la terre,* 11; Outhier, *Journal d'un voyage au nord,* 77–78. Maupertuis said he noticed that the river followed the meridian when he made his initial investigation of the terrain; Outhier remembered that this information came from the rector of the local school, M. Viguelius (Outhier, *Journal d'un voyage au nord,* 83).

67. Outhier, *Journal d'un voyage au nord,* 76, 83.

68. Outhier also referred to the soldiers as "sailors" because of their facility with their canoe-like boats, without which the instruments could not have been transported.

69. This was Anders Hellant, who proved invaluable as much for his interest in astronomy as for his language skills. He later became an astronomer in Uppsala (Outhier, *Journal du voyage au nord,* 84).

Fig. 4. Map of meridian arc measured in Lapland between Torneå and Kittis, with the chain of triangles marked. Maupertuis, *The Figure of the Earth* (London, 1738). Courtesy of William Andrews Clark Library, UCLA.

of people of all ranks and prompted considerable socializing, mediated by Celsius and the two residents of Torneå who could speak French.

The steeple of the Lutheran church provided a convenient starting point for the surveying operations. The academicians would have to scout out other points as they made their way up the river. Until an accessible and well-situated point was found for the northern vertex of the chain of triangles, the success of the process remained uncertain. Leaving the relative comforts of Torneå in early July, the men embarked with their instruments, tents, and provisions. The party included twenty-one soldiers, five servants, eight astronomers, and the interpreter. They spent the next two months in a frenzy of exploration, mountain climbing, construction of markers, observation, checking and calibrating instruments, and calculation. Although they started with a general plan, getting reliable observations required improvised solutions to problems encountered in the field. The group divided into several smaller bands that crisscrossed the landscape, meeting up with each other periodically, checking each other's measurements. They learned as they went along how to make markers by peeling the bark from felled pine trees and building pyramids with the white trunks facing outward, so they would be visible from a distance. The soldiers often had to clear the trees from the summits of hills, to ensure visibility, and they did the work of constructing the signals as well as transporting the instruments (fig. 5). Ultimately, the astronomers

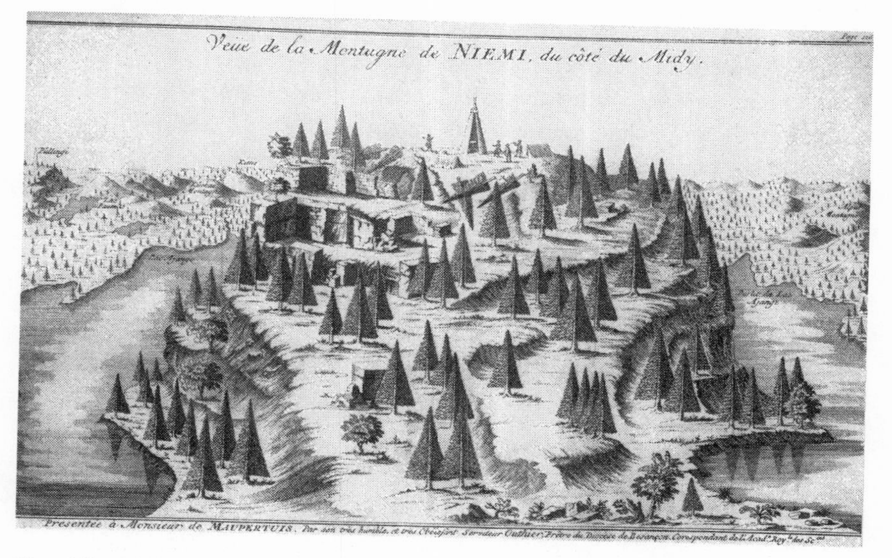


Fig. 5. Niemi Mountain, showing clearing in progress for building observing station. The conical signal constructed of stripped pine trees is visible on top. Réginald Outhier, *Journal d'un voyage au nord* (1744). Courtesy of Special Collections, University of California, Irvine.

established nine geodetic points, in addition to the Torneå church, from each of which they observed as many of the others as possible. From these points and angles, they constructed several interlocking series of triangles that could be checked against each other.

They deferred much of the work of calculating to winter; in the meantime, as the nights became longer in the autumn, they undertook the astronomical observations. No one but Celsius had seen the zenith sector, with its nine-foot telescope, until it arrived in Torneå at the end of August. Although cumbersome, because of the large wooden frame that supported the telescope, the sector was designed to be portable. It promised to provide the critical stellar positions that would establish the endpoints of the chain of triangles. No one knew whether it could deliver on this promise after the rigors of its journey from England. From Torneå, the soldiers transferred it to a small boat for the trip upriver to Kittis, the northernmost mountain of the chain, where they had built a small observatory for it. There was some urgency to accomplishing this phase of the operation before the onset of winter, when camping on the mountain would become impossible. The sector would then have to be dismantled and taken back to Torneå before the river froze, to make the corresponding observations from the southern end of the chain with as little delay as possible. As it happened, the weather cooperated, and the zenith sector functioned extremely well. Alternating teams of three men observed a single star in the constellation Draco over several weeks. They were able to measure the distance of the star from the zenith consistently within two or three seconds of arc, just the degree of precision Maupertuis had hoped to attain. In the same period, the team also worked on a series of experiments with an array of pendula made from different materials. When the weather started to change, in late October, they dismantled the sector and traveled back downstream just before ice made the river impossible to navigate (fig. 6).[70]

Back in Torneå, the Frenchmen and their instruments found lodging. They installed the sector in a small purpose-built observatory and proceeded with the zenith observations over the course of a week in the beginning of November. The next step was to measure the baseline from which the distance between the two endpoints would be calculated. This operation consisted of laying twelve-foot wooden measuring sticks end-to-end over a distance of about five miles. The baseline followed one side of a triangle, along the river about halfway between Torneå and Pello. After considerable debate about the practicality of working outdoors in the short and icy winter days, the enterprise proceeded on the snow-covered river. Working in two teams with measuring devices fabricated for the

70. Maupertuis, *Figure de la terre;* Outhier, *Journal du voyage au nord;* Nordenmark, *Anders Celsius.*

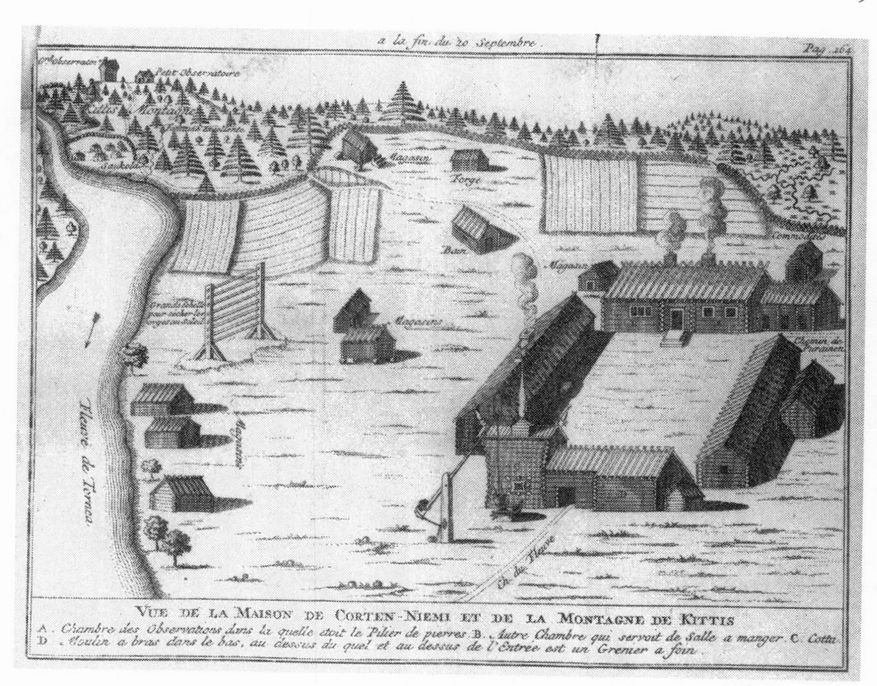


Fig. 6. House in Corton-Niemi where the travelers lodged and performed pendulum experiments. The observatory housing the zenith sector, on top of Kittis Mountain, is visible at upper left. Réginald Outhier, *Journal d'un voyage au nord* (1744). Courtesy of Special Collections, University of California, Irvine.

occasion, the men spent ten days measuring and verifying the length and orientation of their base.

The remainder of the winter was spent in relative comfort in Torneå, calculating, continuing the pendulum experiments, and observing the frequent aurora borealis. Clairaut used the time to finish a paper for the Royal Society.[71] The Frenchmen lived companionably with their Swedish hosts and were much admired by the local populace.[72] When the ice broke up in the spring, they returned to several of their observation points, verifying some angles and remeasuring the direction of the baseline. To increase the certainty of their celestial measurements, Maupertuis

71. Clairaut to Mortimer, 13 Dec.1737, RS, *Letter Book*, 24, fols. 79–81; Clairaut, "Investigationes aliquot, ex quibis probetur terrae figuram," *Philosophical Transactions* 445 (1738), 19–25.

72. Outhier, *Journal d'un voyage au nord*, 179; M to Mme. de Vert[e]illac, 6 April 1737, in *Société des bibliophiles français, Mélanges*.

decided to undertake a second complete set of observations of a different star with the zenith sector. So the instrument was once again disassembled for the journey upriver to the little observatory on Kittis. Further verifications and remeasurements and calculations took up the next two months. By this time, the academicians knew that their results implied a flattened shape for the earth, but they informed no one of this finding in advance of their return to Paris (fig. 7).

## Narratives of Heroic Science

This overview of the Lapland expedition gives a rough sketch of how the Paris Academy team set about resolving, to their own satisfaction, the technical problem of measuring the length of one degree of latitude on the earth's surface. In Lapland, they encountered a physical and social environment that presented certain challenges, but ultimately the terrain, weather, and local residents all contributed to the expeditious completion of their scientific goals. The elements of my story,

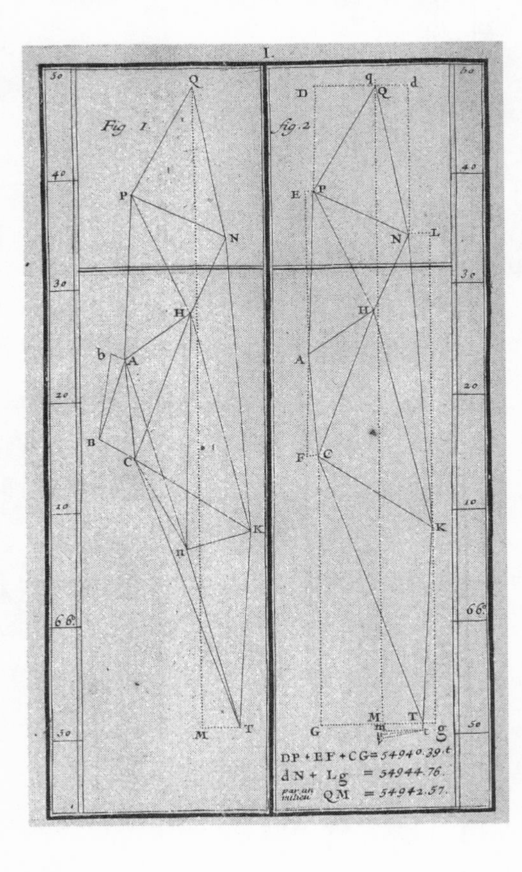


Fig. 7. One of many plates showing one configuration of the chain of triangles, with direction of meridian indicated by dotted line. Calculations were done in as many ways as possible, using all angles measured to form alternate configurations of triangles. Maupertuis, *La figure de la terre* (1738).

which I have purposefully presented as linear and unproblematic, come from the two published first-person accounts by Maupertuis and Outhier.[73] The way these authors crafted their story suggests that the narrative could be used to several different ends. Maupertuis's narrative of the journey—which highlighted the instruments, the measurements, and the calculations—embodied an indirect defense against charges of sloppy technique and theoretical prejudgment.[74] Outhier's book only appeared in 1744, after the controversy had died down. It is an invaluable source for the expedition itself, since it was not written as part of the polemic and it closely follows Outhier's personal journal of events and descriptive details. Apart from these books, only a few letters written from the field survive, some of them published or circulated at the time. There are many things we may wish to know about the actual events in Lapland that we cannot discover in the historical record. But these records tell us a great deal about how Maupertuis's contemporaries came to know of the expedition, the place it occupied in his own life, and how the whole enterprise fit into claims about the practice of science.

Maupertuis told the story, or parts of it, many times for various audiences. His book *La figure de la terre* (1738), containing the narrative account and the associated observations, calculations and diagrams, was published in several editions and soon translated into English and German.[75] He delivered substantial portions of the text to two heavily attended public sessions of the Academy, one just after the expedition's return and the other just before the book's publication the following spring. The press covered these presentations generously, and reviewed the book, which was also widely discussed and admired.[76] The book presented an adventure story, a triumphant quest for knowledge, in the tradition of travel literature. But it was also a detailed technical record of sophisticated science. In fact, this aspect of the account did not go unchallenged. Cassini attacked the Lapland results as soon as the travelers returned, and by the time Maupertuis finished his book several months later, his version of the expedition's technical accomplishments had

73. Manuscript sources do not survive, with few exceptions. Letters and documents pertaining to the provisioning of the expedition are cited in Nordmann, "L'expédition de Celsius et Maupertuis."

74. Maupertuis turned to the English astronomer James Bradley for confirmation of his techniques for using the zenith sector. For explicit attacks on Cassini, he relied on Celsius, a foreigner. See chapter 5 below.

75. There were two editions in 1738, one in Paris and one in Amsterdam; another Amsterdam edition appeared in 1739; English translation, London, 1738; German translation by Samuel König, Zurich, 1741. Versions of the text also appeared in Maupertuis's collected works, starting in 1752.

76. Review in *Journal des sçavans,* October 1738; *Mercure de France,* November 1737, 2461–73. See Chouillet, "Rôle de la presse périodique."

au Cercle Polaire. 81

| Angles observés. | Angles réduits à l'Horison. | Hauteurs. | |
|---|---|---|---|
| *Sur Niwa.* | | | |
| TnK... 87° 44′ 24,″8 | 87° 44′ 19,″4 | T... — 17′ 40″. | Fig. 1. |
| HnK... 73 58 6,5 | 73 58 5,7 | K... + 16 50 | |
| AnK... 95 29 52,8 | 95 29 54,4 | A... + 4 40 | |
| AnH = AnK — HnK | 21 31 48,7 | H... — 0 30 | |
| AnH = 21 32 16,9 | 21 32 16,3 | | |
| AnH est donc . . . . | 21 32 2,5 | | |
| CnH... 31 57 5,2 | 31 57 3,6 | C... + 10 0 | |
| *Sur Kakama.* | | | |
| TKn... 72 37 20,8 | 72 37 27,8 | n... — 22 50 | |
| CKn... 45 50 46,2 | 45 50 44,2 | C... — 4 45 | |
| HKn... 89 36 0,4 | 89 36 2,4 | H... — 5 10 | |
| HKC = nKH — CKn | 43 45 18,2 | | |
| HKC... 43 45 46,8 | 43 45 47,0 | | |
| HKC... 43 45 41,5 | 43 45 41,7 | | |
| HKC est donc . . . . | 43 45 35,6 | | |
| CKT = CKn + nKT | 118 28 12,0 | T... — 24 10 | |
| HKN... 9 41 48,1 | 9 41 47,7 | N... — 8 10 | |
| *Sur Cuitaperi.* | | | |
| | | K... — 6 10 | |
| KCn... 28 14 56,9 | 28 14 54,7 | n... — 19 0 | |
| TCK... 37 9 15,0 | 37 9 12,0 | T... — 24 10 | |
| HCK... 100 9 56,4 | 100 9 56,8 | H... — 2 40 | |
| ACH... 30 56 54,4 | 30 56 53,4 | A... + 5 0 | |

F

Fig. 8. Table of observations taken at three different observation stations, showing reductions of raw observation data to the horizontal, and showing how many observations were made at each point. Maupertuis, *La figure de la terre* (1738).

become part of a lively polemic. The grounds for dispute had as much to do with astronomical practice as with rival physical theories or cosmologies. But the standards of academic comportment forbade direct public attacks on fellow academicians, so the critique of Cassini had to remain implicit, at least for the time being.

When Maupertuis presented the preface to his book publicly in April 1738, the debate about how to interpret the results had already engaged the Academy's attention for some time. Public and Academy were familiar with the story's basic trajectory. Maupertuis used the high level of public interest in the question to justify immediate publication of the account and the observations as a book, rather than waiting for the Academy's annual volume for 1738, which would not appear for several years.[77] In addition to giving a history of geodetic measurements and their utility, the preface outlined a strategy distinguishing this text from other

77. Nevertheless, there was opposition to the rapid publication of Maupertuis's book. Châtelet to Algarotti, 10 January 1738, *Lettres,* 1:206; Souciet to Delisle, 25 September 1738, MS Obs., Delisle papers.

accounts of meridian measurements. Maupertuis claimed to be publishing all relevant observations, taken directly from the notebooks of each observer, "which were all found to conform to each other." Without naming Cassini, he criticized standard practice in "similar works that report such information." These works habitually "corrected" the triangles so that the sum of their angles always added up to exactly 180°; in the case of arc measurements, they published averaged observations, "without giving the observations themselves."[78] Maupertuis implied that his numbers could speak for themselves (fig. 8). Displaying his data openly and exhaustively, he offered them as transparent representations of reality to the reader, who would then be able "to judge the degree of precision that is there, or that is lacking. Finally, he will be able to make corrections himself, as he sees fit, and to compare the different results that corrections made differently than ours would produce."[79] Readers who were likely to examine the numbers this closely made up only a tiny portion of Maupertuis's intended audience, but the capacity of the observations to stand up to such scrutiny, whether real or hypothetical, grounded the rhetoric of precision permeating the text. As we might expect, the published text is in fact far from a transcription of the record books of individual observers. Many measurements were taken in the process of installing and verifying the instruments; these numbers were not published and so remained invisible to the reader. Similarly, observations deemed problematic for any reason did not appear.[80] Instead, the author reiterated the remarkable conformity of the observations and verifications that provided the raw material for key calculations, discreetly reminding the reader of the absence of consistency in Cassini's observations. Maupertuis emphasized his team's obsessive checking and rechecking of key observations just because these observations had already come under attack. "Perhaps we ought to apologize for an exactitude that will appear overly scrupulous to some, as much in our calculations as in the detail of the circumstances of our observations: but we thought it impossible to push exactitude too far in a controversial matter, and one which is of such great importance."[81] In later editions of the book, after the controversy had been resolved, he deleted such justifications

78. Maupertuis, *Figure de la terre,* iii–iv. This was the same critique leveled at Cassini by the review of Poleni's book several years earlier.

79. Ibid., iv.

80. The only manuscript record of such observations I have found is a copy by Le Monnier of his observation notebooks; this manuscript includes a set of observations that were deemed inaccurate because the two observers disagreed too much. MS Obs., A.C. 4.3.

81. "Nous aurions peut-être à excuser une exactitude qui paroîtra trop scrupuleuse à quelques-uns, tant dans nos calculs, que dans le détail des circonstances de nos observations: mais nous avons cru ne pouvoir pousser trop loin cette exactitude dans une matiere qui a été disputée, & qui est d'une si grande importance" (Maupertuis, *Figure de la terre,* xvii).

of the display of raw observations, with their implied challenge to the accuracy and honesty of Cassini's measurements. For the moment, these passages served a clear purpose.

## Heroic Effort in the Name of Science

Without the numbers, there would have been no story. Maupertuis's narrative leads up to the climactic revelation that a degree of latitude at the Arctic circle is fully 1000 *toises* longer than Cassini had projected from his elongated model of the earth, and 377.9 *toises* longer than the average length of a degree in France.[82] (One *toise* was equal to six feet.) As he explained in the beginning, "the arcs of one degree [measured by the stars] will be longer where the earth is flatter. If the earth is flattened towards the poles, a degree of the terrestrial meridian will be longer toward the poles than toward the equator."[83] Much of the book recounts the heroic efforts expended in obtaining that number. Having undertaken the expedition in the public eye, Maupertuis sought to make his results credible to as broad a public as possible. His text had to have something for everybody: utility for the ministers and the king, precision and careful mathematics for the Academy, adventure and spectacle for the literary elite. If we were to separate these features from each other, however, we would lose sight of the interpenetration of motivation, audience, and resources that characterized science in this period.

Let us look at the way Maupertuis told the tale and then at how it was received. In the address to the public session in November 1737, three months after returning to Paris, he framed his story as a quest for the apparently impossible.[84] Skipping over the voyage to Stockholm, he started with a nod to the king of Sweden, whose crucial assistance smoothed the way for the expedition to proceed. The warm welcome afforded them by the French ambassador, Castéja, and the Swedish court reminded the reader of the expedition's status and significance. Even as he gave them his blessing, however, the Swedish king warned of the difficulties of working in the arctic. "The perils they threatened us with in Stockholm did not delay us at all; nor did the kindness of a King, who, in spite of the orders he had given for us [to proceed], repeated to us several times that he watched us leave for such a dangerous undertaking only with distress."[85] The king

82. Ibid., 125. I do not enter into the question of the accuracy of this number by later standards. For a critical reassessment of the measurements, see Delambre, *La grandeur et figure de la terre;* J.-P. Martin, "Maupertuis."

83. Maupertuis, *Figure de la terre,* 9.

84. Read to the public assembly of the Academy, 13 Nov. 1737; reread in closed sessions, 16, 20, 23 November, AS p-v, 1737; published as the preface to Maupertuis, *Figure de la terre.*

85. "Les périls dont on nous menaçoit à Stockholm ne nous retarderent point; ni les bontés d'un

should have known that no talk of danger was going to discourage the travelers at this juncture.

The search for observation sites in the Gulf of Bothnia led to frustration: "We were looking in these islands for what we could not find, and we had to give up hope and abandon them."[86] This is only the first setback the men of science will face in their quest. It sets the tone for the months to come, for whenever they meet an obstacle, ingenuity and perseverance overcome it. When they decide to venture north of Torneå, dangers and difficulties become more palpable, and the narrator hints that the men will be forced to stretch their endurance beyond anything imaginable in Paris. Here is a retrospective sketch of the physical difficulties anticipated when the original plan had to be abandoned:

> We had to perform operations that would be difficult in the most accommodating regions, in the wilderness of a nearly uninhabitable country, in that immense forest that stretches from Torneå up to the North Cape. There were only two ways of penetrating into that wilderness, and we had to use both of them: one by navigating a river full of cataracts, the other by crossing dense forests or profound swamps on foot. Supposing that we were able to penetrate into the country, it would be necessary after the roughest possible walking to scale steep mountains and to clear their summits of trees obstructing the view. It would be necessary to live in this desert with the worst food, exposed to insects so vicious that they force the Lapps and their reindeer to abandon that part of the country in the summer.[87]

Maupertuis was well aware that any mapping expedition faced similar challenges once it left settled territory, even in the provinces of France. But he stressed the extremes of the northern environment to alert his readers (and listeners) to the fact that everything the academicians encountered in Lapland would be that much more difficult, and at the same time more exotic and interesting, than analogous travels in temperate regions. On the one hand, this drama made the account itself more engaging; on the other, it devalued the accomplishments of the domestic cartographers and astronomers. It also brought to the fore the physical effort required to bring the mission to a successful conclusion. All of these themes recur throughout the book.

In addition to the bodily endurance demanded by the setting, the Frenchmen faced the hardship of wondering whether they were attempting the impossible.

---

Roi, qui, malgré les ordres qu'il avoit donnés pour nous, nous répéta plusieurs fois qu'il ne nous voyoit partir qu'avec peine pour une entreprise aussi dangereuse" (Maupertuis, *Figure de la terre,* 7).

86. "Après nous être opinâtrés dans plusieurs navigations à chercher dans ces isles ce que nous n'y pouvions trouver, il fallut perdre l'espérance, et les abandonner" (ibid., 10).

87. *Figure de la terre,* 12.

> Finally we would have to undertake this work without knowing whether it was going to be possible and without being able to ask anyone; without knowing if, after so much trouble, the lack of a single mountain would not ruin the chain of triangles or whether we would find a suitable baseline on the river. If all that worked, we would then have to build observatories on the northernmost of our mountains; we would have to carry there an array of instruments more complete than is found in many European observatories; and we would have to make the most sensitive astronomical observations.[88]

At every opportunity, Maupertuis recalled the fearful and the unpleasant, the painful and the discouraging, as the men doggedly amassed their measurements. Insects pursued them wherever they went: "On the river we had been severely incommoded by large green-headed flies, which drew blood wherever they stung us. We found ourselves on [the mountain] Niwa persecuted by several other kinds even more cruel."[89] They learned from some women pasturing their reindeer nearby how to immerse themselves in the smoke of their campfire to deter the mosquitoes. On the next mountain, they encountered even more vicious insects: "Smoke could not defend us from the insects, more cruel on this mountain than on Niwa. In spite of the great heat, we had to wrap our heads in our reindeer-skin robes and have ourselves covered with a thick rampart of pine branches and whole pine trees, which crushed us, and which did not keep us safe for long."[90] And so it went, up one mountain after another, sleeping on rocky ground and suffering from insect bites and heat. Whenever possible, they took advantage of the river for transportation, but even this had its dangers and discomforts. The boats were lightweight, to enable them to traverse the frequent rapids, and needed skilled oarsmen to keep them from capsizing. "It is a spectacle that seems terrible to those who are not accustomed to it . . . to see in the middle of a cataract, whose noise is terrifying, this frail contraption carried by a torrent of waves, foam, and stones, sometimes tossed into the air, sometimes lost in the torrents. An intrepid Finn steers it with a large oar, while two others ply their oars to free it of the waves coming from behind that are always on the point of submerging it."[91]

The dominant impression the reader takes away from the narrative is one of continual activity. The team split up and regrouped according to the task at hand, carrying instruments and supplies back and forth, often walking all night in the midsummer light. When bad weather enforced temporary inaction, an uncom-

88. Ibid., 12–13.
89. Ibid., 14.
90. Ibid., 16–17.
91. Ibid., 17–18.

fortable restlessness prevailed until they could take their measurements and move on. Some places were particularly difficult of access or uncomfortable for making camp. On Pullingi, the highest and steepest of their mountain observation points, "our sojourn from 31 July to 6 August was as painful as the ascent had been."[92] The persistent flies made the work more difficult:

> We had to clear a forest of good-sized trees; and the flies [mosquitoes] tormented us to the point that our soldiers of the regiment of Westrobothnia, a distinguished company even in Sweden where there are so many valorous soldiers, these men hardened by the most difficult labors were constrained to envelope their faces and to cover themselves with tar. The insects infected whatever we wanted to eat; our food was instantly black with them. Birds of prey were no less famished. They flew incessantly around us to steal morsels of the mutton that had been brought for us.[93]

Here we see the academicians living and working and eating alongside the soldiers, these "men hardened by the roughest work." The collaboration of Finnish peasant/soldier and Parisian man of letters lent an aura of otherworldly authenticity to the enterprise; it also highlighted the resourcefulness of the Frenchmen in adapting to the hardships of the Arctic environment. As the commanders of the expedition, the men of science maintained their superior status, while simultaneously showing themselves the equals of the local soldiers in toughness and manly courage. They had given up the comforts of civilization in the interest of "the progress of science," and their ability to endure hardships with good grace marked them as true heroes. By recounting the trials of life in the wilderness, Maupertuis painted himself, along with his brothers-in-arms, as a man of action and daring. He knew very well that physical suffering, if they could survive it, would enhance the value of their accomplishment.

The landscape, in this story, is always secondary to the accomplishment of the measurements. The men of science making their way through this foreign scene are dedicated, intrepid, and persistent in pursuit of their goal. Eventually they will have to abstract their numbers from the setting that produced them, by reducing all measures to sea level and correcting for atmospheric and other effects. Until their initial quest is complete, however, they are men pushing themselves to their physical limits, not knowing whether they will prevail. In the telling, Maupertuis punctuated the chronological account of their movements with stylistic elements borrowed from travel literature and tales of magical quests. Sometimes, these punctuating moments are disasters, or near-disasters, as when the men watch from a distance as a fire, ignited by the embers of an earlier campfire, consumes one of

92. Ibid., 17–18.
93. Ibid, 22–23.

their laboriously built markers on a mountain peak, along with the surrounding forest.[94] At other times, fortuitous circumstances seemed to conspire in their favor. Such a moment occurred when they arrived at Niemi, another observation point, after a daunting trek through dense and trackless forest and a boat ride across a windy lake. The serenity and beauty of the spot, contrasting with the darkness and difficulty of the approach, made it seem enchanted. The "charm" of the place prefigured the success of the whole expedition.

> The surrounding lakes, and all the difficulties we had to conquer to get to this mountain, made it resemble the enchanted places of fables. . . . We found on one side a pleasant wood where the ground was as even as the paths of a garden; the trees did not hinder walking nor the view of a beautiful lake that bathed the foot of the mountain. . . . On the other side . . . there were rocks . . . that seemed more like walls begun for a palace than the work of nature. We saw, rising from the lake, the mists that the local people call Haltios, and that they take for the guardian spirits of the mountains. . . . This mountain seemed to be inhabited rather by fairies and genies than by bears.[95]

The forest setting (perhaps magically) appeared civilized, recalling gardens and palaces where a man could walk at leisure rather than struggling through underbrush. If such a place seemed incongruous in the rude wilderness of Lapland, it was no more so than the precision of the calculations that the Frenchmen planned to distill from their sufferings.

Whatever the merits of these literary turns, it was crucial that the fabulous or incredible elements of the story not eclipse the empirical. The author never shifted his gaze far from the purposeful collection of data and used the wild or enchanted backdrop as a foil to the meticulous rationality of the whole undertaking. The technical procedures, the array of instruments, and the obsessive attention to detail supported the narrator's scientific credibility, a quality often lacking in classic travel accounts.[96] Whenever possible, two groups of men cross-checked results by measuring the same thing; invariably, they found that their numbers agreed to an astonishing extent. This policy was particularly striking in the measurement of the base, undertaken on the river in the dead of winter. The reader learns that the precision resulting from the flat surface more than compensated for the difficulty of working on the frozen river. The direction of the river relative to the meridian had been determined in the summer; once the river froze, measurement of the baseline (about five miles in length) proceeded by laying carefully calibrated measuring

94. Ibid., 29–30.
95. Ibid., 28.
96. See Adams, *Travelers and Travel Liars.*

sticks end to end. These rods were constructed on the spot, to exact specifications set by the iron *toise* brought for the purpose from Paris. Describing the discomforts suffered by the academicians as they struggled to maintain their standards of precision and accuracy under duress, Maupertuis gave his Parisian audience a graphic and memorable picture of the operation.

> I will say nothing of the hardships nor of the perils of this operation; you may imagine what it is to walk in two feet of snow, carrying heavy measuring sticks, which must be continually set down in the snow and retrieved. All this in a cold so great that when we tried to drink *eau de vie,* the only drink that could be kept liquid, the tongue and lips froze instantly against the cup and could only be torn away bleeding. A cold so great that it froze the fingers of some of us, and continually threatened us with yet greater accidents.[97]

This passage—and others like it—make the reader acutely aware of the bodies of the academicians, as they freeze or sweat or bleed in the service of science. However difficult it might have been to imagine such sensations, the physical suffering of these men made the accomplishment real and substantial to a genteel audience with little direct experience of climates beyond the boundaries of the French provinces. The measurements themselves emerged pristine from this trying process when the results were revealed: two teams measuring the same distance independently came up with numbers that differed by no more than a few inches. "The difference between the measures of the two teams was but four inches [*pouces*] on a total distance of 7406 *toises* 5 feet; exactitude that one would not dare expect and which one would almost not dare to report."[98]

In their winter quarters, while making calculations based on the summer's observations, Maupertuis realized that they had neglected a simple height measurement on Avasaxa, one of their mountain stations. In "a more accommodating country," they could easily have returned to make the measurement, but in the Arctic winter going back promised to be an adventure. "If one imagines a high mountain, covered by rocks hidden by a prodigious quantity of snow, which also covers the cavities that one can fall into, one will hardly believe that it is possible to ascend it." The reader is not surprised to find Maupertuis making the attempt, though, which gives him the opportunity to describe a harrowing journey in a sleigh drawn by reindeer (fig. 9). The animals, "wild and untamable," provided an added element of excitement, as they could not be slowed down and were liable to turn on their passengers and "revenge themselves with kicks." The natives knew how to deal with temperamental reindeer, by hiding under the overturned sleigh:

97. Maupertuis, *Figure de la terre,* 51–52.
98. Ibid., 56.

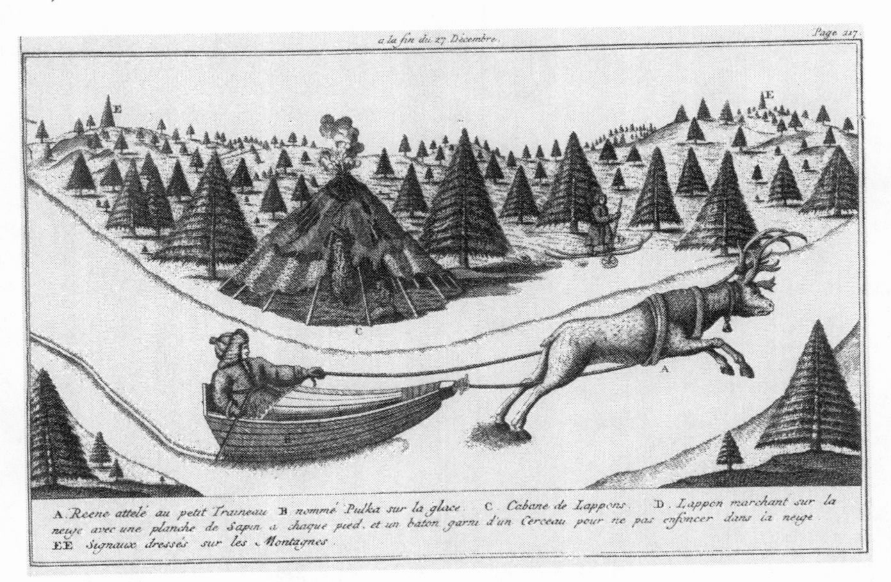


Fig. 9. Exotic transportation used by Frenchmen to scale the mountain Avasaxa. Note the local resident in the background on "long, narrow planks" showing the way. Conical signals are marked E in upper right and upper left. Réginald Outhier, *Journal d'un voyage au nord* (1744). Courtesy of Special Collections, University of California, Irvine.

"As for us, hardly capable of this solution, we would have been killed before being able to get under cover. Our only defense was a small stick that we carried and used like a rudder, to steer the boat [sleigh] and avoid tree trunks. Thus, abandoning myself to the reindeer, I undertook to scale Avasaxa, accompanied by M. Outhier, two Lapp men, one Lapp woman, and M. Brunius, their curate."[99] The journey proceeded without tragic consequences, in spite of the steepness of the ascent and the terrific cold, and the intrepid adventurers returned triumphantly with the altitude measurement in hand.

By this point, it is clear that this is a fable of scientific virtue, in which good fortune blesses the hero time and again, enabling him to overcome the obstacles the helpful king had warned him about at the beginning of the quest. As the author and the protagonist of this tale, Maupertuis recounted the process of assessing dangers of various kinds and then facing up to them. Whereas they met dangers with courage, the men of science met technical challenges with careful calculation and tireless accumulation of detail. The heroic elements of the story

99. Ibid., 54–55.

define a kind of scientific courage and strength, partially grounded in rationality and mathematics, but integrated into a series of physical feats. The author repeatedly reminded his readers of the risks, and the process of careful consideration by which they were evaluated. Physical risks often had to be weighed against the risk of losing observations or losing valuable time, so that the scientific goal was tied to the possibility of bodily harm. But the scientific goal always took precedence. The unavoidable exposure to cold posed one of the greatest challenges even to routine work: "When one would open the door of a warm room, the outside air immediately transformed the vapor in it to snow, making great white swirls; when one went out, the air seemed to tear the chest. We were warned and threatened every moment by increases in cold, by the noise of the wood, of which all the houses were made, cracking. . . . One saw in Torneå people mutilated by the cold: and the inhabitants of such a hard climate sometimes lose an arm or a leg."[100]

Facing dangers and rising to physical challenges, Maupertuis and his companions survived their ordeal unscathed, except for some frostbite. The heroic tale of risk and hardships reinforced their claim to having delivered unprecedented accuracy and precision in their scientific results, against the odds. This rhetorical strategy worked precisely because of the importance of appealing to the judgment of readers beyond the confines of the Academy. In closed sessions, Cassini had mounted a serious challenge to the expedition's results, based on a critique of their instruments and procedures.[101] In the book, Maupertuis never so much as mentioned this challenge, but worked instead to establish his credibility on as many fronts as possible. The book enjoyed a wide success. The admiring reviewer for the *Journal des sçavans* noted that the book is "equally worthy of a good philosopher and a great wit [*un très-bel esprit*]. . . . One senses throughout this work the honorable man [*honnête homme*] and an honorable man who is also amiable." The review repeated the story of the sleigh ride up Avasaxa, and noted approvingly that "other travelers would not have missed the opportunity to insert into the story some little accident. But we find in this account only those that really happened; and if it is not more diverting in this regard, it is certainly not the fault of our courageous philosophers." The author's truthfulness, honesty and courage were reflected in his literary style, in the work's "elegance, vivacity, clarity and precision," which enhanced its claims to truth.[102]

Voltaire, another reader sympathetic to Maupertuis's arguments, also applauded the elements of romance in the book. He described his response effusively, but tellingly, in a letter to the author:

100. Ibid., 59.

101. See chap. 5 below; Terrall, "Representing the Earth's Shape."

102. Review of *La figure de la terre*, *Journal des sçavans* (October 1738): 596–98.

> I have just read a story and a piece of physics more interesting than any novel. . . . Your preface . . . elicits an extreme impatience to follow you to Lapland. As soon as the reader is there with you, he thinks himself in an enchanted fairyland where philosophers are the fairies. . . . In ecstasy and in fear, I follow you across your cataracts and up your mountains of ice. Certainly you know how to paint; only you could be our greatest poet as well as our greatest mathematician. If your operations are worthy of Archimedes and your courage of Christopher Columbus, your description of the snows of Torneå is worthy of Michelangelo.[103]

At least according to Voltaire, the author's literary talents could stimulate visceral and emotional responses in his readers, making Maupertuis an artistic hero (analogous to Michelangelo) as well as a scientific one. However, his status as a stylist and wit in no way detracted from his status as a philosopher, or a truth teller. Voltaire's characterization of the book as fairy tale, or poetry, capable of transporting the urbane reader to the frozen north, reminds us of the book's wide readership. Writing in the heroic vein was in part a play for the sympathy of these readers, including a significant contingent of women. To represent the validation of Newton's theory of gravity as a romance was to claim a cultural significance for science that reached beyond the Academy and beyond its patrons in the government.

Maupertuis carefully conceived and composed his speeches to the Academy and his book to frame the technical achievements of the expedition for broad public consumption. The approval of readers who would appreciate his style and heroism was critical for the consolidation of his polemical position in the dispute with Cassini, and also for his identity as a successful man of science and man of letters. He fascinated influential women of upper-crust Parisian society with his stories about Lapland. The published account, which developed the heroic plot to buttress scientific arguments and results, left out many details of life in the far north, details of interest to a public that traded gossip and curiosities more easily than numbers and diagrams. To this public, centered on a circle of friends and admirers, Maupertuis played the role of the seasoned traveler whose eyewitness accounts of strange phenomena provoked wonder, and sometimes amusement. He used his status as emissary of the Academy to guarantee the accuracy of his reports, while he simultaneously appealed to the curiosity of an elite audience to gain a following for his scientific conclusions.

While he was in Lapland, he corresponded with a number of women in France, although only one of these letters survives.[104] Judging from echoes of its details

103. Voltaire to M, 22 May 1738, Volt. *Corr.,* 89:131.

104. Châtelet continued her correspondence with Maupertuis while he was in Lapland; his letters to her do not survive. She refers to his letters to Mme. de Richelieu, 18 July, 1736, *Lettres*

that show up in other correspondence, the letter and others like it made the rounds of salon society; its author knew it would be circulated and discussed.[105] This particular letter, written to the comtesse de Verteillac, discusses truth and skepticism, as well as various curious details of daily life in the north. Its recipient, who was known in Paris for her intellectual interests and her collection of curiosities, had expressed skepticism about some aspects of Maupertuis's report on local customs and living conditions. He assured her that his reports were trustworthy, however incredible they may have sounded to Parisian ears.

> When one has traveled, Madame, only from one's home to the Tuileries or to the Opera, one has very limited ideas about all the wonderful things there are to see; but just because you have seen nothing, there is no need to doubt the good faith of travelers as honest as ourselves; surely the imagination has no need to make anything up to find strange things here. If I had told you, Madame, that last summer my tent was pitched on a pair of shoes, you would not have believed it; you will believe it when you see them.[106]

He brought the shoes in question home for her collection. Much of the letter concerns the local people, whom he calls "Lapps," and their interactions with the Frenchmen.[107] Although these people and their habits remained strange, Maupertuis maintained an ironic distance from his own place, and that of his companions, in this foreign context. He tried to make his correspondents see that to the locals the visiting Frenchmen themselves were the curiosities:

> They were as surprised to see our faces as we were to see theirs. They had great difficulty in divining what this great instrument was that we took everywhere with us, to which we built temples on the mountaintops, where someone watched over it every night, which we hardly dared to touch, and which we never approached without trembling and often on our knees. The most sensible among them think it is some divinity that we worship, but the freethinkers [*esprits forts*] simply think we are crazy.[108]

---

1:120–121. For letters from Lapland circulating around Paris, see Châtelet to Algarotti, 18 October 1736, *Lettres,* 1:123; Châtelet to M, 1 December 1736, *Lettres,* 1:125.

105. See also letter from Le Monnier to Pont-Charost, extracted in *Mercure de France* (December 1736): 2730–33. This letter includes a description of life in Lapland.

106. M to Mme. de Vert[e]illac, 6 April 1737, in *Mélanges publiés par la Société des bibliophiles français,* 4. (There is no information about the manuscript used to establish the text.) On the comtesse de Verteillac, see Badinter, *Passions intellectuels,* 85.

107. Usually the travelers used this term for the indigenous population of Sami people who lived seminomadically, fishing and herding reindeer; sometimes they used it more broadly for anyone living in the region. On contemporary conceptions of the Sami and the label "Lapp," see Koerner, *Linnaeus,* 56–59.

108. M to Mme. du Vert[e]illac, in *Mélanges publiés par la Société des bibliophiles français,* 7.

When he remarked, "What is truly curious in this country is to see me here," he was responding to his Parisian audience's interest in how Frenchmen could live among people so lacking in the rudiments of civility.[109] When Mme. du Châtelet wrote to him in this unfamiliar setting, she started by asking, "Well, how are you finding the company of the Lapps? By force of [your] imagination have you managed to give them some?"[110] She could not believe that social life was possible among people with no imagination, although Maupertuis made a point of saying that he was quite happy there, and led a convivial life (albeit with the town-dwellers, and not with the indigenous herders). "In the frozen region," he insisted, "there are some very attractive and pleasant people; they sing and dance and do everything that we do in Paris."[111] These accounts of winter diversions in Lapland became part of common knowledge about the expedition.

He played with the notion of becoming naturalized as a northerner, as a "Lapp." The spring was as cold as the worst Parisian winter, he said,

> but I now do not feel the cold any more than a Lapp. I only fear no longer being capable of standing the heat of the next winter in Paris. During these cold spells, which we still have here even though our days are getting to be twenty hours long, our Lapps [*nos Lapons*] do not bother to put up their tents, and sleep on the ground, in the courtyard of the house from which I am writing to you, with no other mattress than the snow. You will not believe it, Madame, but it seems to me that I would do just the same; bodies are so much more docile than minds. If I could chase the bad dreams of the southern countries from my [mind], I would be the happiest Lapp in the world.[112]

However implausible his friends back in France may have found it, Maupertuis was indeed comfortable in this strange and extreme land. Mme. du Châtelet persisted in seeing his voyage more as the result of stubborn bravado than scientific necessity and refused to admit that such unimaginable wilderness could arouse such favorable sentiments. "Is it possible that I must still write to you at the pole? I did not believe that it could be one of those passions that increases with fulfillment [*de ces passions que la jouissance augmente*]." Her own imagination failed her, unable to picture the person she had known in the luxurious surroundings of high society, in the frigid and uncivilized climate of Lapland.[113]

109. Le Monnier had also written home about his ludicrous appearance in native dress: "you would have laughed to see me." (LeMonnier to Pont-Charost, *Mercure de France* (December 1736), 2731.

110. Châtelet to Maupertuis, 18 July 1736, *Lettres,* 1:120.

111. Maupertuis to Mme. de Vert[e]illac, in *Mélanges publiés par la Société des bibliophiles français,* 9.

112. Ibid.

113. Châtelet to M, 1 Dec. 1736, *Lettres* 1:125.

For Maupertuis, life was simpler in Lapland than in Paris. His position was clear, as the respected leader of a willing team, with no one to oppose him or question his judgment.[114] He enjoyed the attentions of his hosts as well as the novelty of the surroundings. Of course, the poignancy of the setting was enhanced by knowing that he would return to the obligations and power struggles of the city and the Academy. Everything had gone remarkably smoothly in Torneå; by the spring, when he was writing to the countess, he was confident that the Academy's interests had been well served, with the enterprise coming to a brilliantly successful conclusion. Clairaut's calculations indicated that their results would be unassailable. All they had to do was to complete the final observations and make the long journey home. As it turned out, the final months in the cold climate, with all the numbers falling into place and the instruments working just as they should, were a pleasant interlude before the storm that broke when the academicians returned to their lives in the city.

114. On Maupertuis's qualities as leader, see Outhier, *Journal d'un voyage au nord,* 44–46.

# 5

# The Polemical Aftermath of the Lapland Expedition

IN THE BATTLE OVER THE SHAPE OF THE EARTH, conceptual issues associated with the labels "Cartesian" and "Newtonian" were embedded in concerns about other features of scientific practice. As the dispute unfolded, cosmological or theoretical positions were no more essential to the dispute than were claims about authority and expertise, the role of analytical mathematics in astronomy and cosmology, the design and use of astronomical instruments, and the appropriate presentation of quantitative results. Personal animosities also entered into the mix, as time went on. The different expeditions (France, Lapland, and eventually Peru) produced irreconcilable numbers, which astronomers and mathematicians deployed for their own ends, both offensively and defensively. After the return of the Lapland expedition, the measurements by all parties became subject to interpretation, skepticism, and reconfiguration. Far from being self-evident, the assessment of precision and error drew on many different intellectual, mathematical, instrumental, political, and textual resources. Both sides then used their alleged exactitude to construct and defend rival scientific programs and practices.

Cassini defended the integrity of his measures of France, and those of the previous generation of Paris astronomers as well, by stressing their adherence to accepted standards of astronomical practice. From this point of view, mathematical brilliance was no substitute for long experience with astronomical observation. Even before the equatorial and polar expeditions had left—in response to the *Journal historique*'s review of Poleni's geodetical proposals—Cassini had asserted his own claim to a specialized astronomer's knowledge that was beyond the reach of a mere mathematician who had never measured a stellar position. Poleni's anonymous reviewer had suggested that a slightly different selection of available data for the latitudes of the endpoints of the French meridian would have led Cassini to the opposite conclusion about the earth's bulge. The effect was small, and very close to the margin of error of the instruments. Cassini explained, for the benefit of "those who do not have a perfect knowledge of the practice of Astronomy," that astronomers must draw their conclusions based on "the choice one must make between different observations, especially when one is using instruments of different structure and size [for the different observations]. Without this discernment, everything will be confused in astronomy, and everyone will be free to doubt all astronomical observations that require some precision, reject the good ones and adopt the less exact ones when they conform more closely to our hypotheses." This indiscriminate questioning posed a terrible danger for astronomy, "against which one cannot take too many precautions."[1] Cassini stood by his interpretation because it rested on his judgment about how to assess the performance of particular instruments and evaluate possible errors. Cassini was defending not only specific observations made by his father, himself, his son, and their associates going back to the 1660s, but also the practice of astronomy and geography as established by Cassini I at the foundation of the Observatory. The defense of the elongated earth was thus the defense of a tradition, of the integrity of many years of observations, and more personally, of a family now in its third generation of practicing astronomy.

There were a few in the Academy for whom the elongated earth represented "Cartesian" celestial physics; there were others, like Mairan, who struggled to find an explanation that would accommodate all the numbers without challenging the observational techniques of either party.[2] For the astronomers and mathematicians returning from the north, their results represented a vindication of Newtonian theory, of English instruments, and of analytic calculational techniques. In the public perception, the travelers were "Newtonian" and represented the new generation of French science, ready to try both theoretical and practical innovations. The travelers agreed, but they saw themselves as doing more than confirming something Newton had already figured out. They argued that they had achieved a degree of precision unprecedented in such observations, and their boasts of intrepidity and stamina were secondary to their claims for the integrity of their numbers and the meticulousness of their procedures. But their observations and the overall design of the scheme were still subject to scrutiny by the Academy, where certification of their validity was by no means a foregone conclusion. The personal stakes involved, and especially the matter of reputation, only heightened the acrimony of the dispute over technical matters.

1. AS p-v, 11 December 1734; J. Cassini, "Réponse aux remarques qui ont été faites dans le Journal Historique de la République des Lettres," *MAS* 1732, 499.

2. Mairan deposited a manuscript with the Academy (apparently not presented formally) that built on his 1720 paper, looking for a model for gravity that would generate a shape to accommodate the measurements in both Lapland and France ("Manuscrit de M. de Mairan," 29 November 1737, AS, pochette de séance 1737). Cassini II defended vortex physics: "De la manière de concilier dans l'hypothese des tourbillons les deux regles de Képler" (*MAS* 1736, 233–43). Privat de Molières also defended vortices vociferously and repeatedly, but he remained a marginal figure in the Academy. On Molières, see Aiton, *Vortex Theory.*

Maupertuis did not anticipate the severity of the attacks on his numbers. A triumphant letter to Bernoulli, written on the return voyage just before reaching France, communicated his elation with the astronomical observations, the surveying operation, the pendulum experiments, and the efficiency of the whole expedition. "I hope that everyone will be content with this work, and that the question about the shape of the earth will be forever settled. I will not speak of the life that we had to lead to reach this happy goal, of the cold, the discomfort, the fatigue, the dangers; all that is past, and of fifteen individuals that I led [in this endeavor], not a single one is dead or ill."[3] Maupertuis had been away from Paris for some eighteen months when he made this optimistic assessment. Apparently, he had forgotten the degree to which Cassini was invested in the contrary result, and underestimated the vigor of his response.[4]

As soon as he returned, Maupertuis and his companions reported their results to the comte de Maurepas and Cardinal Fleury at Versailles; Maurepas presented them to the king and gave them dinner.[5] There were only a few meetings of the Academy before the annual two-month vacation started in September. Although several of the "old academicians" were favorably impressed with the initial informal report, Cassini challenged the Lapland measurements immediately, voicing doubts about measurements made with the vaunted Graham sector.[6] "[Cassini's] main quibble," Maupertuis reported to Celsius, "concerns the fact that we did not turn the sector in both directions to assure against any disturbance that could have arisen in moving it [from one site to another]."[7] This technical objection referred to the standard practice for meridian observations of verifying the placement of the instrument by aligning it with the meridian first in one direction and then physically turning it to check its alignment in the opposite direction.[8] Cassini continued to insist on this point throughout the dispute, charging the Lapland team

3. M to JB, 11 August 1737, BEB.

4. Maupertuis's claims for his results occasioned considerable tension with the older Bernoulli, who accused the travelers of prejudice in favor of the flattened earth. JB to M, 8 May 1735, BEB. See Brown, "From London to Lapland."

5. Outhier, *Journal d'un voyage au nord*, 253.

6. Mairan, for example, told Bernoulli that the measures were done with "infinite precautions," but mentioned Cassini's displeasure (Mairan to JB, 3 September 1737, BEB). "M. de Cassini, comme vous jugez, ne se rend pas, vu la délicatesse des observations" (ibid.).

7. M to Celsius, 6 September 1737, copy in AS, dossier Maupertuis.

8. Cassini described this practice for quadrants, in his response to Poleni ("Réponse aux remarques," *MAS* 1732, 501). Cassini de Thury discussed it further in a paper read to the Academy just after the departure of Maupertuis's expedition. "This method [of reversal] is extremely difficult in practice; nevertheless the exactitude of all observations depends on it" (Cassini de Thury, "Des précautions que l'on doit prendre pour observer le plus exactement qu'il est possible, les hauteurs des étoiles," *MAS* 1736, 203–15).

with using unorthodox observational techniques, although he had had no direct experience with this particular instrument. In response, Maupertuis and his collaborators recounted the measures they had taken to ensure reliable observations. This meant a detailed defense of the design and use of the zenith sector, which Le Monnier set up in his observatory at the Collège Royal in Paris.

Cassini's opening salvo set the tone for the ensuing polemics, which continued for almost three years. The initial exchanges took place behind closed doors in the Academy. As usual, the secretary did not record the uncivil proceedings, but Clairaut and Maupertuis quickly and confidently prepared a written response to Cassini's objections, in order to settle the matter before the Academy adjourned for vacation. Their document suggested that meridian verification was far from a specialized or delicate operation. At issue was the determination of the time a given star crossed the meridian, as observed through the telescope mounted on the zenith sector. "Although one can hardly believe," Clairaut complained, "that six astronomers and mathematicians could be incapable of tracing a meridian, placing an instrument in the right direction and calculating the necessary precision, here are the measures we took for this operation—such a simple one, by the way, that the details have never been demanded of anyone before."[9] He went on to work through the repercussions of a hypothetical error in placement of the sector on the stellar positions measured, concluding that for stars very close to the zenith, such errors would be negligible anyway. Furthermore, while still in Lapland, he and his companions had checked their observed transit times against a calculation of the expected time and found only negligible differences. Turning Cassini's critique back on him, the defense continued, "It is true that there are cases where a mistake in position would be more dangerous, namely those where the stars one is observing are farther from the zenith, like those chosen by Mssrs. Picard and Cassini for their observations."[10]

Maupertuis expected the controversy to die down once Cassini's anger had spent itself. In the beginning of September, he reported back to Celsius that Cassini had "contained his bad humor a bit in the beginning, but he finally ceased to be the master of it, and raised petty objections against us until the meeting yesterday, when he finally shut up."[11] He wrote to Daniel Bernoulli that the expedition had met with "the satisfaction of the court and the Academy," always

9. Manuscript in AS, pochette de séance, 31 August 1737. The paper was delivered at the second meeting after the expedition returned to Paris. The minutes say the piece was read by Maupertuis; the manuscript (preserved separately in the Academy archives) gives Clairaut as the author.

10. Ibid. Clairaut worked out the possible error in position for several stars observed by Picard, used in calculating the length of the degree in France.

11. M to Celsius, 6 September 1737, copy in AS, dossier Maupertuis.

excepting Cassini. "But his discontent does not reflect on the result, and if anything has ever been demonstrated in practical geometry, it is our operation."[12] He also reported on the expedition to James Bradley in England, and asked for observations of certain stars that might corroborate the aberration calculations Clairaut had made.[13] Clairaut also thought the matter settled, when he wrote to the Royal Society, "M. Cassini, who has a great deal to lose by these operations, wanted to make some objections; but we responded immediately in such a way that he held his tongue about it, and our Academy is finally convinced of the truth of the flattening toward the poles."[14]

The public learned of the results of the expedition in the September issue of the *Mercure de France,* which mentioned that the Lapland observations had determined "that the earth is a spheroid flattened at the poles, as Mssrs. Huygens, Newton, and several other great geometers had thought, based on theory."[15] The journalist did not report on the grumbling from Cassini. By the time the Academy reconvened for its public meeting in November, Maupertuis knew that the consensus was not quite so stable. He prepared an account of the Arctic adventure for a large and attentive audience, emphasizing the consistency of the measurements and the hardships endured in the name of precision. His address juxtaposed narrative and numbers, setting the numbers against a backdrop of colorful accounts of the observers' experience. At this point, he was still acting graciously, at least in public. He artfully displaced the abstract issues—particularly the question of Newtonian attraction—onto a discussion of observations, calculations, and error estimates. The *Mercure de France* reported, "The marked interest of the most numerous assembly there has ever been in any academic meeting sufficiently applauds the work of these illustrious voyagers, the finesse and exactitude of their operations, and the clear and elegant manner in which M. de Maupertuis made everyone capable of judging it."[16] Even Réaumur, one of the old guard of the Academy, recounted the success of the speech in the most lavish terms:

> "M. de Maupertuis's report lasted more than an hour and a half, and everyone in the audience found it too brief. Also, it is written soberly as much as pleasantly; the facts

12. M to Daniel Bernoulli, 8 September 1737, copy in BEB; original in Geneva, Société d'histoire, nr. 243.

13. M to James Bradley, 27 September, 1737, British Library, Sloane Papers, Fr. 4285; English translation in Tweedie, *James Stirling,* 77–79.

14. Clairaut to Mortimer, 15 September 1737, RS, *Letter Book,* vol. 24, fols. 23–24.

15. *Mercure de France* (September 1737): 2032. The *Gazette d'Utrecht* had mentioned the return of the expedition, and their support for the Newtonian position, slightly earlier. See Chouillet, "Rôle de la presse périodique."

16. *Mercure de France* (November 1737): 2462.

> that the public likes to hear are distributed in a manner appropriate to hold the attention of those who would be bored by the recital of more technical operations. . . . The gathering of the audience was prodigious; there could be no thought of closing the doors; part of the gallery was filled with those who were not able to enter into the hall. If this account had been printed . . . just after these gentlemen arrived [back in Paris], many of the nasty comments made in the cafés during this vacation would not have been spoken. It revealed that everything was done with the most scrupulous exactitude.[17]

Réaumur particularly appreciated his colleague's finesse in avoiding negative or inflammatory statements, as well as his engaging style. Later, when the controversy became more vicious, the naturalist turned against Maupertuis, but for the time being, he was a valuable supporter.

There the matter might have rested, but for Cassini's continuing objections and Maupertuis's prickly resentment. As the lines of battle were drawn, the academic setting defined the options available to the combatants. Officially, the Academy maintained its habitual posture of decorum and fraternity, polite disagreement rather than vicious combat. In closed sessions, however, Maupertuis accused his adversary of hypocrisy, arguing that Cassini had never objected when Picard omitted to reverse his own sector, nor had Cassini always performed this allegedly essential operation himself. "This paper dishonors him in the Academy; nevertheless it has perhaps only made him more violent in his protests to the ministers," Maupertuis told Celsius.[18]

As Maupertuis went on the offensive, the polemics crossed back and forth over the boundaries between academy and salon, between expert readers of technical memoirs and the readers of literary scandal sheets. "We had hardly returned," Maupertuis complained, "when one saw in all the newspapers that we had done nothing certain or useful; and although I don't suspect Cassini of having personally taken charge of this, one can hardly doubt that it must be one of his emissaries."[19] By complaining of malicious treatment at the hands of his enemies, Maupertuis gained the sympathy of the aristocratic women whose friendship he cultivated assiduously.[20] In a move undoubtedly calculated to stimulate that sympathy, shortly

17. Réaumur to [Bignon], 14 November 1737, Bibliothèque Centrale du Muséum d'histoire naturelle, MS 1998, no. 245 (addressee identified by Badinter, *Les passions intellectuels,* 93).

18. M to Celsius, 31 January 1738, copy in AS, dossier Maupertuis.

19. Maupertuis to Celsius, 31 January 1738, copy in AS, dossier Maupertuis. For accounts of the Lapland expedition in the press, see Chouillet, "Rôle de la presse périodique"; Nordmann, "L'expédition de Maupertuis et Celsius," 94.

20. Hervé, "Les correspondantes de Maupertuis," 755–56; Brunet, *Maupertuis,* 1:60. For complaints, see M to JB II, 23 August 1738, BEB.

after his return to Paris he refused a pension awarded by Cardinal Fleury, the minister of finance, declaring the amount insultingly low. Emilie du Châtelet linked the "mediocre" pensions awarded to the expedition members to "persecution" orchestrated by Cassini and his Jesuit allies, and applauded Maupertuis's decision to divide his pension among his colleagues as "noble disinterestedness," assuring him that everyone's "esteem and consideration" for him had risen because of this action.[21] A month later, however, she was encouraging him to accept the minister's favor, however unworthy the amount, because the continuing stubborn refusal "would make enemies for you of those who should be and who wish to be your admirers."[22]

Cassini doggedly refused to admit the adequacy of the various accuracy checks his opponents had performed. In desperation, Maupertuis wrote to England for confirmation of verification practices for the Graham zenith sector. "There have been great wrangles and disputes in France about this measurement," John Machin informed his fellow mathematician James Stirling. "Cassini has endeavored to bring the exactness of it into Question. . . . So that Mr. Maupertuis was put to the necessity of procuring from England a certificate concerning the construction of Mr. Graham's Instrument, to show that it did not need that sort of verification."[23] The recourse to English expertise can hardly have mollified Cassini. Maupertuis read a translation of the English "certification," probably written by Bradley or Graham, to the Academy in February of 1738. It claimed that turning the instrument would have been useless in detecting any inaccuracy; although normal practice with other instruments, the design of this one made such an operation redundant. "We think that this objection is completely unfounded, that this [reversal] is not at all necessary and that it cannot detract from the truth of the observations, which is absolutely independent of this reversal." Any deformation or misalignment caused by moving and reassembly would have been easily detected through the routine checks performed before each use. Finally, the English expert reiterated Maupertuis's central defense, that the internal agreement of two sets of measurements of different stars, taken six months apart, definitively confirmed the sector's dependability.[24]

21. Châtelet to Algarotti, 10 January 1738, *Lettres,* 1:206,; Châtelet to Maupertuis, 10 January 1738, ibid., 1:208; records of pensions in *Registres du secretaire de la maison du Roy,* Archives nationales, O1/81, fol. 376, 1 November 1737.

22. Châtelet to M, [10 February 1738], *Lettres,* 1:218. Châtelet had discussed this with the duchesse de Saint-Pierre and other friends of Maupertuis; their advice was meant to protect him from ministerial pique.

23. Machin to Stirling, 22 June 1738, in Tweedie, *James Stirling,* 174.

24. AS p-v, 15 February 1738, fols. 34–35 (text of translation of English letter read by Mau-

As we have seen, the zenith sector had been used for English astronomical discoveries closely associated with Newton's celestial mechanics. These associations tainted the instrument for Cassini, prompting his attack on Maupertuis's breach of astronomical practice. But the instrument's link to Bradley was precisely what recommended it to Maupertuis. In the 1720s, Bradley's observations with a Graham zenith sector led him to detect the aberration of starlight, a very small apparent motion of the stars due to the finite speed of light and the orbital motion of the earth. When Maupertuis corrected his own observations for the effects of aberration, he used this newly discovered "English" phenomenon against Cassini, flaunting the up-to-the-minute precision of his results. Clairaut became interested in aberration as a result of the controversy as well. Just when acrimony in the Academy was at its height, Clairaut read several papers on the theory of aberration.[25] Maupertuis's incorporation of Bradley's corrections advertised his adherence to standards of astronomical practice foreign to the French astronomers, standards which aspired to a level of precision only possible with the best instruments. Although the corrections were negligible quantities, they were useful rhetorical weapons just the same. In fact, Maupertuis consulted directly with Bradley to determine corrections for aberration and for the precession of the equinoxes: "M. Bradley very kindly shared with me his latest discoveries on the motions of the stars; and communicated to me the necessary correction to the arcs [we] observed. . . . For greater exactitude, we will use the corrections as he sent them to us, even though they do not differ sensibly from those we have just calculated."[26] He also found that when Picard's numbers were corrected for these apparent motions, using Bradley's tables, the difference between the French degree and that measured in Lapland was even greater than it first appeared.

Maupertuis turned to Celsius, an outsider and a foreigner, for assistance in attacking the French measurements more directly. In encouraging his Swedish colleague to go on the offensive against Cassini, Maupertuis excused his own reluctance: "Collegial scruples [*quelques raisons de compagnie*] have kept me from doing it myself, but I don't know if they will continue to do so much longer."[27] He went so far as to send Celsius a list of page numbers in Cassini's book where he might

pertuis). "De plus le parfaite accord des observations comparées et examinées nous paroit encore une marque suffisante de leur vérité" (ibid.).

25. Maupertuis mentions consulting Bradley in *Figure de la terre,* 123; Clairaut, "De l'aberration apparente des étoiles," AS p-v, 11 December 1737 (published in *MAS* 1737); also Clairaut, "L'aberration des fixes causée par le mouvement de la lumière," AS p-v, 12 February 1738.

26. Maupertuis, *Figure de la terre,* 123. He explained the phenomenon of aberration on 43–45, with an account of Bradley's discovery.

27. M to Celsius, 31 January 1738, copy in AS, dossier Maupertuis.

find ammunition for the attack. Celsius quickly obliged with a scathing pamphlet calling into question the overall reliability of French instruments and denigrating the observational ability of the founder of the Cassini astronomical dynasty, Jean-Dominique, and the selective reporting of data by Jacques Cassini, currently the senior astronomer at the Paris Observatory. He concluded by charging that "Cassini's observations, terrestrial and celestial, in the southern part of France, are sufficiently uncertain that it is impossible to deduce the shape of the earth from them."[28] The astronomer J. N. Delisle, who read the pamphlet in St. Petersburg, objected strenuously to the harshness of the accusations: "I find the conclusion that you draw at the end of your dissertation very rude. . . . Could we not hope for a softening of this position?" But Celsius replied that his opponent had brought the attack on himself: "[He] is surely the aggressor, since he attacked our observations in the Academy and afterwards spread a rumor in the gazettes about the uncertainty of our operations."[29] Cassini rose to the bait and defended himself and his father in a densely argued pamphlet of his own.[30] Celsius and Cassini exchanged accusations primarily in the technical language of astronomy, but the conflict filtered out into the rumor mill, in cafés and salons, and into the press. The harshness of the attack by Celsius aggravated the situation and made the debate a matter of family honor for Cassini. Maupertuis viewed the questions about his measurements as affronts to his honor as well, which ultimately led him to humiliate his adversary. For the moment, however, he refrained from open personal attacks and sought instead to appeal to a wider public by printing his account of the expedition.

Rumors circulated about official interference holding up the publication of Maupertuis's book.[31] As the work of an academician, it did not need an official privilege from the royal censor. But some observers speculated that publication may have been held up because Maupertuis's flattened earth was associated with Newtonianism and hence with Voltaire, whose *Philosophical Letters* had been banned in 1734. At just this moment, Voltaire was trying unsuccessfully to get his *Elements of Newtonian Philosophy* published in Paris.[32] By April 1738, Maupertuis's book was still not out, and he delivered the preface as yet another public address. This was

28. Celsius, *De observationibus pro figura telluris,* 20.

29. Delisle to Celsius, 1 August 1738; Celsius to Delisle, 16 August 1738 (extracts by G. Bigourdan, MS Obs., 1029a).

30. Cassini read his response to the Academy 30 April and 3 May 1738, AS p-v; it was published as *Réponse à la dissertation de M. Celsius* (Paris, 1738).

31. For the rumors, see Châtelet to Algarotti, 10 January 1738, *Letters,* 1:206; Souciet to Delisle, 25 September 1738, MS Obs., Delisle papers.

32. Châtelet to M, [around 10 February] 1738, *Lettres,* 1:216: "Mr. de V. a perdu, non sans re-

the first time he referred disparagingly in public to Cassini's reporting and observing practices, albeit in carefully circumspect language. Under the guise of explaining his decision to publish complete observations, he charged "those who have given us similar works" with keeping crucial information from the public.[33] At the same session, Du Fay read letters from South America by Godin, Bouguer, and La Condamine "of which several proved the correctness of those done in the North."[34] Jacques-Elie Gastelier, the author of regular news reports from Paris to a correspondent in the country, noted at the end of May that printers were working around the clock to get Maupertuis's book out. "For several days now the royal press houses about sixty printers who work day and night with all the secrecy imaginable. They eat in the workshop, and they had mass said there for the holiday, so they would not have to go out."[35] The secrecy surrounding the production certainly heightened expectations for the book, though its contents were hardly scandalous.

When it did finally appear, Maupertuis personally distributed many copies to people all over Europe, with sixteen going to various colleagues in England, including Bradley, Graham, Machin, and other prominent Newtonian mathematicians and astronomers.[36] It was immediately translated into English as well. The English scientific community was well aware of the controversy in France. "It is said that France is not very happy with you," he heard from James Jurin. "That is very surprising. . . . Here everyone heaps praises on you for your skill, ability, extreme exactitude and inviolable fidelity in giving us the observations just as they were made, and not adjusted in favor of any hypothesis. One cannot forget the courage that you Gentlemen have shown on this occasion, a quality very rare in philosophers, but as necessary to them in some encounters as to soldiers. There are many people who gaily engage in battles but who would refuse to confront the ice of Lapland."[37] And the *Mercure de France* reported, "one cannot think of all these obstacles and difficulties without admiring the courage of those who subjected

---

gret, l'espérance de faire imprimer son livre en France, il n'est fait que pour des français . . ." (ibid.). For the tortured publication history of Voltaire's *Elements,* see Walters and Barber, "Introduction."

33. Maupertuis, *Figure de la terre,* iii.

34. *Mercure de France* (April 1738): 737. Letters from South America were also read to Academy sessions on 1 and 5 March (AS p-v).

35. Gastelier, 29 May 1738, in Gastelier, in *Lettres sur les affaires du temps,* 90. Gastelier had heard that it was to sell for 6 *livres,* unbound.

36. M to Mortimer, 25 May 1738. RS Archives, M.M. 20. Maupertuis also sent the book to the crown prince of Prussia, probably at Voltaire's instigation (Frederick to M, 20 June 1738, Koser, *Briefwechsel,* 185.

37. Jurin to M, 5 October 1738, *The Correspondence of James Jurin,* 419. For another testimony to the popularity of Maupertuis's book in England, see Le Blanc, *Lettres d'un françois,* 3:245–46.

themselves to and surmounted them."[38] The numbers, displayed like prizes won by heroic effort, included tables of stellar observations, angles used in the triangulation, observations of the sun for determining the meridian, the linear measurement of the baseline, observations by all five observers for verifying the division of the sector, measurements of gravity with a pendulum, and so on. Calculations and diagrams explaining the complicated chain of inferences and corrections that led from the chain of triangles to the conclusion for the flattened earth figured prominently as well (fig. 10).

Even with the book in print, however, the controversy dragged on. The continued absence of the South American team prolonged the dispute, though their pendulum experiments had already confirmed the variation in gravity that indicated a flattened earth. But as long as the geodetic results remained unfinished, the Academy had to consider the question unsettled. On that score, news from Peru

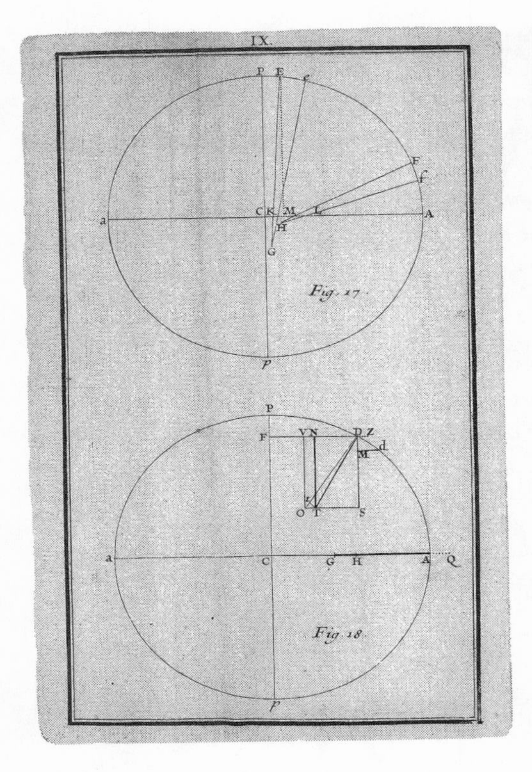


Fig. 10. Diagram showing comparison of lengths of degrees at different latitudes. Note slightly flattened shape of circle. Maupertuis, *La figure de la terre* (1738).

38. *Mercure de France* (November 1737): 2463. Mme. de Graffigny reported that it was not selling very well, "in spite of being accessible to everyone and a part of it can be read like a novel" (Graffigny to Devaux, 27 August [1739], *Correspondance,* 2:122).

was not encouraging, as Maupertuis wrote to Celsius in Uppsala. "After some terrible scenes, they have not been speaking to each other since they last wrote six months ago, and how can such a task be done without speaking to each other? I fear that in the end one of them will die and the two others will come back, one from the east and the other from the west, without having done anything."[39] All parties invoked the missing equatorial data as the definitive arbiter. In the meantime, Maupertuis despaired of convincing his opponents in the Academy: "I am deprived of the advantage I had hoped to get from their measurements, and until their return, which will perhaps never happen . . . M. Cassini's party will hang on and even triumph."[40] He feared that the equatorial expedition might finish its task "so late that everyone will have forgotten that I went to the Arctic Circle."[41]

To escape the fraught atmosphere, Maupertuis left Paris for the calm of Saint-Malo. Where was he to find corroboration for his results, given the apparent disintegration of the equatorial expedition and the impossibility of a return to Lapland? He discussed all this in letters to Celsius, and they toyed with the idea of measuring another degree on the ice of Lake Wättern in Sweden; the arc measurement could then be accomplished with the Graham sector, and they would have another length for comparison with the Lapland degree. Initially, it was to be done in secret, with only local help for the manual labor. "Once the [base] is measured, I will ask M. de Maurepas next spring for the sector, and he will surely not refuse it to me. . . . I would not want to bring the sector with me [initially], nor to start by announcing such an enterprise when we do not know if it will succeed. . . . Until then I would like everything to depend only on you and myself." Maupertuis was sure that he and Celsius could accomplish this before the return of his colleagues from South America and was anxious to have a number "independent of the measurements of Picard and of Godin."[42] He planned to go to Sweden at his own expense and then to surprise his enemies with new and definitive measurements. But in the end, Celsius decided, to Maupertuis's profound disappointment, that the project would be too risky, because the lake did not freeze consistently, and he seemed reluctant to form an alternative plan. "I am sorry to be the only one to feel chagrined at seeing that they are not happy with what we have done, and for that I swear that there is nothing I would not have undertaken."[43] He had

39. M to Celsius, 22 December 1738, copy in AS, dossier Maupertuis. Several letters from the travelers had arrived at the Academy the week before (AS p-v, 17 December 1738).

40. M to Johann II Bernoulli, 21 December 1738, BEB.

41. M to Celsius, 22 December 1738, copy in AS, dossier Maupertuis.

42. M to Celsius, 8 June 1738, copy in AS, dossier Maupertuis.

43. M to Celsius, 22 December 1738, copy in AS, dossier Maupertuis. Celsius's side of the correspondence does not survive.

held out high hopes for Celsius, who had been willing to attack Cassini, and whose residence in Sweden would have made a secret expedition possible; this rebuke marked the end of their correspondence.

## Visitors from Lapland

Maupertuis had hoped to dispel his dissatisfaction with a trip to Sweden; instead, he went to Basel and then on to Cirey to visit Châtelet and Voltaire, accompanied by the younger Johann Bernoulli. But just before he left Paris, two unexpected visitors caused him some embarrassment. The arrival in France of two young women from Torneå caused a sensation in high society. In and of itself, this is a trivial incident, but it gives some insight into the way gossip made its way into print, and how Maupertuis's public image encompassed both the heroic and the ridiculous. It was well known in Paris that the Frenchmen had been free with their affections during the long Arctic winter. Maupertuis had even included at the end of his letter to the comtesse de Verteillac a "song" he had written in honor of a local maiden.[44] These verses made the rounds of Parisian society, prompting comments and gossip for years. Their author was not particularly embarrassed by this; in July 1739, Maupertuis was regaling Graffigny with the same poem: "Maupertuis forced on me some couplets he wrote in Torneå for a Finnish woman he was in love with," she noted with distaste.[45] As for the actual women who appeared in Paris in November of 1738, very little about them can be traced back to reliable sources. It was commonly said that the women (always referred to as *lappones*) had been brought to Paris by Maupertuis, as souvenirs of his voyage. According to Jérôme Lalande, writing many years later, Maupertuis "brought a *lappone* (or two) back with him to Paris," and F. M. Grimm recalled that "he claimed also to have conceived a violent passion for a young Lapp woman whom he

44. This had circulated in manuscript; it also appeared appended to the anonymous *Anecdotes physiques et morales,* a satire on the Lapland expedition. (The authorship of this book is uncertain, but there seems no reason not to attribute the poem to Maupertuis.) On the fictitious publication date and the doubtful attribution of the *Anecdotes* to Maupertuis, see Beeson, *Maupertuis,* 141–43.

45. Graffigny to Devaux, 21 July 1739, *Correspondance,* 2:68. The songs had a remarkably long life. La Beaumelle asked Maupertuis for a copy of them many years later: "Je vous prie de m'envoyer vos chansons sur Christine, mesdemoiselles de Montolieu me persécutent pour les avoir" (La Beaumelle to M, 2 June 1758, in LS, 249). The song was performed shortly thereafter: "Je vous remercie de votre chanson lapone que Mlles de Montolieu, qui la chantent ce soir, attendaient avec beaucoup d'impatience" (La Beaumelle (in Nîmes) to M, 12 June 1758, in LS, 251); and "[Madame du Boccage] a trouvé votre chanson lapone fort jolie, elle ne la connaissait point" (La Beaumelle to M, 14 June 1758, in LS, 253).

brought to France, and who [later] died there."[46] Voltaire could not resist the story either, when he apostrophized Maupertuis in verse:

> Ramenez des climats soumis aux 3 couronnes
> Vos perches, vos secteurs, et surtout deux Lapones.
>
> [Bring back from the climes governed by Sweden
> Your rulers, your sectors, and above all two Lapp women.][47]

In fact, although the women had come from Torneå, they were the daughters of the Swedish family Plänstrom, where Camus and Herbelot, the expedition's artist, had lodged for the winter.[48] Of bourgeois Swedish parentage (their father was apparently a merchant), they were as unlike the indigenous Sami reindeer herders as were the Frenchmen.

Maupertuis himself left little record of his reaction to their arrival in Paris. A letter to Celsius in early 1738 mentioned that he had left in Stockholm "two poor young ladies who do not want to return to Torneå, and to whom I would very much like to be of service and to be good for something." He identified one of them, whom he linked with Le Monnier, as "Mlle. Forsstrom," which may plausibly be read as "Plänstrom," the name given by Outhier to the household where Camus and Herbelot stayed.[49] Maupertuis's remark suggests that the young women traveled to Stockholm with the returning expedition, though this is not confirmed elsewhere; the judicious Abbé Outhier never mentioned the women. However they got to Stockholm, it is hard to imagine that Maupertuis wished them to come to Paris, where he would have been in some measure responsible for them and where they would have disrupted his social life considerably. How they traveled to Paris, and on whose initiative, remains a mystery; they arrived more

46. Lalande, *Bibliographie astronomique,* 407; "Il prétendait aussi avoir conçu une passion violente pour une jeune Lapone qu'il avait amenée en France, et qui y est morte" (Grimm, *Correspondance littéraire,* 7:149 (7 November 1766)). Lalande (1732–1807) knew Maupertuis in the 1750s in Berlin, but he had no firsthand knowledge of the story. All secondary accounts subsequently repeated Lalande's version, including Brunet, *Maupertuis,* Velluz, *Maupertuis,* and Beeson, *Maupertuis.*

47. Voltaire, *Quatrième discours de la modération en tout,* in *Oeuvres,* ed. Beuchot, 12:72–73. The poem was revised in the 1752 version, after Voltaire fell out with Maupertuis, to include these lines. The original version (1738) was laudatory rather than satirical.

48. Outhier, *Journal d'un voyage au nord,* 160.

49. M to Celsius, January 1738, copy in AS, dossier Maupertuis. "J'ay laisse à Stockholm 2 pauvres demoiselles qui ne veulent pas retourner à Tornea et à qui je voudrois bien rendre service et etre bon à quelque chose" (original autograph letter in Uppsala; I have seen only a copy in another hand). Des Essarts (recounting a much later court case) gives her maiden name as P***, consonant with Outhier's identification (*Causes célèbres et intéressantes,* 7:74).

than a year after the return of the expedition. They seem to have been remarkably determined and resourceful, even though they had little or no financial means. One of the earliest published versions of their story goes as follows: "Two young *lappones,* who are said to be daughters of the judge of Torneå, arrived recently. Our astronomers from the Academy converted them [to Catholicism] and procured for them the means to get to Sweden and to embark from there for France. They were received upon their arrival at Rouen by the representatives of our astronomical apostles, who brought them here. They were initially living with a young Arab, daughter of a rich African merchant, who escaped from her home with a portion of his treasure, to follow a Parisian slave, from whom she received the first lessons in Christianity."[50] This nicely conflates two stories of young women escaping from strange lands and strange religions, and shows that Lapland could participate in the contemporary taste for the exotic, but the account does not otherwise inspire confidence in the accuracy of the details. Such stories inspired embellishment as they circulated, especially since Maupertuis already had a reputation as a libertine man-about-town, equally happy to consort with duchesses and their maids.[51]

The first trace of the story in the correspondence of Maupertuis's circle comes from Mme. de Graffigny, then visiting Voltaire and Châtelet at their country retreat in Cirey, where she had heard a bowdlerized version in which the secretary of the expedition played the role of faithless lover: "The secretary of M. Clairaut, one of the voyagers to the pole, made love to a Lapp woman; he promised her marriage and left without keeping his word. The young lady has just arrived in Paris with her sister, to follow her lover. They landed at the home of M. Clairaut, who is sheltering them, although [he is] not very rich. The fiancé does not want to marry, and the young lady does not want to return." Clairaut had told Voltaire that he was trying to find a convent where they could live, at least temporarily. Meanwhile, Graffigny went on, "All Paris is going to his house to see these Lapp women. My God, how could one possibly be a Lapp?"[52] The young women found themselves on display as curiosities who had crossed over from one world to another, perhaps not quite realizing what they would find in Paris. Graffigny's expression of wonder and incomprehension captures the fascination, mingled with

50. La Barre de Beaumarchais, *Amusemens littéraires,* 3:44.

51. For one such rumor, see the following piece of gossip: "Il avait affecté une grande amitié pour la femme-de-chambre de madame la duchesse d'Aiguillon, qu'il voyait beaucoup; mais si l'on n'avait jamais dit dans le salon de madame d'Aiguillon, que Maupertuis était monté à l'entresol de demaoiselle Julie, je crois que sa liaison avec mademoiselle Julie aurait peu duré" (Grimm, *Correspondance littéraire,* 7:179 (7 November 1766).

52. Graffigny heard this from Voltaire, to whom Clairaut had written, in a letter now lost: "Le secrétaire de M. Clairaut, l'un des voyageurs aux poles, a fait l'amour à une Lapone; il lui a promis le

a bit of horror, they provoked in their viewers. Their ties to the mathematicians made them that much more of a curiosity. Maupertuis had regaled his public with stories of French gallantry thriving in Lapland, bringing civilization to the frozen outlands, and here was living proof.

Maupertuis mentioned the women in letters to his sister, written from Cirey in January 1739. He did not use their names, but at least referred to them as "young Swedish ladies," and displaced responsibility for the seduction onto the expedition artist, M. Herbelot.

> Our artist had made one of them sacred promises of marriage and told so many lies about his wealth, saying he was very rich and could make her a great lady, that this poor girl came to find him, and her sister, counting on sharing this imaginary fortune, accompanied her. Not only is this artist as poor as a church mouse, but he was unfaithful and married someone else. The poor girl was so hurt by this misfortune that she fell very ill and was not out of danger when I left Paris. Everyone was touched by her trouble and we have been led to hope that if [the ladies] change their religion, the court could do something for them. In the meantime, all the company of the polar circle have contributed [to a fund for their support] and we have placed them in a Paris convent to learn French and develop a taste, if possible, for our religion. I was quite astonished by the desire of these young ladies to come to France, not knowing at all that the youngest was coming there to marry the artist, nor anything about the promises he had made. It is quite embarrassing for us to have to find situations for two young ladies.[53]

---

mariage, et est parti sans tenir sa parole. La demoiselle vient d'arriver à Paris avec une soeur à elle, pour suivre son amant. Elles sont débarquées chez Mr Clairaut, qui les heberge, quoique tres médiocrement riche. L'épouseur ne veut point épouser, et la demoiselle ne veut point s'en retourner" (Graffigny to Devaux, 18 December 1738, *Correspondance* 1:231). Voltaire and Mme. du Châtelet wrote sympathetically to Maupertuis and contributed to a fund for the support of the young women (Voltaire to M, 20 December 1738, Volt., *Corr.* 89:452–54; Châtelet to M, 28 December 1738, *Lettres,* 1:284–85).

53. "Notre Dessinateur avoit fait à l'une des promesses de mariage si sacrées & tant de mensonges sur sa fortune, se disant fort riche la pouvant rendre grande dame, que cette pauvre fille l'est venue trouver & que sa soeur comptant partager cette fortune imaginaire, l'a accompagnée. Non seulement ce dessinateur est gueux comme un rat, mail il étoit un infidèle et en a epousé une autre. La pauvre fille a été si sensible à ce malheur qu'elle en est tombée fort malade et n'étoit pas encore hors de peril quand je suis partie de Paris. Tout le monde a été touché de son infortune et on nous a fait esperer que si elles changeoient de religion, la Cour pourroit faire quelque chose pour elles. toute la compagnie du cercle polaire en attendant s'est cottisée et nous les avons mises dans une communauté de Paris pour apprendre le françois, et prendre du goût, s'il est possible, pour notre religion. J'etois assés etonné de l'envie qu'avoient ces Dlles. de venir en france ne sachant point que la cadette y venoit pour epouser le dessinateur ny les promesses qu'il lui avoit faites. Cela est assés embarassant pour nous d'avoir des deux d.lles à placer" (M to Marie Magon (his sister), 15 January 1739, AS Fonds Maupertuis).

He thought of sending them to Saint-Malo to live with his sister, where they "could amuse a philosopher who would withdraw from the world to St.-Elier [the family estate]," but decided that such an arrangement would provoke too much gossip in the provincial town.[54] Instead, he invoked his connections to Maurepas and Fleury, and arranged small pensions for the sisters once they renounced Lutheranism. The elder entered a convent permanently, and the younger, Elisabeth, was protected and housed by the duchesse d'Aiguillon, who may also have been Maupertuis's lover at this time. Eventually, with a generous dowry supplied by the duchess, Elisabeth married a down-at-heels nobleman who mistreated and subsequently abandoned her, attempting to make off with the dowry. Her short-lived fame did her little good in the end, though she did succeed in getting out of Torneå.[55]

In Torneå, the sisters had provided welcome companionship for the travelers, and they no doubt seemed appealingly exotic. Once in Paris, they could hardly compete with the fashionable ladies of Maupertuis's acquaintance. None of the stories describes them as individuals, but only as *les lappones.* Until a series of lawsuits brought many years later by Elisabeth against her reprobate (and possibly bigamous) husband, they had no voice of their own at all. Shunted aside by the men they had known in the distant north, pitied by the Parisian women, they became part of the myth of the Lapland expedition. Voltaire immortalized them in *Micromegas* a few years later: "The geometers take their quadrants, their sectors, and two Lapp girls, and descend onto the fingers of the Sirian."[56]

54. Ibid. His sister did take in the unwed mother of Maupertuis's illegitimate son; the son, Philippe, was sent to China as a boy and returned to France after the death of his father, in 1760. Diderot reported on the arrival in Paris of Maupertuis's natural son, as a "curiosity" (Diderot to Sophie Volland, 28 October 1760, *Correspondance de Diderot,* 3:205). La Condamine gives Philippe's baptismal date as 1742 (La Condamine to JBII, 1 August 1766, BEB).

55. Nordenmark, *Andreas Celsius;* Des Essarts, *Causes célèbres et intéressantes,* "XLIVe cause," 7:49–114. Maupertuis would have seen Elisabeth before her marriage, at the country estate of the duchesse d'Aiguillon, where he spent a month in August 1741, but nothing is known of their relations (M to JBII, 1 August 1741, BEB, in which he announced his impending visit to the duchess). Châtelet provides details on the marriage: "Vous savez peut-être que la Lapone vient d'épouser un homme qui a dix mille livres de rentes et qui lui en assure sept. Elle a bien fait assurément de quitter Tornea" (Châtelet to JBII, 12 December 1745, *Lettres,* 2:146). Des Essarts gives a date for the marriage in 1743. According to his account of the court records, the husband's accusations sent Elisabeth to prison for four years; subsequently she sued him for the recovery of her possessions. The court proceedings between them dragged out for years (Des Essarts, *Causes célèbres et intéressantes,* "XLIVe cause," 7:74).

56. "Les geometres prennent leurs quarts de cercle, leurs secteurs, deux filles laponnes, et descendent sur les doigts du Sirien" (Voltaire, *Micromegas,* in Beuchot, ed., *Oeuvres* 33:182).

## Further Adventures

In the summer of 1739, Maurepas approved a proposal by Cassini de Thury (Cassini III) to redo some meridian measurements in Provence made years earlier by his father and grandfather. He had new instruments for the occasion, including a six-foot sector by Langlois, the premier French instrument-maker (fig. 11).[57] While the younger Cassini went south, Maupertuis, restless and frustrated with the unsettled state of affairs, asked the minister for permission to go north to Amiens with the Graham zenith sector to remeasure the angular separation of the endpoints of Picard's base, from which Cassini II had calculated all his distances. According to Graffigny, the minister initially refused to sanction the expedition, even when Maupertuis proposed to go at his own expense. "The Academy is engaged in a civil war worse than one can say," she reported.[58] However, by August, Maupertuis was in the field with the sector and the other members of his old team: Camus, Clairaut, and Le Monnier.[59] Working on Cassini territory this time, they streamlined their operation by limiting it to stellar observations. They modified their verification procedures to take account of Cassini's objections and carefully recorded stellar positions taken with the instrument in opposite directions.[60] Two of Picard's markers, the cathedral in Amiens and the southern tower of Notre Dame in Paris, were conveniently located on the meridian almost exactly one degree apart. This allowed Maupertuis's team to compare their arc measurements directly to the seventeenth-century measures by Picard, using the same procedure they had used for the endpoints of the Lapland arc. The short and manageable expedition was ideally suited to the ongoing polemic, as it was designed to demonstrate the accuracy of the sector and to revise Picard's measure, as well as to keep the question in the public eye.

In December, Maupertuis presented to the Academy a corrected length for the arc from Paris to Amiens, based on the new observations. Even though the length on the ground had not been checked, since they did not want to survey a whole chain of triangles, the correction to the astronomical observations was sufficient

57. Langlois incorporated a micrometer in the eyepiece of the telescope of a 52-degree sector, a feature lacking in the earlier French instruments (Cassini de Thury, *La méridienne de l'Observatoire royal de Paris,* 8). See also Wolf, *Histoire de l'Observatoire de Paris,* 187–88.

58. Graffigny to Devaux, 22 June, 1739, *Correspondance,* 2:7. Maurepas did finally approve the expedition (Maupertuis, *Degré du méridien,* ii).

59. Gastelier mentions their departure from Paris on 11 August (*Lettres sur quelques affaires du temps,* 20 August 1739, 276).

60. Maupertuis, *Degré du méridien,* xxiv–xxvii.

Fig. 11. Zenith sector by Claude Langlois, Paris instrument maker, for Cassini de Thury's triangulation of the Paris meridian, 1739. Cassini de Thury, *La méridienne de l'Observatoire Royal de Paris* (1744).

to confirm the flattening of the earth when compared to the Lapland numbers.[61] Maupertuis took the opportunity to produce another book, with elegant engravings of the zenith sector and detailed diagrams of its parts (Fig. 3). This time the instrument, rather than the mathematicians, occupied center stage. The polemic remained implicit in the narrative and visual descriptions of the instrument, in the record of observations and in the new length for Picard's arc. Maupertuis did not question Picard's acumen as an observer, and even relied on the old triangulations for the terrestrial length, but showed that his predecessor had been led astray by inadequate instruments. "Since [the arc] is the part of the operation where negligence would be the most dangerous, and where the least imperfection in the instruments would expose it to more considerable errors, it is there that we made all our efforts; and it is there that we could have some advantage over M. Picard, by the excellence of our sector, the same one that we used in Lapland."[62] Picard's

61. AS p-v, 5 December 1739. No text of the speech is given in the minutes, nor was it published in the *Mémoires*; it appeared as the first chapter of *Degré du méridien.*

62. Maupertuis, *Degré du méridien,* iii–iv.

book, originally published in 1671, was reprinted in the same volume, with the original vignettes and plates depicting his instruments. Apart from stretching the book to a reasonable length for an octavo volume (without it, Maupertuis's account of the Amiens expedition ran to no more than 56 small pages), the addition of Picard's text made the new work look like the sanctioned heir to the tradition of astronomy going back to the first years of the Paris Observatory. The very structure of the book co-opted the assent of Picard, one of the founding fathers of French astronomy, for the flattened earth, especially since it did not question his triangulations, but simply corrected his astronomical observations. "And since the correction that we are making to such celebrated measurements implies quite large errors in them, we would make it only fearfully [*en tremblant*] without all the care that we have brought to it. And it is yet another piece of good fortune for us that the nearness and the immobility of the endpoints of our measured arc make it easy to verify at any time."[63] At the end of the volume, Le Monnier included his observations of the aberration of starlight, made in Paris over the course of two years with the same instrument.

Maupertuis read his account of the operation to the Academy in December, but the Amiens expedition seems to have had little effect on the dispute in the Academy, presumably because the results were not yet in for the southern portion of the meridian, where Cassini de Thury was working.[64] His pride suffering, Maupertuis stayed away from meetings for weeks at a time. He found a much more friendly reception in the homes of the social elite. He confided to the younger Johann Bernoulli, who became a close friend in this period. "Since I do not have the approval of the Academy, I am going to take advantage of this time to wander around and divert myself."[65] Before and after the trip to Amiens, he was a frequent guest at the tables of the duchesse de Richelieu, the duchesse de Saint-Pierre, and the duchesse d'Aiguillon, where he charmed the numerous company with his gallantry and often outrageous wit. Mme. de Graffigny saw him at Mme. de Richelieu's townhouse almost daily, along with Voltaire. Madame du Châtelet was back in Paris in the fall of 1739 as well, once again importuning him for rendezvous and resenting the time he spent with other women.[66] His public persona

63. Ibid., vi. One of the problems with redoing old measurements was that exact location of markers used years earlier was often difficult; in this case, the cathedral spires were easy to sight, and they were indisputably the same points measured by Picard.

64. "Les observations faites à Amiens avec le secteur qui a servi à celles de Lapponie" (AS p-v, 5 December 1739); "Exposition de l'Opération faite pour vérifier le Degré entre Paris et Amiens," in Maupertuis, *Degré du méridien*, i–vii.

65. M to JBII, 21 December 1738, BEB.

66. "Madame de Richelieu compte que vous souperez avec elle aujourd'hui. Vous etes engagé depuis huit jours chez Mme. de Saint-Pierre demain, mais je veux vous voir toute la journée et si vous

as an elegant, if eccentric wit, and seducer of aristocratic ladies took the "civil war" in the Academy out into *le monde,* where the elite who made the reputations of writers and conversationalists took his side. As a result, his contributions to the scientific dispute, whether in the Academy or outside, were noticed, read, and discussed by that portion of the public who saw his commitment to the polemics as a defense of the "cause" of Newtonianism.

During the Academy's vacation in 1739, he set off for Fontainebleau, where the royal court spent two months every year.[67] There, he was able to reconcile with the ministers, especially Maurepas and Fleury, who had been distressed at the contention in the Academy. His presence at court was known around town, where people took note of his ties to the center of power. In the words of the duchesse de Saint-Pierre, one of his many admirers, "It's a new life for you, in the tumult of the great people at court, becoming a gambler, and even a hunter. What a life for a mathematician! . . . You are extremely *à la mode.*" She recognized that such a situation could work to his advantage and advised him to make the most of it. "Think of your affairs and profit from your long stay. . . . You will have the advantage of being on better and better terms with the minister because you will have frequent opportunities to see him."[68] And his success at court did pay off. As soon as he was back in Paris, the news spread that he had a new honorific position that came with a pension of 3000 *livres.* Maurepas, minister of both the navy and the Academy, created a post especially for Maupertuis, charging him with the task of "perfecting navigation." Unlike the pension he had scorned earlier, this one, being unique as well as munificent, brought honor and distinction along with a substantial income. "It is a position [*place*] created expressly for me by M. de Maurepas. . . . Everything was done in the most gracious way for me. It makes me very happy and quite comfortable."[69] He saw this mark of favor not only as recompense for service performed, but as a vindication of his conduct and his accomplishment. "This position," he crowed to his old mentor Bernoulli, "does me so much honor that nothing remains to be desired, other than to . . . merit the confidence which the minister has in me. The way in which it was done added greatly to its value. It makes all the professors of hydrography [at the new Ecole

ne le voulez pas, vous etes un ingrat" (Châtelet to M., [n.d.] September 1739, *Lettres,* 1:378). Mme. de Graffigny, who hated Châtelet, noted "C'est elle qui tiraille, agasse, fait des train a la Puce [Maupertuis] devant tout le monde" (Graffigny to Devaux, [9 November 1739], *Correspondance,* 2:242).

67. He was at court from early October through mid-November. See M to JBII, 9 November 1739, BEB.

68. Duchesse de Saint-Pierre to M, 29 October 1739 (BN, n.a.f. 10 398).

69. M to JBII, 28 December 1739. BEB. He also reported his new pension to his sister (M to Mme. Magon, 28 November 1739, AS Fonds Maupertuis).

de la Marine] subordinate to me; I am charged with examining their theory [*science*] and practice."[70] And Mme. de Graffigny remarked, "We must hope that he will be content, although I believe that to be impossible. Some people torment themselves with saying that they are miserable. He said that he wanted honor, glory, and money. He has all that, since he was feted at Fontainebleau by the king, by Mme. de Mailly [the king's favorite] and all the court, as one might expect. This post can give him a lot of glory, and a thousand crowns [3000 *livres*] in addition are plenty honorable [honnêtes]."[71] There were those who viewed this mark of favor cynically, however. Abbé Le Blanc, perennially looking for patronage for himself, resented Maupertuis's success:

> M. de Maupertuis, for whom the voyage to the north was worth a pension of only twelve hundred *livres* has stayed in Fontainebleau the whole time the court was in residence there. He played his guitar at the toilette of duchesses and at the suppers of ministers. They have paid him with a position without responsibilities [*un emploi sans fonction*] that was created just to give him 1000 crowns more than he already had. Here one gets everything when one is vaunted by women, and it is worth more to amuse men than to be useful to them.[72]

Le Blanc associated court life with feminized corruption and idle frivolity rather than productive work. But he missed the crucial point: that Maupertuis was playing both ends of the apparent opposition between utility and amusement. For the mathematician turned explorer and then part-time courtier, the court was just one more avenue to exploit in the quest for standing and reputation. The expedition to Lapland opened up this possibility, by offering both amusement and the promise of utility. Maurepas, although he may have enjoyed listening to the guitar, also had a very real interest in improving the practice of both navigation and cartography, which explains his willingness to devote substantial resources to geodesy.[73]

Most of those resources went to the mapping team, for which Cassini de Thury and Nicolas-Louis de la Caille worked on the meridian in Provence in 1739. At the

70. M to JBI, 11 January 1740, BEB.

71. Graffigny to Devaux, 29 November 1739, *Correspondance*, 2:255.

72. Le Blanc to Bouhier, 13 January 1740, Volt., *Corr.*, 91:89. Le Blanc spent many years seeking patronage. For his many unsuccessful attempts at getting a seat in the Académie française, see Monod-Cassidy, *Un voyageur-philosophe*. In his next letter to Bouhier, Le Blanc noted, "Whatever they call this position, I would happily say to Maupertuis, 'Take the title for yourself, and leave me the income'" (Le Blanc to Bouhier, 29 January 1740, Volt., *Corr*, 91:89).

73. Maupertuis's pension was increased again several years later to 7000 *livres* (M to JBII, 21 May 1740, BEB).

age of twenty-five, Thury was by this time well versed in the field techniques of the astronomer and geographer, having participated in numerous expeditions, and was poised to mediate between the affronted parties in the Academy. He was beginning to recognize that the length of the degree along the Paris meridian had not been determined with sufficient precision, in spite of the best efforts of the astronomers. He could not capitulate without the tacit acquiescence of his father, however, and remained circumspect about the ongoing verifications. As early as spring 1739, even before Maupertuis took the Graham sector to Amiens, Thury prepared the way for admitting errors in the old meridian measurements. In spite of all precautions, he told the Academy, "it is extremely difficult to avoid little errors, which when multiplied may produce quite considerable ones over an extent as large as this line, which is almost 500,000 *toises* in length." He quite rightly pointed to the difficulties inherent in measuring such a distance, but also judiciously referred to the Lapland measures as having "all the exactitude that one could wish." These numbers, he said, when compared to the older measurements along the Paris meridian, indicate that "degrees [of latitude] increase in length as they approach the poles and we are persuaded that they will be confirmed by those currently underway in Peru."[74] Given Cassini II's continuing hostility to Maupertuis, Clairaut and Le Monnier, this was a conciliatory gesture. Still, it was not exactly a capitulation. Thury outlined his program for further refinements of the Paris meridian measurements, incorporating new practices in direct response to criticisms of his father's work. To reduce the potential for error, all three angles of each triangle were to be observed directly, no triangles would contain angles of less than 30 degrees, and as many lengths as possible would be measured on the ground (fig. 12).[75]

One year later, in April 1740, Thury formally announced to the Academy's public meeting that corrections to measurements in France confirmed the conclusion supported by the Lapland expedition. Maupertuis, who had difficulty forgiving Cassini II for his stubborn resistance to this conclusion, reported to Johann II Bernoulli: "The Cassinis have publicly recanted on the elongation of the earth and have confessed to having found it flattened in the sixth operation that they have just finished. They are a bit the talk of the town and well deserve it."[76]

74. Cassini de Thury, "Sur les opérations géometriques faites en France dans les années 1737 et 1738," *MAS* 1739, 129–30 (read on 8 April, public session). The manuscript text was copied into the minutes.

75. Ibid., 130.

76. "Les Cassinis se sont rétractés publiquement sur l'allongement de la Terre et ont confessé l'avoir trouvée applatie dans la sixieme opération qu'ils viennent de faire. Ils sont un peu la fable de la ville et le meritent bien" (M to JBII, 2 May 1740, BEB).

Fig. 12. Cassini de Thury's zenith sector in use in the French countryside. The recumbent astronomer is observing the passage of a star while his assistant notes the position of a pointer on the limb of the instrument with a micrometer. Cassini de Thury, *La méridienne de l'Observatoire Royal de Paris* (Paris, 1744).

Actually, Thury did not "recant"; he carefully avoided any reference to the dispute between Cassini II and the partisans of the flattened earth. Thury even used Maupertuis's formula for determining the ratio of the earth's axis to its diameter, tacitly acknowledging the authority and utility of a mathematics never used by Cassini II.[77] Not only was the son more mathematically adept than his father, he was also more rhetorically skillful. He acquiesced to the flattened earth, but made it peripheral to the central problem of moving forward with the map of France that had been in the works for many years. He went on to show how a new generation of French instruments had enabled an unprecedented level of precision, giving the map project renewed credibility. Acknowledging the virtues of Maupertuis's "very exact instruments," he described the Observatory's new Langlois sector carefully and justified its design features (such as its large 52-degree arc length), subtly suggesting that it was more versatile that Graham's instrument

77. Cassini de Thury, "Observations astronomiques, géographiques, et physiques faites en diverses provinces de la France en 1739 & 1740," AS p-v, 1740. This address was not published in the *MAS*; much of it appeared in the preface to Cassini de Thury, *La méridienne de l'Observatoire Royal de Paris.* Calculations from Maupertuis's formula are on 113–14.

without being any less accurate.[78] Then, he went a step further and used the phenomenon of aberration to excuse the inaccuracy of previous measurements. The first two generations of Cassinis could not have achieved results beyond the capability of their instruments, nor could they have factored in corrections for aberration that were not yet known.[79] Given the rhetorical use Maupertuis had made of aberration, and also given the very small size of the correction, this was a clever move. By validating an English discovery that had antagonized the older generation of Paris astronomers, Thury co-opted one of Maupertuis's weapons. In so doing, he attempted to reconcile the antagonistic visions of the local cartographic project and the global geodetic expeditions.

## Satire and Subterfuge

With his gracious capitulation on the geodetic question, Thury strengthened his own position in the continuing cartographic project.[80] The Academy's attention was no longer focussed on geodesy. Although reports continued to trickle back slowly from South America, the work there was not completed until 1744. Clairaut continued his mathematical work on the theory of the earth's shape, but Maupertuis turned to new work in mechanics and astronomy.[81] However, he simultaneously pursued several strategies to keep the geodesy controversy in the public eye, through carefully orchestrated publication ventures. The first of these, a literary hoax that he must have had in mind for some time, appeared in early 1740, when the dispute with Cassini was winding down. It had the rather bland title *A Disinterested Examination of the Different Works on the Shape of the Earth,* with a fictitious place of publication (Oldenbourg) and date (1738). Included in the same volume was *Examen des trois dissertations que M. Desaguliers a publié sur la figure de la terre. . .,* a detailed examination of Desaguliers's much earlier criticism of Cassini II's derivation of the elongated earth (Fig. 13).[82]

78. Cassini de Thury, *La méridienne de l'Observatoire Royal de Paris,* 11–12; for a full description of instrument and its use, see ibid., lxxi–lxxiv.

79. Joseph Delambre later argued that aberration was just a "subterfuge" on Cassini de Thury's part, missing the rhetorical cleverness of this move (Delambre, *Grandeur et figure,* 66).

80. See Pelletier, *La carte de Cassini.*

81. Maupertuis, "Loi du repos," *MAS* 1740; "Traité de loxodromie," presented 1742, published in *MAS* 1744.

82. Maupertuis, *Examen désintéressé des différentes ouvrages qui ont été faits pour déterminer la figure de la terre* (Oldenbourg, 1738 [1740]). Although it appeared before Cassini III's public capitulation, the discussion of the volume's authorship and intent continued in the press for many months, and revived when a new edition appeared a few years later (in 1742 or 1743, with a fictitious date of 1741). On the *Examen,* see Beeson, *Maupertuis,* 122–31, and Badinter, *Les Passions intellectuels,* 135–41.

As Montesquieu wrote to Martin Folkes in England, "A very well written book has just appeared here, *A Disinterested Examination*. . . . The author seems to be a wise and reserved man, who does not utter inanities."[83] The book purported to summarize the debate about the shape of the earth without taking sides. The satire is so understated as to have been invisible to many of its readers; it was the talk of Paris, not because of any novel or outrageous claims, but simply because the author's sympathies were difficult to read. Debate about its authorship continued for several years, fueled by discussion in journals and then by a second edition a few years later. Mairan praised the book lavishly for its accuracy and measured tone. "Whatever [the identity of the author] may be, this book is a credit to whoever wrote it. It is in the hands of everyone in Paris, including the ladies, who also

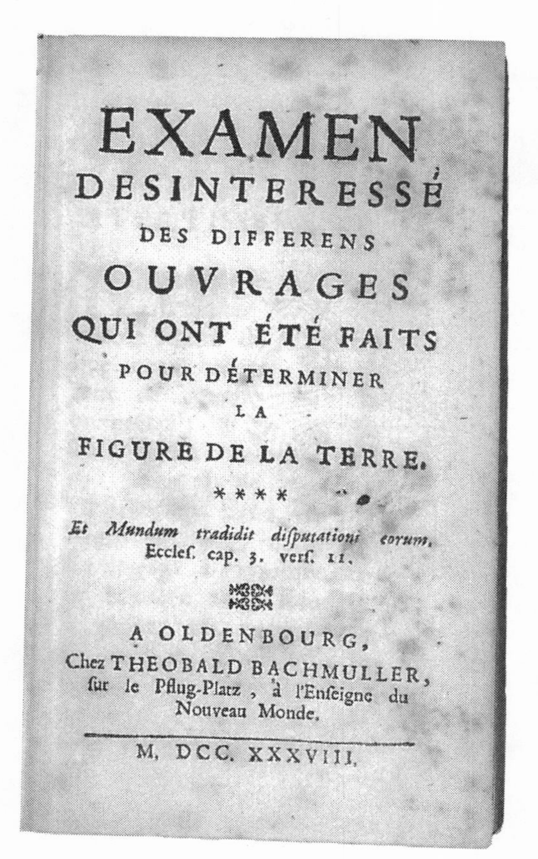

EXAMEN
DESINTERESSÉ
DES DIFFERENS
OUVRAGES
QUI ONT ÉTÉ FAITS
POUR DÉTERMINER
LA
FIGURE DE LA TERRE.
****
*Et Mundum tradidit diſputationi eorum,*
Eccleſ. cap. 3. verſ. 11.

A OLDENBOURG,
Chez THEOBALD BACHMULLER,
ſur le Pflug-Platz, à l'Enſeigne du
Nouveau Monde.

M, DCC. XXXVIII.

Fig. 13. Title page of Maupertuis's anonymous *Examen désintéressé des differens ouvrages qui ont été faits pour déterminer la figure de la terre*, with fictional publisher and date. Courtesy of William Andrews Clark Library, UCLA.

83. Montesquieu to Folkes, 17 February 1740, in Montesquieu, *Correspondance*, 347.

dabble in wanting to know the shape of the planet they walk upon."[84] To the younger Johann Bernoulli, Maupertuis reported, "The little book you speak of is creating a great sensation in Paris and for the last three months has been exercising everyone's efforts to uncover [the secret]. It has been successively attributed to your father, Fontenelle, Cassini, Mairan, and finally to me; and no one has been able yet to be sure if it is for or against the Cassinis."[85] What are we to make of a book, written by Maupertuis, that many of his contemporaries read without recognizing it as an attack on Cassini? Why did a book filled with technical discussion of astronomical observations and instruments create a stir in literary and scientific circles in Paris, and even across Europe?[86]

The book starts by summarizing the work of the different expeditions. After five pages on the Lapland expedition, there follow twenty-eight pages on the measurements of Picard and the Cassinis. The catalogue of their results carries the burden of the attack. Although qualitatively they all implied an elongated earth, each set of measurements gave a different degree of elongation, undercutting any claim for precision. After listing the numbers, the author comments: "Who would not admire [the fact that] on a baseline of 7246 *toises*, 2 *pieds*, after 48 triangles, there were only 3 *toises* of error? The corrections having diminished this error still further, who will not recognize that we can count on the whole distance between Paris and Collioure having been measured with an astonishing exactitude?"[87] But the unarticulated question remained: why, with all this precision, had all the expeditions come up with different numbers in different years?

Maupertuis worked hard to keep his authorship secret, actively encouraging speculation about the book's proper attribution. He persistently denied his responsibility to Johann II Bernoulli, while enlisting his help in getting the controversy into the Swiss press. Bernoulli proved a rather gullible ally; he only realized much later that his friend had cavalierly deceived him. An anonymous letter (probably by Maupertuis) to Bernoulli was printed in the *Journal helvétique,* in July 1740, suggesting the senior Bernoulli as the author, and concluding, "Some people nevertheless suspect that this book was written maliciously against the party that holds for the elongation of the earth." In his response, published by the same jour-

84. Mairan to Jallabert, 29 February 1740, quoted in Badinter, *Les Passions intellectuels,* 136. Mairan was favorably impressed by the harsh treatment of Desaguliers's challenge to the French.

85. M to Johann II Bernoulli, 3 April 1740, BEB. Mairan mentioned rumors about Maupertuis as a possible author (Mairan to Jallabert, 29 February 1740, quoted in Badinter, *Les Passions intellectuels,* 136).

86. Maupertuis himself sent copies to Folkes, Jurin, Bradley, and Desaguliers in England. See M to Folkes, 12 February 1740, in Rigaud, ed., *Correspondence of Scientific Men,* 2:357–59.

87. Maupertuis, *Examen désintéressé* (1st ed.), 39.

nal, Bernoulli included excerpts from a letter of Mairan to the senior Bernoulli, and denied his father's involvement. The Swiss became interested because of the attribution to Bernoulli, but none of them saw the joke.[88] Those who trafficked in wordplay were less obtuse; Voltaire saw the irony and suspected Maupertuis immediately: "Plenty of skill is devoted to acquitting the Cassinis, but as for reasons in their favor I hardly see any. This book proves only that they must have been badly mistaken."[89] Algarotti also got the point, but almost ruined the game in Switzerland, where he discussed it with Johann Samuel König, an admirer of Maupertuis who had just translated *The Shape of the Earth* into German.[90] Maupertuis wrote to Algarotti: "I am annoyed that you have kept my translator König from refuting the *Disinterested Examination*, that would have been a great joke for me; and I do not want anyone to have any certainty about the author of this book. . . . You have seen in König an incomprehensible contrast of thickness and subtlety; but I think thickness predominates in all those who out of friendship for me would like to refute the *Examen*."[91] König had written a preface to his translation, defending Maupertuis from the attacks in the *Examen*, but had been warned just in time to stop the printer from including it. "One must think like you," König wrote abashedly to Maupertuis, "to dare to write against oneself. This method is as new as the discovery of the shape of the earth itself and will surely cause no less of an uproar. It is the real way to make fun of your adversaries, although a bit at the expense of everyone."[92]

The confusion about the text's point of view was played out in the press in conjectures about the identity of the author.[93] Anonymity, and stubborn denial of his

88. These maneuvers were discussed in several letters to Johann II Bernoulli (M to JB II, 3 April, 21 May, 29 June, and 9 August 1740, BEB). See also "Lettre à M. Jean Bernoulli, Docteur en Droit . . . sur la figure de la terre," *Journal helvétique* (July 1740); "Réponse de Jean Bernoulli fils à la lettre anonime sur la figure de la terre . . . ," *Journal helvétique* (September 1740) (extracts in AS Fonds Maupertuis). Châtelet told Bernoulli that Maupertuis was the author (Châtelet to JBII, 30 June 1740, *Lettres*, 2:19). See also Châtelet to M, 23 December 1740, *Lettres*, 2:36. On the very lively correspondence among Swiss savants, see Badinter, *Passions intellectuels*, 136–38.

89. Voltaire to 'sGravesande, 29 February 1741, Volt., *Corr.*, 91:116–17.

90. König, a former student of Bernoulli, had met Maupertuis in Basel. The German translation (*Figur der Erden*) was published in Zurich in 1741.

91. M to Algarotti, 18 February 1741, in Algarotti, *Opere del conte Algarotti*, 16:182.

92. König to M, 11 February 1741, in LS, 117. M followed up with another to Algarotti: "pourquoi avez vous dit à König que j'étois l'auteur de *l'examen*? je le lui nie tout ouvertement, et lui dis que vous avez voulu apparement vous divertir. Il prétend cependant qu'il avoit déjà deviné l'auteur; mais ce n'est pas l'homme le plus fin du siecle" ( M to Algarotti, 26 February 1741, in Algarotti, *Opere del conte Algarotti*, 16:184). König, "not the sharpest man of the century," tangled with Maupertuis later in Berlin (see chap. 9).

93. *Journal des sçavans* (March 1740): 153–58.

authorship even to intimate friends, allowed Maupertuis to create a minor literary phenomenon. Like many anonymous books, it came out in a second edition (dated 1741, but printed considerably later, probably in 1743) prefaced by a "History of this book" that hinted broadly at the author's true identity, and then repeated rumors that claimed the text for the other side. "There were people in Paris who sought the place of publication of the *Examen* with as much care as others had employed to discover the shape of the earth," the preface boasted. "This book itself became an object of interest for that type of connoisseur who prides himself on knowing everything about literature, from the ink to the style. . . . People in high society took a different interest in it, since it speaks continually of the savants with whom they live, whom they esteem or scorn."[94] The false attributions served ironically as evidence that the text could not possibly be hostile to the Cassinis. Quoting positive responses to the first edition by Cassini II himself, Fontenelle, and Mairan, the preface went on, "Surely no one will believe that [these men] lack sufficient knowledge [*science*] and intelligence [*lumières*] to be mistaken about what favors or destroys their own position. The common man may be an Oedipus in judging that which concerns him personally, but are we to believe that the most enlightened men in the kingdom of France are deluded and take for praise what are only ironies?"[95] This new preface played on readers' interest in stories of authorship and publication of controversial books. It also dropped clues about how to read the book.[96]

The manipulation of anonymity allowed Maupertuis to display his cleverness to an audience that appreciated (and knew how to read) such authorial contortions. Tricking not just Cassini, who had no facility for wit, but the elegant and eloquent Fontenelle as well, was a subversive claim to literary aplomb. Some of his friends thought his pursuit of this line to be in poor taste, especially after the question of the shape of the earth had been settled in the Academy.[97] But more was at stake than the approbation of the official scientific elite. He wanted personal revenge on Cassini, certainly, but he also wanted to shift the ground of debate from the Academy to the press, the wider Republic of Letters, and conversations in drawing rooms. This move made the scientific quarrel into a literary one, drawing

94. Maupertuis, "Histoire du livre," *Examen désintéressé des différentes ouvrages qui ont été faits pour déterminer la figure de la terre,* 2nd. ed. Amsterdam [Leipzig], 1741 [1743?], n.p. Maupertuis mentioned that it had been printed in Leipzig (M to JBII, 22 August 1743, BEB).

95. Ibid.

96. Johann II Bernoulli was still answering questions about the book's authorship in 1743 (JBII to Beck, September 1743, BEB).

97. See, e.g., Châtelet's comment: "Je ne sais si ce n'est pas là une plaisanterie un peu trop poussé" (Châtelet to JBII, 30 June 1740, *Lettres* 2:19).

on a long tradition of such exchanges going back to the dispute between the self-styled Ancients and Moderns at the end of the seventeenth century.[98] Maupertuis could not have written a satirical attack on a fellow academician under his own name without risking his position in the Academy. He also knew that the full impact of the satire would only become apparent once he had been unmasked, after he had tricked his enemies into approving of the book. As one reviewer noted in retrospect, "These were the equivocal civilities of a panegyrist in disguise."[99] Although Maupertuis never acknowledged his authorship in print, many people had recognized him by the time the second edition appeared; any reader who did not already know the truth would have solved the puzzle from the hints in the preface.[100] Without his academic credentials, which he could not display openly in this context, Maupertuis could not have been convincing as a witty author of "ironies" (as he called them). He wanted to defend his credibility as an astronomer and mathematician, but he also set out to establish credibility in a different social context.

As Maupertuis's authorship became known, Cassini and his partisans looked especially foolish for having failed to recognize the hoax. Various comments became barbed in light of the attribution. For example, the text played down the importance of Bradley's discovery of aberration, disallowing it as a correction to the Lapland measures, because "this theory is not yet accepted by everyone, and we do not believe in hurrying to accept a discovery that has only been witnessed by a single author, which does not agree with the observations made by French astronomers, and which is founded on the successive motion of light (which the ablest astronomers still doubt)."[101] Given the Cassinis' public retreat from their previous conclusion, the defense of their original results became a denunciation. Disavowing any ulterior motives in his discussion of Cassini's results, the author

98. DeJean, *Ancients against Moderns.* Maupertuis's first mentor in Paris, Houdar de la Motte, had been a major player in this dispute, and was revered by his protegés as the model of the "modern" man of letters. His polemics with the representatives of the ancients, especially the classicist Anne Dacier, took the form of pamphlets and small books. For a contemporary account of literary disputes through history, see Irailh, *Querelles littéraires.*

99. Desfontaines, *Jugemens de quelques écrits de ce tems* (Avignon, 1744), I:359 (review of *Anecdotes physiques et morales*).

100. Second edition falsely dated 1741; actual publication was more likely in 1742, or even 1743. Maupertuis referred to it as a "new edition" in August 1743 (M to JBII, 22 August 1743, BEB). By this time, Johann II Bernoulli recognized Maupertuis's authorship: "It is obvious that the whole book was only written to put Cassini in an even worse light . . . when the flattening of the earth at the poles would be verified" (JBII to Jacob Christoph Beck, September 1743, BEB).

101. Maupertuis, *Examen désintéressé,* 1st ed., 9–10. The "successive motion" refers to the Newtonian theory of light.

remarked, "The calculation of the errors that M. Cassini would have had to commit if the earth were oblate is an arithmetical calculation; it cannot be made arbitrarily, any more than 2 times 2 can be said to equal 5."[102] The hypothesis of a flattened earth, since it generates numbers very different from Cassini's results, leads to the "absurd" conclusion that Cassini "is the clumsiest and the most unfortunate astronomer in the world."[103] But new numbers supplied to the Academy by Cassini's own son (incorporating corrections due to aberration) had already destabilized the alleged absurdity of this conclusion . Maupertuis also denied that the Cassinis had changed their minds. "[Y]ou have been deluded when you were told that [M. Cassini and his son] gave in to the flattening. However strong the spirit of conciliation and modesty might be, it would never make one prefer a single alien operation to five one had done oneself; one would have to abandon one's glory and justice for that."[104] Anonymous satire was only one of many tools Maupertuis put to use in the service of his self-promotion. Through a complex subterfuge—planting letters in the press, duping his closest friends and allies, taking on multiple voices, backdating the publication, and masking the printer—he explored the possibilities afforded by print for making people talk. This was not common practice for a man of science, although the strategies were recognizable from other genres, where frivolity and viciousness were no strangers to the literary marketplace.

## The Portrait

As another piece of his energetic campaign to remind the public of his accomplishment, Maupertuis engaged a famous portrait painter to paint him in the triumphant pose described above in chapter 1. He sat for the portrait in the autumn of 1739, just before leaving Paris for Fontainebleau.[105] Robert Tournières (also known as Levrac-Tournières) was at this time an acclaimed member of the Academy

102. Maupertuis, "Histoire du livre," in *Examen désintéressé,* 2nd ed., unpaginated.

103. Ibid.

104. "Je ne vous parle point de M. Cassini, ni de M. son Fils, main on vous a trompé lorsqu'on vous a dit qu'ils s'étaient rendus à l'aplatissement. . . . Quel que soit l'esprit de conciliation et de modestie, il ne serait jamais aller à faire préférer une seule opération étrangère à cinq opérations qu'on a faites soi'meme; il faudroit abandonner sa gloire et la justice pour cela" (Maupertuis, *Examen,* 2nd edition, 53–56). Privately, Maupertuis could be much more vicious about Cassini; in another satire, circulated only among friends, he asserted that the Cassini operations were no more reliable than throwing dice ([Maupertuis], *Lettre d'un horloger anglois à un astronome de Pékin* (n.p., 1740), 13–14). Only four copies were printed (Lalande, *Bibliographie astronomique,* 407). The text was reprinted with notes and introduction by Beeson, "Lettre d'un horloger anglois."

105. "On travaille à mon portrait et dés quil sera fait M. votre pere l'aura, il me fait trop dhoneur de le vouloir pour que je differe a le luy envoyer" (M to JBII, [October, 1739], BEB). This referred

of Painting and Sculpture who had made his reputation by painting powerful nobles and government ministers during the Regency.[106] He was well acquainted with Maupertuis's family, having painted a group portrait of them in 1715. (See fig. 1 in chapter 2.) Tournières had a flourishing career in Paris, with a studio on the Rue Sainte-Anne, near where Maupertuis lived. His portraits of great men like the chancellor d'Aguesseau and Pontchartrain, minister of the navy, were displayed in periodic exhibitions in the Louvre. (Maupertuis's portrait was exhibited in the Salon of 1741, where it hung on public view for a month.)[107] The choice of artist thus lent a luster to the painting apart from the actual execution. Maupertuis was claiming a place in the company of illustrious political figures and aristocrats who formed the bulk of Tournières's clientele. It was a flamboyant act of self-advertisement, but also an assertion of nobility.

It was highly unusual to depict a man of science so elaborately, especially in a setting other than his study. Gastelier considered it newsworthy, informing his correspondent that Maupertuis "had himself painted dressed like a Lapp, holding between his hands a globe that he presses and that flattens toward the poles. This indicates his opinion on the shape of a Dutch cheese that the famous Newton gives to our globe."[108] The picture was the topic of conversation and jokes, and soon appeared, embellished with a flattering quatrain by Voltaire, in an engraving by Jean Daullé, which Maupertuis distributed widely to friends and allies across Europe. The image was printed in several different forms over the next few years (fig. 14).[109]

---

to a copy rather than the original; Maupertuis sent Bernoulli a painting in early 1740. Mme. d'Aiguillon owned a portrait, possibly the original; La Condamine bought a copy in 1765: "On a reconnu la touche de Tourniere[s], sous les yeux duquel cette copie a été faite et à laquelle il a mis la main. je l'ai aussi comparé à l'original de Mde. la Dsse. d'Aiguillon, et il n'y a pas de difference sensible" (La Condamine to JBII, 13 August 1765, BEB).

106. Tournières was christened Robert Levrac, and took the name from his place of birth when he moved to Paris. For his biography, see Bataille, "Tournières."

107. "M. Moreau de Maupertuis, pensionnaire de l'Académie Royale des Sciences, en habit de Lapon," Livret de Salon, 1741; also *Mercure de France* (September 1741). A copy of the portrait (possibly, but not necessarily, the original) hangs in the municipal museum of Saint-Malo. The portrait may have been commissioned by Maupertuis's father, who had done business with Tournières before.

108. Gastelier, 28 January 1740, in *Lettres sur les affaires du temps,* 346.

109. The Daullé engraving was commissioned by the marquis de Locmaria, a Breton nobleman and acquaintance of Maupertuis (and probably of his father). Locmaria solicited the verses from Voltaire. See Voltaire to Locmaria, 17 July 1741, Volt., *Corr.,* 92:70. See also Graffigny to Devaux, 27 November 1742, *Correspondance,* 3:465. The engraving was exhibited in the Salon of 1743. Daullé also engraved a bust from the portrait in 1755, for the frontispiece to Maupertuis's *Oeuvres;* the portrait was engraved again by J. Haid (BN, Estampes). Several versions in different sizes survive (Archives of Académie française; BJ). Other copies were made in oils: one went to Johann

Fig. 14. Engraving by Jean Daullé of Tournières portrait of Maupertuis. Courtesy of Owen Gingerich.

In the picture, Maupertuis stands in a window, gesturing to the scene behind him, where we see the huts and hills of Lapland, with the crucial observation signals smoking on the peak of a snow-covered mountain. Other mementos of the journey lie on the windowsill: a diagram of triangulations and the fur leggings worn against the rigors of the arctic winter. Another image beneath the window shows the fur-clad traveler bundled into a boat-like sledge, navigating with a stick as an energetic reindeer pulls him through the snow, as described in *The Shape of the Earth.*[110] In the foreground, a reindeer-skin robe drapes over the window ledge to show its ornamentation to best advantage. The embroidery and ermine adorning the robe give an impression of luxury, but the garment is supposed to be the customary attire of the residents of Lapland, a souvenir of the frozen wilds, like the fur hat. Maupertuis wears the costume like a trophy earned in the name of science; he displays the globe itself as a trophy too. The flattened globe stands in for Newtonian theory as well as for the power of precision instruments and the promised supremacy of French navigation. The Frenchman, it seems, has appropriated the Sami (or perhaps Finnish) costume and English science for his king and himself.

The coherence of the portrait masked conflict and contention, and rhetorically asserted control, both of results and of personal identity. The self-assured posture of the man in the picture ignored the attacks he and his results had sustained, claiming the viewer's attention and approbation. The public would have recognized the image itself as part of the polemic. Nollet, for example, reported to a correspondent in Geneva that

> Maupertuis . . . is declaring open war on anyone who does not proclaim at the top of his lungs, "The earth is flattened!" Never did Don Quixote make such a row to defend his Dulcinea. He has just had himself painted as a Lapp [*en Lapon*], one hand on a globe that he flattens with an air of assurance. He is giving out copies of it, and I expect that he will have it engraved. I would have wished that they had put in the background his five collaborators. . . . [T]hey are taking so little part in the dispute that it makes it look like they had little to do with the work that initiated it.[111]

Nollet, who ridiculed Maupertuis's arrogance without denying the reality of the flattened earth, noticed the remaking of the collaborative effort into the triumph of

---

Bernoulli and one to Frederick II of Prussia (now in Berlin, Staatliche Schlösser und Gärten: see *Friedrich der Grosse: Ausstellung des Geheimen Staatsarchivs Preussischer Kulturbesitz anlässlich des 200. Todestage König Friedrichs II. von Preussen,* Berlin 1986, 137).

110. This is based on a detail of an elaborate plate, presumably drawn by Herbelot, published in Outhier's *Journal du voyage au nord* (see chapter 4 above). The image of the reindeer-drawn sledge also appeared as a vignette in *La figure de la terre* (1738).

111. Nollet to Jean Jallabert, 2 January 1740, in Benguigui, *Théories électriques du XVIIIe siècle,* 98.

one individual. But this was just the point. Maupertuis felt abandoned by his collaborators, who had not defended their work vociferously enough, and by this time had appropriated the conflict for himself.[112] Although his flamboyance made him an easy target for ridicule, Maupertuis excited admiration as well, and his fame spread beyond the boundaries of France. The empress of Russia offered him a pension, in recognition of his accomplishments; the new king of Prussia, Frederick II, flattered him with an offer to come to Berlin.[113]

## Geography, Geodesy, and Utility

In the early 1740s, Maupertuis tailored much of his writing to the same literary and fashionable public who admired his performance at public sessions of the Academy and who read his *Shape of the Earth.* He gave very few mathematical papers to the Academy in this period, but wrote numerous books that drew on his geodetical work to consolidate his public stature, while simultaneously keeping alive the polemic with Cassini. He released the books strategically, sometimes anonymously, often publishing revised or updated versions as conditions changed. All of his publication ventures reflect a sensitivity to the uses of different forms and styles for refining an identity as a public figure equally at home in academy and salon. With the added recognition of the special pension for improving navigation, he was also conscious of playing to his patrons in the government. For all of these audiences, he exploited the related themes of utility and progress.

The geodetic expeditions had interested the government because of their potential payoff for navigation and cartography. The ministers imagined that these applied sciences would serve commercial and military ends, with improved transport and mapping systems within France and more accurate navigation techniques for long sea voyages. From the academic end, the initial inspiration for determining the shape of the earth was not utilitarian, as we have seen, although the academicians spoke the language of utility when trying to impress their sponsors in the government. There was always a tension in geodesy between observational precision and theory, because of the complex role played by Newtonian theory in the

112. On his resentment of Clairaut's passivity, see M to JBII, 17 December 1740, BEB.

113. For the offer from Empress Elisabeth, see Maupertuis's letter to Johann II Bernoulli: "Come j'ay actuellemt environ 7000# de pension du Roy et que j'ay cru voir qu'on trouveroit mauvais que j'acceptasse des pensions d'autres souverains, j'ay pris le party de remercier, mais de la maniere que j'ay crue plus capable de faire connaitre à l'imperatrice mon respect et ma reconnaissance" (M to JBII, 21 May 1740, BEB). Frederick first invited Maupertuis to Berlin in June: "Vous avez montré la figure de la terre au monde: montrez aussi à un roi combien il est doux de posséder un homme tel que vous" (FII to M, [June 1740], Koser, *Briefwechsel,* 185–86).

challenge to Cassini II's conclusion. Determining the earth's shape could have been pursued as a strictly empirical question, but in fact the wider connotations were never entirely absent from the dispute. In France, these connotations included the application of analytic techniques to astronomical and geographical problems, techniques that were not Newtonian as such, but which Clairaut, Maupertuis, and others were also applying to mechanics.

Maupertuis, in his quest to upstage the astronomers, played the utility card as a polemical tool. He was not a cartographer, and had no intention of doing the painstaking kind of work on which Cassini de Thury was building his own reputation; nor did he ever manage to translate his equations into practical rules for navigators. But he took care to explain to his readers why they (and the government) should care about the niceties of geodetical and astronomical practice. In Maupertuis's book on the Amiens expedition, for example, he noted, "Independent of the utility of this measure for determining the shape of the earth, it will also be of considerable utility for the particular geography of France." In other words, he proposed to help the cartographers in spite of themselves. But he also claimed for his own work a usefulness that went beyond the "particular utility" of better roads and canals, or the discovery of easier sea routes to distant ports; the shape of the earth carried with it a "general utility for all people and for all times."[114] This is the more grandiose knowledge, grounded in the generality of analytical mathematics, that Maupertuis was offering to the King of France and that he contrasted with the systematic geographic measurements that aspired to no more than producing detailed maps.

But geodesy made its strongest utilitarian claims in the potential advantages for navigation. Maupertuis framed the report of his expedition in just these terms: "It is consequential for navigators not to think themselves to be sailing on one of these spheres when they are actually on the other. . . . On routes of 100 degrees of longitude, one would err by more than 2 degrees, if in sailing on the spheroid of Newton, one thought oneself on that of [Cassini]. And how many ships have not perished for smaller mistakes!"[115] The hyperbole served a double rhetorical function, to assert that his new numbers were accurate enough to improve determination of coordinates at sea and to imply that Cassini's numbers would lead sailors astray. He developed this line in another book, originally published anonymously in 1740, devoted to the "cause" of the flattened earth.[116] The *Elements*

114. Maupertuis, *Mesure du degré,* vi; idem, *La figure de la terre,* xv.
115. Ibid,, xii.
116. Châtelet recognized his authorship immediately (Châtelet to M, 22 October [1740],

*of Geography*, written for a nonspecialist audience, retold the history of the geodetic measurements, but spelled out the ramifications for the practice of navigation. Now that we know that the earth deviates from the perfectly spherical, he argued, sailors and travelers need a new geography, and such a science would be useful in turn for astronomy. The author noted somewhat facetiously that such work *must* be useful; otherwise, the government would not have supported it. "Those who govern can protect the sciences even in their useless speculations, . . . but they do not order considerable enterprises unless the state can derive from them more substantial advantages. And though savants can give their time to frivolous things when they are in their studies, they are not permitted to cross oceans and to risk their lives and those of others, except for discoveries whose utility justifies their perils and their pains."[117] With this, he implied that sixty years of royal support for the mapping project produced inaccurate measurements; his own expedition had redeemed the state's investment.

To demonstrate the practical importance of knowing the length of a degree, he devised a table to compare graphically the lengths of degrees at different latitudes and longitudes, according to the rival measurements, in a section titled "The Peril of Navigators." The disparity between the lengths of "M. Cassini's degrees" and "M. Maupertuis's degrees," tabulated in a third column, showed the magnitude of error a navigator would be making if he used the wrong model (fig. 15). The relevance of the difference jumped out at the reader, even though the table gave no means of deciding which set of numbers was correct. "It is true that all those who would have avoided shipwreck by one of the columns of this table, would have perished if they had followed the other. It may seem that its use is as dangerous as it is helpful. But it is up to the prudence of the navigator to know how to decide for the elongated or the flattened earth."[118] As in the *Disinterested Examination,* the anonymous author pretended to leave the choice up to the reader. "If the least utility becomes of greater value when applied to a great multitude, what price shall we put on a discovery which considerably diminishes danger for the innumerable number of men who risk their fortune and their lives on the sea?"[119] He left the reader to draw the obvious conclusion, relying on the skill of the mathematician

---

*Lettres,* 2:32–33). She read the *Elements of Geography* as "the second volume of the *Disinterested Examination,*" in other words, as a continuation of Maupertuis's attack on the Cassinis.

117. [Maupertuis], *Elemens de géographie* (Paris, 1740), art. XVIII, quoted from idem., *Oeuvres,* 3:6–7. Maupertuis considered this comment to be potentially offensive to his patrons in the government; hence the anonymous publication (M to JBII, 17 December 1740, BEB).

118. Maupertuis, *Elemens de géographie* (1740 ed.), 134; also in idem, *Ouvrages divers* (1744), 59. The table appeared in the first three editions; it was left out of the revised version of the book included in the 1756 edition of the *Oeuvres.*

119. *Elemens de géographie* (1740), 134.

DE GEOGRAPHIE.

TABLE

DES DEGRÉS DE LATITUDE.

| Latit. du Lieu. | Degrés ſuivant M. Caſſini. | Degrés ſuivant Mrs. Clairaut, Camus, le Monnier & moi. | Différences. |
|---|---|---|---|
| 0 | 58020 Toiſ. | 56625 Toiſ. | + 1395 T. |
| 5 | 58007 | 56630 | 1377 |
| 10 | 57969 | 56655 | 1314 |
| 15 | 57906 | 56690 | 1215 |
| 20 | 57819 | 56740 | 1079 |
| 25 | 57709 | 56800 | 909 |
| 30 | 57580 | 56865 | 715 |
| 35 | 57437 | 56945 | 492 |
| 40 | 57285 | 57025 | 260 |
| 45 | 57130 | 57110 | 20 |
| 50 | 56975 | 57195 | — 220 T. |
| 55 | 56825 | 57275 | 455 |
| 60 | 56683 | 57350 | 667 |
| 65 | 56555 | 57420 | 865 |
| 70 | 56444 | 57480 | 936 |
| 75 | 56355 | 57530 | 1175 |
| 80 | 56287 | 57565 | 1278 |
| 85 | 56243 | 57585 | 1342 |
| 90 | 56225 | 57595 | 1370 |
| | *Axe de la Terre* 6579368 Toiſ. | *Axe de la Terre* 6525600 Toiſ. | 53768 Toiſes. |

Fig. 15. Table comparing lengths of degrees on earth's surface, extrapolated from Cassini's measurements in France and the Lapland measurements, with the disparity clearly noted. In the original (anonymous) edition, Maupertuis referred to himself in the third person; here he acknowledged his authorship in the column headed "Degrees according to Mssrs. Clairaut, Camus, Le Monnier, and myself." Maupertuis, *Elémens de géographie, Ouvrages divers* (Amsterdam, 1744). Courtesy of History Division, Biomedical Library, UCLA.

who had made the calculations for the comparison. When the *Elements of Geography* came out under Maupertuis's name in 1742, approved by the Academy, the *Histoire* noted that the chart was "very useful for practice, even though the theory is based on a delicate calculation that M. de Maupertuis suppresses because he restricted himself to giving only what would be accessible to the greatest number of readers."[120]

Maupertuis was well aware of the difficulties of turning mathematics and astronomy into applications of value to mariners. Given that rules for new navigational practice were not immediately forthcoming, what meaning does utility have here? Is it just an idle promise made by the recipients of royal patronage in order to justify their pensions? The point of contact between the academic expert and the navigator remained elusive.[121] But the rhetoric of utility referred also to the advancement of astronomy and geography as sciences. In *Discourse on the Parallax of the*

120. [Mairan], "*Elemens de géographie* de M. de Maupertuis," *HAS* 1742, 114–15.
121. He attacked this problem in *Astronomie nautique* (1743), with only limited success.

*Moon* (1741), Maupertuis developed the notion that knowing the shape of the earth "is useful for the perfection of the theory of the moon, which is today the most important thing that remains to be discovered in astronomy. The determination of longitudes at sea depends on this theory."[122] Practical rules, "as simple as those drawn from the spherical shape of the earth," could translate the abstruse knowledge of savants into calculating techniques for ship's captains. Determining longitude at sea was not a problem he could pretend to solve because of the difficulty of keeping time on board ship. But he did argue that a theory of the moon's motion, based on parallax observations, was integrally tied to accurate knowledge of the shape of the earth. Ultimately, the utility arguments were tied back to Newtonian physics, through the equations used to derive the earth's shape once the latitude measurements were known. This was the result from Newton's *Principia* that Maupertuis had demonstrated analytically in 1735. His readers did not have to follow the demonstration, of course, but the reference back to the rigorous mathematics legitimated the result and reiterated his ties to Newtonianism, now fashionable in France, largely because of his own efforts.

## Conversations

In addition to his challenges to Observatory practice, we have seen how Maupertuis pursued a variety of legitimization tactics, addressing new audiences and multiplying the kinds of texts in which esoteric technical issues were discussed. Much of the academic opposition to the *Messieurs du nord* was expressed in private and made its way into rumors and gossip.[123] Maupertuis raised the stakes by taking the dispute to the public and asking for a different sort of validation. Without abandoning the academic context completely, he forced the dispute outward from the closed ground of the Academy to the literate society beyond. This society delighted in scandal, but Maupertuis's success showed that it would also entertain arguments couched in quantitative terms. They didn't necessarily follow the details; nevertheless, he derived his reputation in part from his knack for explaining philosophical and technical matters engagingly. He managed the dispute by moving between genteel society and academy, between France and England and Lapland, between anonymity and publicity. His opponents did not cross so easily between contexts and constituencies. In transposing the argument into forms

122. Maupertuis, *Discours sur la parallaxe de la lune* (1741), in *Oeuvres,* 4:216.

123. Brunet names Réaumur and Mairan as Cassini's strongest allies in the Academy, although Réaumur was still favorably disposed to Maupertuis in November 1738 (Brunet, *Maupertuis,* 1:60); Maupertuis lists Bragelogne, Fontenelle, and de Molières as his other opponents (*Examen désintéressé,* 2nd edition, 53–56).

other than the academic memoir, Maupertuis contradicted Cassini's notion of astronomical practice as much as when he transgressed against the norms of stellar observation by failing to reverse his zenith sector.

Cassini de Thury, almost a generation younger than Maupertuis, found it easier than his father did to admit the advantages of new instruments and new mathematics. He continued to defend his father's conduct, however. A manuscript in Thury's hand dated November 1742 from Avignon, recounts a discussion in the home of a provincial "learned lady," where the author was asked his opinion of the controversy about the earth's shape. Framed as a letter (to an unnamed recipient), it is not clear who wrote it originally. It may well be fictional, but it articulates the key questions of style and comportment that characterized the whole dispute. Whether or not it recorded an actual conversation, it shows Maupertuis's strategies from a different vantage point, that of a provincial reader; its setting suggests that such discussions would not be out of place even in drawing rooms distant from Paris.[124]

In the manuscript, the hostess asks her guest how the king's astronomers could possibly have made such errors, while Maupertuis, a mathematician working in more difficult physical conditions, "determined and settled the true shape of the earth." She is quite knowledgeable, having read three of Maupertuis's books, and she understands the relation between theory (physics) and astronomy, at least to some degree.[125] Her guest defends Cassini II by saying that the instruments used to measure the Paris meridian were neither stable enough nor precise enough for the task. The zenith sector, because of its novel design, made everything easy for the Lapland expedition. Anyway, he continued, Cassini had not set out to find the shape of the earth and knew that the difference he had found from south to north was not significant enough to decide the question. Maupertuis, on the other hand, promised more than he could deliver, since he had found several different numbers for the ratio of the axis to the diameter, depending on which numbers he compared. "In fact, we are impatiently awaiting his final word on the subject, and so are the navigators who have in their hands a table where this author has marked the danger that ignorance of the true shape of the earth could throw them into.

124. [Cassini de Thury?], "Lettre sur la figure de la terre," 1 November 1742, Obs. MS, B.5.4, vol. X. The manuscript has been inserted in a manuscript copy of Jacques Cassini, *De la grandeur et figure de la terre* (published 1720). It refers to Thury in the third person; he may have copied it from a letter written by someone else, or he may have been experimenting with the genre. In 1742, the Academy was in recess until 14 November, so Thury could have been in Avignon in early November.

125. The three books probably would have been *La figure de la terre* (1738), *Elemens de géographie* (1740), and *Degré du méridien de Paris* (1740).

Our gentlemen of the Observatory are a bit more reserved, they respect the public and scorn the so-called honor of furnishing libraries with repetitious books which undercut each other." Whoever the author of this letter may be, he clearly recognized Maupertuis's literary strategy, without approving of it.

Cassini's defender objects to Maupertuis's presumption, and to the use of print to appeal to readers like the hostess. "Impatient to see his name on a work that decides such a famous question," he sought to preempt the latest conclusions of Thury, then in the process of verifying the meridian. Thury, the letter suggests, always acted honorably in his public references to Maupertuis and remained impartial in his research. Maupertuis, on the other hand, knew ahead of time what he wanted to find, and only remeasured Picard's arc in Amiens to counter ongoing revisions by Thury to the more southerly degrees. The hostess is well informed on the technical matters, but also insists that Maupertuis is "the most admirable man in the world; . . . his works are the delight of savants and his conversation that of people of wit [*les gens d'esprit*]." By the end, the narrator implies that his interlocutor is not an appropriately serious reader: "The lady was on the point of raising another argument when they announced the guests she had invited for supper, since things here are as they are in Paris: learned ladies split their time between the sciences and entertainment; for dinner they gather together men of letters, for supper the pleasant fashionable people." The hostess objects at one point that Cassini unfairly questioned the accuracy of the zenith sector, "seducing the public who are bluffed by the name of Cassini into believing," provoking her guest to respond that Cassini had been attacked by Celsius "in an indecent manner, since he accused the elder M. Cassini of incapacity for observation." This vicious work "gave rise to all the speeches that M. de Maupertuis attributed to Cassini's passion.[126]" Whatever its provenance, the letter captures the combination of personal animus and technical challenge behind the conflict, and indicates how people like the provincial lady might have perceived the debate. She is presented as Maupertuis's ideal reader, eager to evaluate the rhetorical moves of the antagonists. In the eyes of the author of the letter, association with such readers implied a kind of weakness, no doubt linked to the frivolity of the "learned lady."

## Conclusion

Without denying the philosophical differences separating the two sides of this dispute, we can see how they were also contesting academic and disciplinary territory. When he gave priority to new mathematical skills and English instruments over accepted astronomical training practices and locally crafted equipment,

126. [Cassini de Thury?] "Lettre sur la figure de la terre."

Maupertuis questioned the value of the Paris astronomer's way of life, established by Jean-Dominique Cassini sixty years earlier. The young mathematicians in the Academy also initiated a new form of geodetic practice involving long-distance travel, substantial expense, new instruments, and a variety of mathematical tools. When the results of this geodesy were not unanimously accepted, Maupertuis brought into play a range of strategies to justify it, including the co-optation of the utility argument from the astronomers engaged in the very visible and practical project of producing an accurate map of France.

Maupertuis and Cassini differed radically in their reasons for being interested in the problem of the shape of the earth in the first place. For Maupertuis, the problem lent itself to the kind of mathematical analysis he wanted to promote in the Paris Academy; it also provided a graphic illustration of the Newtonian theory of universal gravitation; and finally, it supplied him with the means to make himself visible beyond the confines of the academy. His commitment to the natural philosophy of gravity cannot be separated either from his desire to spearhead a distinctively French Newtonian physics, of which geodesy would be a part, or from his personal ambition. For the Cassinis, on the other hand, the shape of the earth had emerged as a byproduct of cartography, a project they had inherited from the first generation of Observatory astronomers. In the course of the dispute, Cassini II was forced to defend practices that had been accepted as standard for decades. The elongated earth came to represent those practices and the integrity of the Observatory astronomers.

In making universal gravitation respectable in France, Maupertuis and his academic cohort changed the rules governing the use of mathematics in observational astronomy and geography. They used their explorations of Leibnizian analysis to wrench the problem of the earth's shape out of the cartographic context and to give it new meaning. While Jacques Cassini, following in his father's footsteps, had relied primarily on the empirical data of positional stellar astronomy and surveying, the new generation of mathematicians applied their analytic methods to mechanics, hydrodynamics, lunar motion and tidal theory, as well as to geodesy. For the problem of the earth's shape, this meant challenging existing observational data that contradicted the results of their calculations. In 1733, Fontenelle complained, "Clairaut made some reflections [on Cassini's measurements] taken from transcendental geometry, which nevertheless had nothing to do with it."[127] By 1744, when Thury published his account of the remeasurement of the meridian, he incorporated into his observational record just such mathemat-

127. Fontenelle, "Sur la description du parallele de Paris, ou de sa tangente," *HAS* 1733, 59; Clairaut, "Détermination géométrique de la perpendiculaire à la méridienne tracé par M. Cassini; avec plusieurs méthodes d'en tirer la grandeur et la figure de la terre," *MAS*, 1733: 406–17.

ics as Fontenelle had dismissed as irrelevant. In so doing, he tried to retain for the Observatory, of which he would soon become the first official director, control of geographic measures within the realm of France, subject to new degrees of precision with the advent of new instruments. He was implicitly admitting that the mathematical program of Maupertuis and Clairaut had found a place in the practice of French cartography and astronomy.

# 6

# Beyond Newton and on to Berlin

MAUPERTUIS INCORPORATED HIS LAPLAND EXPERIENCE into his public identity and his subsequent work to a much greater extent than did any of his collaborators. At every opportunity, he brought up the success of the expedition and the utility of the measurements. Throughout the years of conflict with Cassini, as we have seen, he was lionized by high society and by the "philosophical" elite, for whom English natural philosophy signified enlightenment. Voltaire contributed as much as anyone to Maupertuis's image as a daring Newtonian, not only by singing his praises, but also by calling on his specialized knowledge and by decrying the "persecution" perpetrated by his enemies. So, for example, when Voltaire set out to answer critics of his own *Elements of Newtonian Philosophy* in 1739, he published his defense as an open letter to Maupertuis, broadcasting his alliance with the controversial man of science. "After thanking you for the lessons on Newtonian philosophy I have received from you, allow me to address to you the ideas that are the fruit of your instructions."[1]

This was all very well. But Maupertuis's ambitions did not end with flattening the earth. He wanted to be more than a Newtonian, however fashionable that identity might be. In the 1740s, he continued to court the genteel audience of *la cour et la ville* with a series of small, elegantly produced books. But he also went back to mathematical physics, to reassert his academic credentials. Without the supervision of Bernoulli, his mathematics became less sophisticated, but his claims for generality and his own originality became more insistent. His physics became more philosophical, or metaphysical, as he drew out the implications of his results in a way that had not been possible for geodesy or astronomy. In so doing, he emphasized not just problem solutions, but a distinctive approach to mechanics.

His ambition to break new ground led him to equilibrium problems, an outgrowth of his earlier work on the shapes of rotating fluid bodies. In 1739, he consulted the senior Johann Bernoulli for the last time, asking him to review a manuscript on the shape of the earth analyzed in terms of forces acting on the oceans.[2]

1. "Lettre de M. de Voltaire à M. de Maupertuis," *Bibliothèque française* (1739), reprinted in Besterman et al., eds. *Complete Works of Voltaire,* 15:697–718.

2. M to JB, 21 March 1739, BEB.

Bernoulli had not been happy with the results of the expedition, and the polemics that followed, but Maupertuis had just visited in Basel and smoothed the elder man's ruffled feathers. In spite of their renewed friendship, the exchange of letters about the new paper was not entirely satisfactory. Bernoulli showed frustration with his former pupil's mathematical approach, and Maupertuis resented his mentor's lack of appreciation for his insight about the problem.[3] As a result of the critical comments, he put the paper aside. When he got around to presenting something to the Academy, almost a year later, he had transformed the work into a question about the equilibrium of a system of bodies subject to attracting forces, rather than the shape assumed by a fluid mass.[4] This paper had passed the scrutiny not of the elder Bernoulli, but of his son Daniel, by this time an illustrious mathematician in his own right, who pronounced it "very beautiful and well demonstrated throughout."[5] The paper posited a "law of rest" for masses subject to attractive forces. For a system of bodies attracted by forces acting as a power **n** of the distance to the centers of force, "in order for the bodies to remain at rest, the sum of the products of each mass by the intensity of its force and by the power (n+1) of its distance to the center of force (which we can call the sum of the forces of rest) must be a maximum or minimum." Maupertuis knew that Johann Bernoulli, some years earlier, had defined a principle of virtual work, to represent equilibrium in terms of a balance of forces multiplied by elements of distance (infinitesimal displacements). Without using Bernoulli's terminology, Maupertuis's law of rest built directly on this principle.[6] Maupertuis argued that he had articulated a principle similar in intent, but far more economical, and therefore more elegant, than those of his predecessors: "One deduces from our principle with two lines of calculation the whole section V of the *Mechanics* of Varignon where he took 87 pages *in quarto* to demonstrate several useful and curious propositions that are nothing but very particular cases of those [propositions] we shall give [here]."[7] The published version omitted the disparaging reference to Varignon,

3. Bernoulli was polite, but completely rewrote Maupertuis's solution, regarding it as ill-conceived and obscure: "J'ai taché de donner à la solution de votre question un autre tour plus naturel, plus intelligible et tiré immediatement des premiers principes de statique" (JB to M, 12 April 1739, BEB). Maupertuis responded with further questions, M to JB, 19 April 1739. Bernoulli's response to this is not preserved.

4. Maupertuis, "Loi du repos des corps," *MAS* 1740, 170–76, reprinted in *Oeuvres* 4:43–64.

5. Daniel Bernoulli to M, 26 January 1740, AS Fonds Maupertuis.

6. Maupertuis, "Loi du repos," *Oeuvres,* 4:48. Bernoulli discussed virtual velocities and the principle of virtual work in J. Bernoulli, *Discours sur les loix de la communication du mouvement,* 23–26. For the history of virtual work and virtual velocities, see Hankins, *Jean d'Alembert,* 199–202.

7. Maupertuis, "Loi du repos des corps," text in AS-pv, 20 February 1740. A week later, he presented one of these cases, showing that the principle governing the descent of the center of grav-

but stressed the same point: "One acquires immediately by this theorem the solution of several problems of mechanics which have in the past arrested able mathematicians, and for which they have given only particular solutions, that cost them much trouble and time."[8]

Maupertuis took the opportunity to reflect on the utility of general principles for the practice of physics, a concern on which he would expend considerable energy in the years to come. These principles lie along a spectrum of generality between axioms, which need no demonstration, and particular instances, which can be tested empirically. Such principles cannot be demonstrated rigorously: "Nevertheless their certainty is so great that some mathematicians do not hesitate to make them the foundations of their theories, and use them every day to solve problems that would cost them much more trouble without these principles."[9] Mathematics (in the broadest sense, including mechanics) depends for the success of its solutions on the solid foundation of its principles. These general principles, though lacking strict demonstrative rigor, function as mental economy principles, since they facilitate the mathematician's journey from axioms to particulars. They are, he says, "Shelters for minds exhausted or lost in their researches."[10] This paper does not seem to have elicited much comment at the time; Fontenelle did not know what to make of it and did not write an extract for the *Histoire.* It gained significance only in light of later work on the principle of least action.

Although he did not follow up on this immediately, the law of rest and the quest for general principles took Maupertuis back into abstract mechanics, but in a more ambitious way than he had attempted before. He certainly drew on the insights and techniques learned from Bernoulli as he thought about how to venture beyond direct commentary on Newton. His subsequent development of the perspective adumbrated in the law of rest indicates that he consulted Leibniz, as well as Descartes and Fermat, in thinking about the relation of metaphysics to physics, and some of these questions may well have been addressed in his scientific correspondence with Châtelet.[11] When he next wrote about mechanics, he brought

ity of a system is a corollary of the general principle. (This remained unpublished.) AS-pv, 27 February 1740.

8. Maupertuis, "Lois du repos," in *Oeuvres* 4:56.

9. Ibid., 4:46.

10. Ibid., 4:47.

11. In 1738 and 1739, they exchanged letters about gravity, *vis viva* and light, as Châtelet worked out her own approach to dynamics. Maupertuis's letters do not survive, so his views must be extrapolated from her comments and questions. She asked him to explain his 1732 paper on attraction, and especially God's reason for preferring the inverse-square law over all others. They also discussed collisions, forces, conservation, and laws of motion. (Châtelet to M, 2 Feb., 10 Feb., 30 April,

metaphysics to the fore, in a way that was highly unusual in the context of the Paris Academy.

## The Principle of Least Action and the Refraction of Light

In 1744, after a hiatus of several years, in which he devoted himself to other literary genres, Maupertuis presented a paper on the refraction of light, a completely new subject for him.[12] Clairaut had shown how Newtonian attraction could be applied to refraction several years before.[13] Maupertuis, leaving the difficult mathematics of attractive forces to his younger and more brilliant colleague, used refraction to demonstrate the usefulness of metaphysical principles, "those laws to which Nature herself seems to have been subjected by a superior Intelligence, who, in the production of His effects, makes her always proceed in the simplest manner."[14] In unimpeded propagation and reflection, light follows the path of least time and least distance, as any macroscopic body would do. In order to extend the economy principle to the case of refraction, he noted that "action," defined for the purpose as the product of the speed of light and the path length, could serve as nature's minimum "expenditure" (*dépense*). The title of the paper, "The harmony of different laws of nature that seemed incompatible until now," referred to the applicability of the minimum principle to rectilinear propagation, reflection, and refraction, but it indicated Maupertuis's ambition for the principle, which he would subsequently extend to all of mechanics. As it turned out, this was his last effort for the Paris Academy, and it must have startled his colleagues to hear him talking about how physics reveals the intentions of God.

The argument throughout assumed the corpuscularity of light, in a Newtonian vein. Light, being corpuscular, ought to obey the laws of mechanics. Could existing explanations for refraction be amended to allow optical and mechanical phenomena to follow the same laws? Rectilinear propagation and reflection presented no deterrent to such a project because they were strictly analogous to inertial motion and elastic rebound off a surface, respectively. "When light passes from one medium into another, the phenomena are completely different from

9 May, 21 May, 22 May, 21 June, 7 July, 1 September, 3 September, 1738, and 20 [January], 1739, in *Lettres,* 1:212–13, 215–18, 220–21, 224–28, 230–39, 242–43, 252–59, 310–12). See also Janik, "Searching for the Metaphysics of Science"; Terrall, "Emilie du Châtelet and the Gendering of Science."

12. Maupertuis, "Accord de différentes loix de la nature, qui avoient jusqu'ici paru incompatible," *MAS* 1744, 417–26, reprinted in *Oeuvres,* 4:1–23. He presented the paper on 15 April, and re-read it to the public meeting three days later (AS p-v 1744).

13. Clairaut, "Sur les explications Cartésienne et Newtonienne de la réfraction de la lumière," *MAS* 1739, 259–75.

14. Maupertuis, "Accord des lois," *Oeuvres* 4:8–9.

those of a ball traversing different media; and however we try to explain refraction, we find difficulties that have not yet been overcome."[15] There must be a way to understand and explain refraction, he suggested, to excise this anomaly and to highlight the fundamental similarity between light corpuscles and macroscopic bodies. If light is corpuscular and obeys a minimum principle, larger bodies ought to follow suit. In pursuing this parallel, Maupertuis's assumption of a Newtonian corpuscular theory of light ultimately led him to a decidedly non-Newtonian mechanics.

Maupertuis situated his idea relative to the work of his predecessors: Fermat, Descartes, Leibniz, and Newton. Fermat, assuming that the speed of light decreases with the density of the medium, had said that the sine law minimized time; for Leibniz (who assumed that light moves *faster* in denser media) the law minimized "resistance." Newton also argued that velocity increases in direct proportion to density, but analyzed light as a stream of material particles attracted by the media through which they travel.[16] They all agreed on the mathematical description of refraction, whereby the sines of the angles of incidence and refraction stand in a fixed proportion to each other. But what is the physics underlying the mathematics? By understanding the behavior of light in terms of a new economy principle, Maupertuis maintained that he had resolved the whole controversy, consistent with a corpuscular theory of light. He adopted Newton's suggestion that the attraction exerted by the medium on light corpuscles causes refraction, but he substituted his own metaphysics to go with this claim, deducing the sine law from the metaphysical principle rather than analyzing the forces between matter and light to get the general law. The economy principle led him to talk about not only *how* light behaves, but *why*.

Maupertuis set out to vindicate Newton's physics with a Leibnizian metaphysical argument, although he couched it in terms of an attack on "the edifice that Fermat built." Time is only minimized if light slows down when it moves to a denser medium, so siding with Newton on this point meant amending the principle of least time. Calling on the Leibnizian principle of sufficient reason,

15. Ibid., 7.

16. Newton speculated on the forces between matter and light in "Queries" appended to the second edition of *Opticks* (1717). On Fermat and Leibniz, see Sabra, *Theories of Light from Descartes to Newton*, 116–58; on Newton, see ibid., 231–50. Maupertuis misunderstood Leibniz's position on the speed of light in different media, and represented him as agreeing with Fermat. In a note appended to the version of "Accord des lois" published in *Oeuvres* (4:23), he explained that in 1744 he had only known Leibniz's work on refraction through reading Mairan, "Suite des recherches . . . sur la réflexion des corps," *MAS* 1723 (making Mairan responsible for his mistake). This was all sorted out by Euler; his clarification appeared in Maupertuis, *Oeuvres*, 4:23–28, but it is not relevant for the purposes of the argument here.

Maupertuis argued that there would be no reason for light to choose time rather than distance as the minimum quantity in the case of refraction. Rectilinear propagation and reflection minimize both time and path length; why should refraction be different? "What preference could [light] have for time over distance?"[17] This language of preference and choice recalled Maupertuis's speculations, back in 1732, about God's reasons for choosing an inverse-square law of gravity over all other possibilities. In that case, he had based his argument on symmetry and consistency, a different kind of efficiency. Here, God prefers a world functioning economically, where all changes or motions cost the least "expenditure." In both cases, the metaphysical language stood out starkly from the other papers in the Academy's *Mémoires.*[18]

In challenging Fermat and Leibniz, both of whom also argued from final causes, Maupertuis sought a minimum that would apply to the motion of light in all situations, and ultimately to the mechanics of collisions. He defined "action" as the product of speed and distance and tried minimizing it.[19] To check this mathematically against the empirical sine law, he calculated the action (as a simple product) for each segment of the light's path, and minimized the sum of the two quantities, finding the point where the light ray must bend in order to get from A to B with the least expenditure of action (fig. 16). (This was essentially the procedure Fermat had followed in minimizing time.) The result does indeed coincide with the sine law, and the same quantity can easily be seen to be minimized for rectilinear propagation and reflection, bringing all three laws under the same general principle. The mathematics in this paper was exceedingly simple, not even requiring integration because he took the two distances as arbitrary, but not infinitesimal, lengths. The simplicity of the mathematics contributed to the rhetorical point about the generality and utility of metaphysical principles.

## Final Causes and Scientific Method

The principle of least action, like Fermat's principle of least time, made the behavior of light teleological. Each corpuscle of light must somehow "know" where it will end up in order to choose its path, and the calculation of its motion takes into account both the starting point and the endpoint. Maupertuis made explicit the methodological role of final causes in physics, as well as the interdependence of efficient and final causes in nature.

17. Maupertuis, "Accord des lois," *Oeuvres* 4:16.

18. On metaphysics in the Academy, see Terrall, "Metaphysics, Mathematics."

19. Leibniz also used term "action." For light, Leibniz defined action to be the product of resistance and distance, claiming that resistance is minimized. Brunet, *Etude historique sur le principe de la moindre action,*12–13; Guéroult, *Dynamique et métaphysique leibniziennes.*

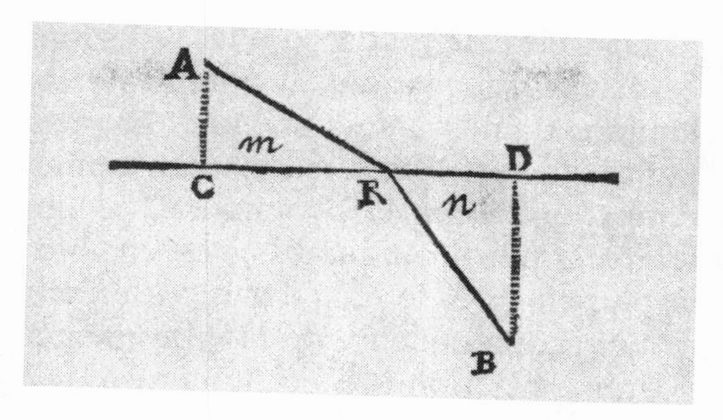


Fig. 16. Refraction of light, showing path from A to B. Maupertuis, "Accord des différentes loix de la nature," *MAS* 1744.

> There is no doubt that all things are regulated by a supreme Being who, even as he imprinted on matter the forces that denote his power, destined it to execute effects that mark his wisdom; and the harmony between these two attributes is so perfect that undoubtedly all natural effects could be deduced from each taken separately. A blind and necessary mechanics follows the designs of the most enlightened and most free Intelligence. If our mind were vast enough, it would see the causes of physical effects equally by calculating the properties of bodies and by investigating that which is most fitting to carry out these effects.[20]

The Leibnizian overtones of this passage are unmistakable. Leibniz designated the realm of forces and efficient causation as the "kingdom of power," and final causes as evidence of God's wisdom. The interplay of these divine attributes governs everything about the universe.[21] Maupertuis presented his global principle as testimony to God's wisdom and as the destiny fulfilled by the action of mechanical forces. Isolated mechanical interactions might appear "blind and necessary," but considered in metaphysical context, they become part of "the designs of the most enlightened and free Intelligence."

Maupertuis claimed to be doing several things with the principle of least action: revealing truths about God; using these truths to solve problems; and understanding the causes behind these problems. Physical problems are, at least ideally, soluble in either of two ways—through calculation of the mechanical forces causing observable effects or through consideration of global principles, those divine purposes that impose conditions on phenomena. In practice, given the limitations

20. Maupertuis, "Accord des lois," *Oeuvres* 4:21.

21. In Leibniz's dynamics, all events are caused by the kingdom of power or efficient causes and the kingdom of wisdom or final causes. . . . These two kingdoms everywhere interpenetrate

of human understanding, Maupertuis insisted that physics must follow *both* paths. Mechanics alone doesn't take us far enough, while teleology can be misleading, since "we can be mistaken about the quantity we should regard as nature's expenditure [*dépense*] in the production of her effects."[22] According to Maupertuis, this was exactly the mistake both Fermat and Leibniz made when investigating refraction. "Let us calculate the motions of bodies, but let us also consult the designs of the Intelligence that causes them to move."[23]

Although Maupertuis's programmatic statement about proper methods in physics came out of the refraction problem, he did not wish to limit it to optics.[24] The passage quoted above, for example, refers to "matter," "forces," "bodies," and "mechanics," although it was discussing the behavior of light. Forces ("properties of bodies") can be measured by their effects, but action, the result of calculating, cannot have the same kind of concrete locus in matter. Since action depends on the initial and final points of a path, it belongs to an event rather than a body. We might say it has a metaphysical reality, although it cannot be physically localized in a body. This is as it should be, since the case of minimum action expresses the metaphysical economy principle, which in turn reflects God's wisdom. Action cannot be measured directly, the way moving force can; it is not located in the physical world per se, but in God's calculations for the physical world. By demonstrating the power and generality of extremum principles, Maupertuis was putting on the table an approach to mechanics markedly distinct from anything else being done by his colleagues in Paris. No record of their reaction survives, and no one picked up on it as a new direction for research. For that, Maupertuis had to wait for assistance from Euler, in Berlin (see chapter 9 below).

## Overtures from Berlin

In 1740, Crown Prince Frederick (later known as Frederick the Great) inherited the Prussian throne. The new king fancied himself a philosopher and a poet, in the French vein, and immediately set to work inviting men of letters from France and Italy to his court. His ambitions included not only extending the boundaries of his political domain but also creating a locus of literary, scientific,

without confusing or disturbing each other's laws." Leibniz, *Specimen dynamicum*, in *Selections*, 132–33.

22. Maupertuis, "Accord des lois," 22.

23. Ibid.

24. There was a substantial tradition linking optics with mechanics. Newton discussed refraction as a corollary to his treatment of the motion of a mass through a resistant medium (*Principia*, book I, proposition 93). Descartes sometimes treated light by analogy to objects like tennis balls, although technically he considered it a pressure pulse in the aether.

and artistic activity in Berlin. Given his strongly francophile tastes, and his scorn for German language and literature, he planned to do this by importing the best minds—who would also be the best wits—to his court. As part of this projected cultural boom, he wanted to patronize a vibrant academy of letters and sciences to rival those of France and England. Although Leibniz had founded an academy in Berlin in 1700, it had fallen on hard times under the disdain of Frederick's father.[25] Voltaire, whom Frederick considered the epitome of the man of letters, had vouched for Maupertuis as early as 1738: "A man like him would establish in Berlin . . . an academy of sciences that would outdo the Parisian one."[26] With his public visibility and reputation for taking philosophical risks, and also because of his ties to Voltaire, Maupertuis seemed the perfect candidate for the task of putting the Berlin Academy on the map of the Republic of Letters. Frederick invited him to do so in the most flattering terms: "My heart and my inclination aroused in me, from the moment I assumed the throne, the desire to have you here, so that you should give the Berlin Academy the form that only you can give it. So come, come and graft the slip of the sciences onto this seedling, so that it may flourish."[27] Shortly thereafter, the invitation was repeated in the same horticultural terms: "I am working at grafting the arts onto a wild and foreign root stock; I need your help. It is up to you to know whether the work of extending and rooting the sciences in these climates would not be as glorious for you as that of teaching the human race about the form of the continent it cultivates."[28]

The invitation was soon the talk of the town in Paris; by August 1740 the *Gazette d'Utrecht* had printed the text of Frederick's invitation for all to read.[29] Frederick's letters reached Maupertuis at an opportune time, when he was feeling unappreciated in the Paris Academy and was looking for another adventure. There is no question that he was honored by such personal attention from a monarch. But however appealing the proposition, he did not think of leaving France permanently at this juncture, or even of accepting foreign patronage. In any case, Frederick did not mention terms. He did ask Maupertuis to meet with him at his castle in Wesel, just across the Dutch border; the Prussian ambassador requested

25. On Leibniz's vision for the Berlin Academy, see Ramati, "Harmony at a Distance."

26. Voltaire to FII, [1 July 1738], Volt. *Corr.*, 93:282–84. Maupertuis had exchanged letters with Frederick himself in 1738, when he sent the prince a copy of *La figure de la terre* (Koser, *Briefwechsel*, 185).

27. "Venez enter sur ce sauvageon la greffe des sciences, afin qu'il fleurisse" (Frederick to M, [June] 1740, Koser, *Briefwechsel*, 185).

28. "Je travaille à inoculer les arts sur une tige étrangère et sauvage, votre secours m'est necessaire" (FII to M, 14 July 1740, Koser, *Briefwechsel*, 185).

29. Gastelier, 15 August 1740, *Lettres sur les affaires du temps*, 440–41; the passage is transcribed from the *Gazette*.

that the French release their subject temporarily, which they graciously did. As always, Maupertuis was intensely conscious of the honor in play, from both sides of the Rhine: "I received a very flattering furlough from M. le Cardinal [Fleury] and M. de Maurepas, both of them telling me that they were only letting me leave for the honor that this brought to me and the Academy, and on my promise to return, and a thousand other most obliging things."[30] Once in the Prussian king's presence, he basked in the radiance of royal favor. "I do not yet know at all what he will do with me," he reported to Johann II Bernoulli. "He showers me with favors, often does me the honor of supping with him either with the whole court or at his private table. Nevertheless, [the French ministers] have only loaned me to him for a time, and I have given my word to return to France. That was done in a very agreeable way, and very glorious for me, as much on the part of the ministers of France as on that of Prussia."[31]

Frederick imagined that recruiting men of letters and mathematicians would be more or less like recruiting foot soldiers. Writing to Voltaire in June of 1740, just after contacting Maupertuis, he noted cheerfully:

> I have started by increasing the state's forces by sixteen battalions, five squadrons of hussars and one squadron of bodyguards. I have established the foundations of our new academy. I have acquired [Christian] Wolff, Maupertuis, Algarotti. I am waiting to hear from 'sGravesande, Vaucanson and Euler. I have established a new college for commerce and manufacturing; I am engaging painters and sculptors.[32]

As far as the Academy was concerned, this was a gross overstatement. Maupertuis had agreed to come to Germany to work on a plan for the Academy, but he had no intention of staying in Berlin. As for the others, Wolff refused categorically to come to Berlin at all, preferring a professorship in Halle; Algarotti came to Frederick's court for brief periods but had little to do with academic affairs; 'sGravesande and Vaucanson demurred, as did many others who were invited subsequently. Of the illustrious men mentioned in the letter, only Euler actually moved to Berlin, and he was dismayed at the state of affairs for some time after he got there.[33]

30. M to JBII, 13 August 1740, BEB. Before leaving France, he stopped at Compiègne "to pay court to the ministers" (M to JBII, 9 August 1740, BEB).

31. M to JBII, 2 September 1740, BEB.

32. FII to Voltaire, 27 June 1740, Volt. *Corr.,* 91:221–24. See also Voltaire to M, 7 July 1740, Volt., *Corr.,* 91:239, where he reports that the king spoke of Maupertuis as one of his "acquisitions."

33. See Euler to Delisle, 21 July 1742, MS Obs.. Delisle Papers, for the poor state of the Berlin Observatory.

Maupertuis left Paris for Wesel and Berlin in August 1740, amid much gossip about the Prussian king's plans for him. Mme. de Graffigny reported that he had gone to Prussia "to set up an Academy."[34] In the short interlude before Frederick decided to invade Silesia (in December), Maupertuis and Algarotti enjoyed the company of the king, briefly joined by Voltaire. Frederick offered Maupertuis a grandiose pension of 12000 *livres,* which he refused, but the prospects for an academy remained uncertain.[35] "In spite of all the pleasure of traveling with the king," he told Bernoulli, "I very much want to get to Berlin and to see what direction the Academy will take."[36] He even made a vigorous effort to convince the Bernoulli brothers, Daniel and Johann II, to accept positions in Berlin.[37] Maupertuis found the king "very amiable, and full of great projects for all sorts of things" but made no long-term commitment to him, conscious of the promise to return to Paris and well aware that Frederick had not set his own course yet.[38] "Nevertheless I do not plan to settle in Berlin. I have even refused all proposals for a fixed establishment, however advantageous it might be for me, but I hope to live there for a time and to return occasionally."[39] It seems that he expected to hold an honorific position in Prussia as advisor to the king, without giving up his academic status and pensions in France.

Undoubtedly, the king's attentions fed his substantial vanity, and he relished his place in the inner circle of the Prussian court, even if the company was not as

34. Gastelier, 15 August 1740, *Lettres sur les affaires du temps,* 440. Graffigny to Devaux, 16 August 1740, *Correspondance* 2:434. Maupertuis's departure was also reported in the newspapers.

35. For the pension offer, see M to JBII, 17 December 1740, BEB; he reported his good fortune to his sister (14 January 1741, AS Fonds Maupertuis). It was immediately discussed across Europe as well: Voltaire to Theiriot, 24 November 1740, Volt. *Corr.,* 91:360–61; Gastelier, 5 November 1740, *Lettres sur les affaires du temps,* 472. Gastelier noted wryly that Frederick could probably ill afford such a lavish expenditure "étant dans le cas, par sa situation, de jouer un rôle où il pourrait avoir besoin de son argent et de son application pour tout autre chose que cela" (Gastelier, ibid., 472–73).

36. M to JBII, 20 September 1740, BEB.

37. "Malgré la maniere equivoque dont vous m'avés parlé sur l'Etablissement que j'envisageois pour vous à Berlin, je n'ay pas laissé de vous proposer au Roy et je crois que vous pourrés avoir une condition trés agreable et tres avantageuse" (M (from Rheinsberg) to JBII, 15 November 1740, BEB). As a matter of protocol, Maupertuis also invited Bernoulli senior, knowing he would not leave Basel (M to JB, 15 November 1740, BEB); he hoped that tensions between Johann II and his father might prompt the younger man to accept the invitation.

38. M (from Wesel) to JBII, 2 September 1740, BEB. Frederick, for his part, described Maupertuis as "joli garçon, aimable en compagnie, cependant de cent piques inférieur à Algarotti" (Frederick to C. E. Jordan, 2 September 1740, in Frederick of Prussia, *Oeuvres,* 17:67).

39. "Cependant je ne compte point me fixer à Berlin. J'ay mesme refusé toute proposition d'établissement fix quelqu'avantageuse qu'elle eut pu etre, mais j'espere y demeurer quelque tems et y revenir quelque fois" (M to JBII, 15 November 1740, BEB).

brilliant as at home. "The favors the king and all the royal family have bestowed on me increase daily, and the way they treat me would satisfy any man ambitious for honors. I often have the honor of dining with the queens, who are charming princesses and full of kindness."[40] But not long after the royal retinue arrived in Berlin, the king was off to war in Silesia, where he hoped to enlarge his territory and his revenues. The war, which quickly made Frederick's reputation as a ruthless military strategist, took priority over his philosophical aspirations. While Frederick went to battle, Maupertuis stayed in Berlin, working on various writing projects and wondering what to do next. Given the king's initial enthusiasm, Maupertuis could not help but experience a letdown at the shift in state priorities. He asked the king to look over his plan for the renewed academy and especially to approve his suggestions for recruits, so that things could move forward.[41] In the meantime, he sought permission to leave Berlin, either to return to France or to take a longer journey to Iceland, where he had always wanted to go. As he wrote to Algarotti, who had gone to Turin on a diplomatic mission for Frederick,

> The king was here for a few days and has gone back to Silesia: I do not know yet what I will do; but I think that I will profit from this time to make some journey; and now is the time for a trip to Iceland, or never. It will be quite different to do it myself, than it would have been to have done it with you, as we had planned. But there you are, a minister of state, and me an adventurer who has just retailed his merchandise ineptly [*mal à propos*]."[42]

But soon the king, bored with the lack of suitable company at the front, summoned his French guest with a letter of a single sentence: "Come here, you are awaited with impatience."[43] A few weeks after Maupertuis joined the Prussian camp, Frederick's troops engaged the Austrians at Mollwitz in a confusing and devastating battle. The actual sequence of events with respect to Maupertuis is impossible to know with certainty, since there are many conflicting versions of the tale, and the details were twisted and embellished with each retelling. Even Maupertuis could not have known just what was happening in the chaos of the moment. In retrospect, Frederick downplayed his own confusion during the battle

40. M. to Marie Magon (his sister), 14 January 1741, AS Fond Maupertuis.

41. M to FII, 13 January 1741, Koser, *Briefwechsel,* 187. Samuel Formey, later the secretary of the Berlin Academy, recalled meeting Maupertuis in this period (Formey, *Souvenirs d'un citoyen,* 1:210–11).

42. M to Algarotti, 26 February 1741, in Algarotti, *Opere,* 16:185.

43. FII (from Schweidnitz) to M, 17 March 1741, Koser, *Briefwechsel,* 187. Frederick also wrote to his trusted friend Charles Etienne Jordan, asking him to travel with Maupertuis to a nearby town and await further instructions. Subsequent stories about Maupertuis's insistence on traveling to the battle front are unfounded; he was answering a summons from the king.

and refused responsibility for his guest's adventures. One version has it that Maupertuis refused to stay safely behind the lines with the baggage carriages and insisted on obtaining a horse in order to follow the king, or at least to observe the action. "As he is known for having been a cavalry captain in France, he thought he ought to request to serve His Prussian Majesty on this occasion, and he had the honor to be employed as adjutant . . . of this Prince."[44] At some point they were separated, the king retreating to one flank, under the impression that his army had lost the battle, and the savant losing his horse. When the smoke cleared, some hours later, the Austrians had surrendered, but Maupertuis was nowhere to be found. Frederick could learn nothing for some days; the Frenchman was presumed dead, but eventually it turned out that he had been taken prisoner.[45] Stories differ as to whether he was robbed by Austrian soldiers or by local peasants, how much they took, and how he communicated with his captors, but in the end someone realized that he was not a Prussian officer, and he was taken to the Austrian general, Count von Neipperg, who knew an international celebrity when he saw one. The general had him transported under his protection to Vienna, armed with letters of recommendation.[46]

In Maupertuis's telling of the story, he deterred his captors from their initial plan to go to "some fortress in Hungary" and convinced them to get him to the Hapsburg court in Vienna instead, where he knew he would find friendly faces.[47] Once at court, he was on familiar ground and found that he could charm the Austrians as easily as the Prussians. The Archduchess Maria Theresa (not yet the empress) treated him with every kindness, offered him compensation (which he refused), and enjoyed his company. News of the capture and the encounter with the Viennese court circulated quickly throughout Europe in letters and conversation.

44. Gastelier, 4 May 1741, *Lettres sur les affaires du temps,* 552–53.

45. The rumors of his death reached France immediately. Mme. du Châtelet heard an early version of the story via a letter written by the French ambassador to Berlin, Valory, according to which Maupertuis had been murdered by Silesian peasants (Châtelet to JBII, 28 April 1741, *Lettres,* 2:49; Châtelet to d'Argental, 2 May 1741, *Lettres* 2:49, 51–53). Before she finished the letter to d'Argental, she had news that Maupertuis had appeared in Vienna. She wrote to him there, where she hoped he would see one of her cousins (Châtelet to M, 2 May 1741, *Lettres* 2:53–54). Mme. de Graffigny heard the rumor of his death too, but knew by 30 April that he was in Vienna (Graffigny to Devaux, 30 April 1741, *Correspondance,* 3:187). News of his capture appeared in the *Gazette de France* (6 May, 208) and in the *Mercure de France* (May 1741, 1023); see also Gastelier, 4 May 1741, *Lettres sur quelques affaires du temps,* 553.

46. This last detail is mentioned by the duc de Luynes, *Mémoires,* 3:388. The duke's account was the version familiar at Versailles.

47. M to Algarotti, 10 June 1741, Algarotti, *Opere,* 16:189. In Vienna, Maupertuis stayed with Prince Joseph von Lichtenstein, former ambassador to Paris, where Maupertuis had known him at court (Graffigny to Devaux, 7 May 1741, *Correspondance,* III, 196).

In early May, it was rumored in Paris that the prisoner would only be freed in an exchange with Prussia. This turned out to be false, like many wartime rumors. He was sent back to Berlin with every courtesy, and no question of requiring an exchange. A few weeks later, the news from Vienna reported Maupertuis's audience with the archduke, who had presented him with a gold watch.[48]

Although he had been treated courteously in Vienna, and even given the opportunity to display his wit, Maupertuis was acutely aware that some would see the whole episode as amusing, even ridiculous. Certainly his encounter with the soldiers had been humiliating. Voltaire wrote sympathetically to him in Vienna, but dressed up the story with satirical touches behind his back. "He was robbed by peasants in that cursed Black Forest where he was doing penitence, like Don Quixote. They left him completely naked; some hussars, one of whom spoke some French, took pity on him. . . . They gave him a dirty shirt and took him to Count Neipperg; all this happened two days before the battle. Count Neipperg loaned him fifty crowns, after which he took the road to Vienna, as a prisoner on his own recognizance."[49] Voltaire invented virtually every detail, making the story into an episode worthy of *Candide.* Nor could Frederick resist making fun of Maupertuis for getting himself captured. The king sent his version of the events, in verse, to Voltaire.[50] In due course, these lines made the rounds in Paris, as Frederick knew they would.

By the time he got back to Berlin, betrayed and disillusioned, Maupertuis took his leave of Frederick as quickly as possible and returned to France, not sure what sort of reception he would find there. He confided in Algarotti, with whom he had hoped to travel on less distressing adventures: "The journey to Lapland and to Iceland would have been rosy compared to the one I just made with the army, where as you will have heard I ended up being captured and sent to Vienna. . . . Will we lead the life you speak of one day, and when will it be? . . . [W]e are not wise enough [to do so]: the disordered state of our hearts makes us run after idle dreams, and to sacrifice to them anything more real."[51] Only a few weeks later, en route to Paris, could he bring himself to expand on his reactions to the adventure: "I will not tell you about the humiliating way in which, for three weeks, I followed

48. Gastelier, 25 May 1741, *Lettres sur les affaires du temps,* 563.

49. Voltaire to Valory, 2 May 1741, Volt. *Corr.,* 92:12–13.

50. Frederick to Voltaire, 14 May 1741, Volt. *Corr.,* 92:24–26.

51. "Le voyage de Laponie et d'Islande auroit été des roses aupres de celui que j'ai fait a l'armée, ou comme vous aurez su j'ai fini par etre pris et envoyé a Vienne. . . .Menerons nous un jour la vie dont vous me parlez, et quand sera-ce? je suis bien persuade comme vous que c'est ce qu'il y auroit de mieux a faire. Mais nous ne sommes pas assez sages: le déreglement de nos coeurs nous fait courir apres des chimeres, et leur sacrifier ce qu'il y a de plus réel" (M to Algarotti, 18 May 1741, in Algarotti, *Opere,* 16:187).

an army where the king had brought me, nor of my chagrin at being taken for want of a horse, and of help; nor of the misery that I experienced during the time I was a prisoner; all that is too mortifying for me to be able to repeat it." Worse still was the news that the king, or at least his advisors, viewed his demand to be taken to Vienna as a crime.

> I had been called by the king to come found an Academy; I stayed in his court almost a year, during which I always tried not to deserve to be dishonored: I have ended by being badly used [*prostitué*] in his army and taken by the hussars, and after the hope of a great honor I am returning to France loaded down with ridicule and degradation. . . . I tell you, I am already trembling for my arrival in France where my enemies will have plenty to crow about.[52]

Maupertuis represented himself as a man with many enemies, just waiting to jump on him at the first opportunity. When he arrived back in Paris, however, in spite of the jokes at his expense, he was well received, both officially and socially. Gastelier noted that Cardinal Fleury greeted him as one academician to another: "His Eminence told him that he should be like the pigeon in the fable, who having had the curiosity to travel, returned so disgusted with traveling that he no longer wanted to get away."[53]

Nevertheless, Maupertuis was not entirely satisfied, especially since he had been led to believe that turning down Frederick's offer would translate into improved prospects at home. "All of that has evaporated, and I look ridiculous for having gone and come back."[54] Algarotti, back in Berlin, took on the role of intermediary between Maupertuis and Frederick, who wished to restore friendly relations. However much he had been insulted, Maupertuis seems to have decided to keep his options open for the long term, and he instructed his friend in Berlin to assure the king of his respect and to insist that he did not expect to be remunerated for his trouble.[55] In spite of his frustrations, he was also able to make the most of his adventure, and even to turn it to his advantage. The watch obtained in

52. "Voila, mon cher Algarotti, un abregé de mon histoire. j'avois été appellé par le Roi pour venir fonder une Académie; j'ai demeuré dans sa cour pendant près d'un an, pendant lequel j'ai toujours taché de ne pas meriter qu'on m'avilit: cela finit par être prostitué dans son armée, et pris par les houzards: et apres l'esperance d'un grand honneur je m'en retourne en France chargé de ridicule et d'avilissement. . . . Je vous avoue que je tremble déjà pour mon arrivée en France où mes ennemis vont avoir bien de quoi triompher" (M (from Frankfurt) to Algarotti, 10 June 1741, in Algarotti, *Opere,* 16:190.

53. Gastelier, 29 June 1741, *Lettres sur les affaires du temps,* 579.

54. M to Algarotti, 16 July 1741, in Algarotti, *Opere* 16:195.

55. Maupertuis had to disavow the misguided attempts of Mme. d'Aiguillon to obtain compensation for him from Frederick: "Je crois qu'elle s'est mêlée fort mal à propos de mes affaires, et

Vienna became a conversation piece: "He is showing everywhere a watch that the grand duke gave him, saying 'The hussars brought me your watch, but I am keeping it and I beg you to take mine; it is set with diamonds.'"[56] And accounts of his gallant repartee with Maria Theresa became part of the record as well, buttressing his image as a man of the world.[57] "I have been overwhelmed with such a great number of visits since my arrival that I have not had a moment to myself. At court and in town [*la cour et la ville*], everyone has been interested in my adventure, and they have shown me that I would have trouble finding elsewhere what I have here."[58]

Overall, though he regretted the indignity of his capture, it did little harm to his reputation in Paris, once he was on the scene to tell his own story. When Frederick renewed his invitation, Maupertuis put him off, preferring to solidify his position in Paris.[59] In any case, he was not at all certain that anything would happen with the Berlin Academy, in spite of the king's assurances. Frederick's cavalier attitude toward his ordeal at the hands of the Austrians only reinforced his reluctance to put himself back into an uncertain position that would risk offending his French patrons. Returning to the Paris Academy was now a matter of fulfilling obligations to those ministers. True, he had quite consciously alienated key men in the Academy, not only Cassini, but Fontenelle, Mairan, and most recently, Réaumur. Undoubtedly, the second edition of the *Examen désintéressé* fueled this animosity, especially since the Cassinis had openly retreated from their earlier position. But in spite of this, Maupertuis still held a respected place in the Academy, now as a senior member, and put considerable effort into his responsibilities. As director in 1742, and a prize commissioner the following year, he was at the center

de vous les recommender d'une maniere qui me choque fort. . . . Quant à des récompenses je ne m'en suis jamais proposé d'autre que celle de lui [Frederick] être utile" (M to Algarotti, 28 June 1741, in Algarotti, *Opere*, 16:192–93).

56. Graffigny to Devaux, 11 July 1741, *Correspondance*, 3:242. According to a slightly different version, the stolen watch had been made by George Graham; the Grand Duke returned to him another fancier watch by the same maker (Formey, *Souvenirs d'un citoyen*, 1:214).

57. Gastelier, 3 August 1741, *Lettres sur les affaires du temps*, 598. La Beaumelle tells the same story (*Vie de Maupertuis*, 69–70), as does Formey (*Souvenirs d'un citoyen*, 1:214). "La reine lui demanda s'il était vrai que la reine de Prusse fût la plus belle princesse du monde. M. de Maupertuis répondit: 'Je l'avais cru ainsi, Madame, jusqu'à aujourd'hui'" (La Beaumelle, *Vie de Maupertuis*, 70).

58. M to JBII, 1 August 1741, BEB.

59. "Jay receu ces semaines passées une invitation de la part du Roy de Prusse pour retourner à Berlin. j'ay eu lh[onneu]r de repondre à S.M. que rien ne me seroit si agreable que de retourner luy offrir mes services, mais quelle scavoit que je n'etois pas libre, et que javois icy differens emplois que je ne pouvois quitter, cette année encor moins que les autres etant à la teste de lAcademie" (M to JBII, 27 November 1741, BEB).

of academic business. And he was able to ally himself with the new younger members, especially the Abbé de Gua and d'Alembert.[60]

## Navigation

For the next few years, Maupertuis juggled two or three different personas, expressed publicly in different sorts of books. He simultaneously promoted himself as an expert on the application of astronomy and analysis to navigation, and as man of letters at his ease in the most elite drawing rooms and country houses. In his first public performance after his return from Prussia, he read a piece on a method for finding longitude at sea using the parallax of the moon. This was part of his new work on navigation, which he carefully tied to his earlier geodetical work. On the one hand, he claimed to be devising a practical method for the use of pilots, limited to naked-eye observations (because telescopes are difficult to use on a moving ship). On the other hand, he used the occasion to argue once again for the utility of knowing the actual shape of the earth, not just for geography and navigation, but for "the perfection of the theory of the moon, which is today the most important thing that remains to be discovered in astronomy."[61] To be useful for sailing ships, knowledge also had to be useful for the "perfection" of astronomy. In particular, the parallax of the moon (its exact position relative to the fixed stars, as observed from two different points on the earth) cannot be known accurately without knowing the shape of the earth. This perspective allowed him to build on his own method for calculating the ratio of the earth's axis to its diameter. It also made the connection between the practical matter of determining positions at sea and the notoriously difficult question of the motion of the moon, a question that would occupy others in the Academy for years to come.[62]

The work on navigation culminated in two books, *Discourse on the Parallax of the Moon* (1741) and *Nautical Astronomy* (1743). Although not written directly for navigators, these books fulfilled the obligation to Maurepas to address the problem

60. On the shifting balance in the Academy with retirements and new elections, see Badinter, *Les passions intellectuelles,* chapter 1. For his academic duties, M to JB II, 6 April 1742, BEB (for tasks associated with directorate); and 2 November 1743, BEB (for the onerous work of reading manuscripts submitted for the annual prize).

61. Maupertuis, *Discours sur la parallxe de la lune,* 1st ed. (1741), in *Oeuvres* 4, quotation on 216. The preface was delivered at a public meeting of the Academy, 15 November 1741 (AS-pv).

62. Notably Clairaut and d'Alembert; see Hankins, *Jean d'Alembert,* 32–36. The astronomer Le Monnier, Maupertuis's associate from the Lapland expedition, had been observing the moon for years and was also interested in its applications to longitude determination. He presented a paper on the subject at the same meeting, 15 November 1741 (AS p-v).

of improving navigation techniques. Of limited practical value, they addressed the relation between the elite science pursued in the Academy and the practice of captains and pilots. "I thought that by reducing the problem to observations one can do in a vessel with as much precision as in a fixed observatory, I would have a method that would give latitude at sea as exactly as it could be given on land."[63] But Maupertuis also revealed his own predilections, when he told the reader that he wanted to liberate navigation from geometry, by basing it on algebraic analysis.[64] As a result, the work remained a display of mathematical insight applied to ideas for observation techniques that were too abstract for the use of non-mathematicians.

## Man of Letters

Along with these attempts at defining an analytic astronomy, Maupertuis pursued several other publication projects for a different audience. He was by this time something of a public figure. He often spoke at the semiannual public meetings of the Academy; the periodical press covered his honors and adventures; literary journals reviewed his books; and he frequented the best company, in public and in private. His supplementary pensions and connections to the aristocracy and the government gave him extra marks of distinction over and above his pension as mathematician in the Academy. Although he was already known as an author, and as a Newtonian, he made a push to show that his literary production was commensurate with his prominent position. He did this, in the first instance, by publishing new versions of the *Discourse on the Shapes of the Heavenly Bodies* (originally published in 1732) and the *Elements of Geography* (originally published anonymously in 1740). Both appeared in elegantly printed editions in 1742, reminding the public of the author's role in settling the question of the earth's shape, in a clear and eloquent style. The *Discourse,* no longer framed as an exposition of the Cartesian and Newtonian "systems," displayed a confident command of astronomical phenomena for genteel readers. He collapsed the material on Descartes into a single chapter and expanded the discussion of Newton to modify the guarded tone of the original. Newton's greatest accomplishment, he now explained, "was to have discovered an attractive force spread throughout all the particles of matter, which acts in inverse proportion to the square of their distances [from one another]."[65] The new edition also added a frontispiece—representing the stars as centers of gravitational force and light, each with its orbiting planets and comets—for an elegant

63. Maupertuis, *Astronomie nautique,* in *Oeuvres,* 4:79.

64. Ibid., 4:91.

65. Maupertuis, *Discours sur les différentes figures des astres,* in *Oeuvres,* 1:138. On the revisions, see Beeson, *Maupertuis,* 147–50.

volume printed on heavy paper with plenty of white space on each page (fig. 17). But the greatest difference between the two editions was that the mathematical problems of rotating fluid bodies no longer appeared at all. All remaining equations and diagrams, the preface reassured the reader, had been segregated in an appendix "so that this book can be read easily by everybody." The text stood independent of the calculations, which were nevertheless there for "those who want to take the trouble."[66] The mathematics signaled the author's special knowledge, which were not accessible to all readers but were displayed nevertheless, marked off from the body of the text by a smaller typeface. Of course,

Fig. 17. Frontispiece, *Discours sur les différentes figures des astres,* 2nd ed. (Paris, 1742). Note the prominence of cometary orbits. Newton had singled out comets as troublesome for Descartes's vortex mechanics. Courtesy of Special Collections, Young Research Library, UCLA.

66. Maupertuis, *Discours sur les différentes figures des astres,* 2nd ed. (1742). The subtitle referring to a comparison of Cartesian and Newtonian systems was deleted. Idem, *Elemens de géographie,* 2nd ed. (1742). These books were printed by the publisher Martin in the same format, and were sometimes bound (and presumably sold) together. In some copies, the frontispiece was printed as a blue-green mezzotint; many of the existing copies are lavishly bound as well.

the cultural resonances of a defense of Newtonian gravity had shifted radically in the ten years since the book was first published—in large part due to the efforts of the author, as his audience knew. Now, instead of carefully crafting a judicious presentation of an unfamiliar and vaguely suspect theory, he took for granted the mutual attraction of matter as the basis for physical and mathematical explanations of the universe. In this context, however, the book also reminded readers of Maupertuis's prescience in recognizing attraction as a viable concept in advance of his colleagues, and his public.

The *Elements of Geography* similarly served to show how all the effort to measure the shape of the earth had been well spent, given the implications for navigation and geography. In both these cases, Maupertuis adapted existing books to enhance his current status and his increasingly avid ambition to gain recognition as a man of letters. Elegant production values emphasized the recently confirmed respectability of Newtonian science, translated into the French context. The broadly "curious" and interested public he addressed was an elite group; to call this writing popularization would miss the point that it served to cement an image of literary finesse coupled with arcane scientific knowledge.

Meanwhile, a comet visible from Paris in March 1742 provided an occasion to venture farther into the territory of genteel literature. Maupertuis's anonymous *Letter on the Comet* drew on the conventions of gallantry used to such brilliant effect by Fontenelle years earlier. "You wished, Madame," the little book begins, "that I should speak to you of the comet that is today the subject of all the conversations of Paris; and I take all your desires as my commands."[67] Comets crossed the boundary dividing polite conversation and gossip from the discourse of Academy and Observatory. They could be seen as wonders or portents by "the common man," as curiosities by elegant ladies, as intractable calculational challenges by mathematicians, or as poorly understood celestial phenomena by astronomers. Comets also represented both the triumph and the limitations of Newtonian natural philosophy, since using the theoretical model of central forces to calculate exact orbits for particular comets was a tricky business, as Maupertuis knew from his own experience.

Maupertuis explained to his gentle reader what enlightened modern astronomy, authorized by the best analytical mathematics and the best telescopes, had to offer a public anxious to understand the appearance and possible consequences of comets.[68] His tone is both authoritative (he commands the technical knowledge

67. [Maupertuis], *Lettre sur la comète qui paroissoit en 1742*, Paris, 1742. Translated by Esther Burney in [Charles Burney], *An Essay towards a History of the principal Comets that have appeared since the year 1742* (London: T. Bedeet, 1769).

68. Schaffer has shown how cometary science in this period cannot be separated from public

to understand the erratic behavior of comets) and entertaining (he unrolls fanciful scenarios for cometary encounters with the earth). No longer the province of the astrologer, comets are subject to physical law, and any dangers associated with them must be assessed rationally. Such an assessment requires the support of mathematics, for which even the most open-minded reader needs the assistance of the expert. Although the gallant author framed his arguments with the conceit of flirtatious didacticism, he devoted most of the book to a clear summary of the current state of knowledge about comets, including an overview of the dynamics of the Newtonian solar system and Halley's heroic calculations of cometary orbits. It evolved into a set piece on the analytical power of Newtonian cosmology. Along the way, the extremely elliptical orbits of comets undermined any cosmology built on swirling vortices carrying planets around the sun.

But nothing in Newton's physics, the book pointed out, precludes the earth from crossing paths with an errant comet, and the reader might wonder about the possible consequences of such an encounter. An errant comet might knock the earth out of its orbit, for instance, changing conditions for its inhabitants. Or a comet might "rob us of our moon," or even make the earth its unwilling satellite. None of this should leave the enlightened reader in mortal terror; the text suggested that she should temper speculation about possible futures with a calculation of their likelihood.

> Everything shows us that comets could bring deadly changes to our earth and to the whole economy of the heavens. . . . But we are right to feel safe. . . . However terrible thunder might be, its descent is not much to be feared by each individual man, because of the small space he occupies in the area where the lightning could strike. The tiny place that we occupy in the immense expanse where these great events happen, annihilates the risk for us, although it does not change the nature of the danger.[69]

In other words, the intelligent reader would recognize that comet-induced catastrophe, although physically possible, was also improbable enough to become interesting and amusing rather than terrifying. Such a collision might even have beneficial effects, such as altering the angle of the earth's axis to induce a perpetual spring, or being captured by the earth's gravity to become a second moon. Taken as a whole, the book artfully combined such lighthearted speculations with serious discussion of the most up-to-date astronomical knowledge. The rational cosmology underlying the speculations about possibilities was that of Newton, shorn of its mathematics.

---

fears about the dangers associated with comets. See Schaffer, "Authorized Prophets." On popular and learned perceptions of comets, see Genuth, *Comets, Popular Culture and the Birth of Modern Cosmology.*

69. Maupertuis, *Lettre sur la comète,* in *Oeuvres,* 3:247–48.

Although there was nothing particularly risky about the argument of the book, anonymous publication allowed its author to adopt a style and tone distinct from his academic voice. But he hung onto his expertise nevertheless, by keeping his other identity present to the reader in a slightly disguised form. The book advertised his place in that elite society through the intimate terms of his reference to his female companion and reader. A second edition (also 1742) pointed directly to the author's purposeful use of gallant style, by adding a preface "by the bookseller" discounting criticisms on that account. "Persons of good taste were shocked by some evidence of a kind of quite insipid gallantry in several places," but the author refused to change anything. Knowledgeable readers would recognize the authority of a man of science speaking to an educated public, carefully figured as female in the person of the lady addressed in the "letter." They might even pick up on hints about the author's identity, as when he explained the theory of the origin of the rings of Saturn as captured cometary debris. "It was explained in the *Discourse on the Shapes of Heavenly Bodies* how a planet could appropriate the tail [of a comet]; and without being swamped by it, nor breathing in the bad fumes, form from it a kind of ring or an arch suspended around itself."[70]

Unlike the *Examen désintéressé,* the book did not remain anonymous for long. At least two reviewers identified Maupertuis by name, describing him as "a Philosopher who is known no less as a man of wit [*homme d'esprit*] than as a profound mathematician."[71] These reviews noted the gallant style of the book and applauded it as an authoritative compendium of current knowledge about comets.[72] But not all readers were so happy with the author's posture. The *Critique of the Letter on the Comet, or Letter to a Nine-Year-Old Young Lady* objected violently to a review that had compared Maupertuis favorably to Fontenelle. The barely masked author of the *Critique,* Gilles Basset des Rosiers, viewed Maupertuis's book as a sacrilegious perversion and co-optation of Fontenelle's style. The *Critique* is interesting for the parallelism of its rabid attacks on the literary form and on the Newtonian content of the *Letter,* while it also participates ironically, albeit clumsily, in the same discourse by addressing a young girl not yet part of the adult world of irony and sexual intrigue.

> The address [to the lady] is the only thing that links it to [Fontenelle's] *Worlds.* . . . In the new work, there are only English names, that you could not pronounce without giving yourself a sore throat, and that your ears could not hear without being

70. Ibid., 3:251.

71. *Journal des sçavans* (June 1742): 351.

72. Desfontaines called Maupertuis "the ingenious rival of the author of *The Plurality of Worlds*" (*Observations sur les ecrits modernes,* 31:135).

> wounded. It is nothing but cones, ovals, parabolas, conic sections. . . . I tell you, Mademoiselle, it is frightening. I don't believe there is in France a single woman tough enough to read this *Letter* and learned enough to understand it, unless it be Madame the marquise du Châtelet.[73]

Maupertuis's choice of style and genre for his discussion of comets once again enabled him to make a splash on the literary scene. The *Critique* only made it more visible; the two were read together.[74] One of his aims, as with his satire of Cassini, was to get himself talked about, to be the subject of gossip and speculation, as well as admiration. The appreciation of the target audience could only enhance the authority of the "*savant astronome.*"

While he was addressing the genteel public with these various works, old and new, he also consciously held back from publishing a book he had had in hand for some time, waiting for the right moment. A version of this eventually appeared in 1750 as the *Essay on Cosmology,* which used the principle of least action, construed metaphysically, to prove God's existence. It was this theological component that made it appear a risky venture, although the argument was neither atheist nor materialist. It was, rather, an attack on the argument from design, where God's existence was proved from the wonders of nature. As early as the winter of 1741, he had written to Algarotti from Berlin that he had completed a work on "the world and its fortunes," which he intended to publish as soon as possible. "I tell you, however, that I am a bit afraid for the success of this work, which is just the thing to scandalize the fainthearted [*les faibles*]."[75] This may have been quite different from the text eventually published, since there is no indication that the principle of least action had been fully articulated mathematically in 1741, but the manuscript must have treated metaphysics and cosmology.[76] Châtelet knew about it: "If I was ever curious about anything, it is your cosmology. The parallax of the moon is more interesting for the astronomers, but as for us other earthly people, I would like the cosmology much more, and I am indignant about not seeing it."[77] Back in Paris

73. [Gilles Basset des Rosiers], *Critique de la Lettre sur la comète* (Paris, 1742), 30–31. The objectionable reviewer was Desfontaines (cited in previous note). Basset des Rosiers was a philosophy professor in a Jesuit college in Paris.

74. Graffigny sent the two books to Devaux, who had heard of them in Lunéville. Graffigny to Devaux, 7 October 1742, *Correspondance,* 3:390, 393 n.17. The Bibliothèque de l'Institut de France holds a volume in which *Lettre sur la comète* is bound together with Basset's *Critique,* and the *Anecdotes physiques et morales* (attributed to Maupertuis).

75. M to Algarotti, 18 February 1741, in Algarotti, *Opere,* 16:183.

76. Maupertuis notes that the followers of Christian Wolff would not like what he says about space and time; see M to Algarotti, ibid.

77. Châtelet to M, 8 August 1741, *Lettres,* 2:62. Voltaire also asked for a copy (Voltaire to M, 10 August 1741, Volt. *Corr.,* 92:95).

after his adventure in Vienna, he claimed to be on the verge of printing it once again. "The novelty of my ideas may attract criticism, but I will try not to make quarrels out of them; and I prefer that small minds remain mired in scandal, than to take the trouble to see them well refuted."[78] But he thought better of provoking another scandal and kept the manuscript away from public scrutiny for the time being.

The ultimate affirmation of literary status would be election to the Académie française, that society of "immortals" of French literature whose selection was heavily influenced by behind-the-scenes lobbying in the most exclusive circles. Limited to forty members, each elected for his lifetime, the Academy voted to fill vacant seats, often under the influence of one or another faction in the government. Elections were generally of great interest in fashionable society, where many of the well-connected hostesses pushed for their favorites.[79] Men of science only rarely attained positions in both academies. Fontenelle had been a member of the literary academy since 1691; Mairan was elected in early 1743, just at the time when Maupertuis was working to build up his own literary reputation.[80] Given that he had provoked the animosity of these two men in the course of his conflict with Cassini, any campaign for Maupertuis was likely to be an uphill battle. But when the Abbé de Saint-Pierre died in May of 1743, releasing his seat in the Académie française, several of Maupertuis's supporters suggested that he put himself forward. He may well have thought of it himself, especially since Voltaire had tried unsuccessfully to obtain the previous slot, vacated just a few months earlier. Unlike the Academy of Sciences, the literary academy was embroiled directly in politics, and elections often brought out all sorts of rivalries, gossip, and slander. Candidates had to seek the votes of the members, but they also had to be approved by the crown, and key individuals at court had de facto veto power. Mme. Geoffrin, well connected to the Academy and the court, thought Maupertuis had a good chance, especially since the comte de Maurepas, the minister with oversight of all the royal academies, was his ally.[81] Maupertuis also had personal friends in the Academy itself, notably Montesquieu and President Hénault, who lobbied vigorously on his behalf. In the weeks leading up to the election, rumors swirled around

78. M to Algarotti, 28 June 1741, in Algarotti, *Opere,* 16:191–92.

79. See Badinter, *Passions intellectuels,* for the lobbying efforts.

80. Another dual member was Maupertuis's old friend from the Café Gradot, the Abbé Terrasson (1670–1750). Subsequently, several pensioned members of the Academy of Sciences were elected to the Académie française: Buffon (1753); d'Alembert (1754); and Condorcet (1781). Voltaire was only elected in 1746; he had even more enemies than did Maupertuis.

81. Mme. Geoffrin informed Martin Folkes just after the death of Saint-Pierre that Maupertuis would fill his slot (Geoffrin to Folkes, 17 May, [1743], in Brown, "Madame Geoffrin and Martin Folkes," 233). La Beaumelle says Maurepas and Montesquieu encouraged him to go for it, against

the salons and the halls of power. When he went to see the Bishop of Mirepoix, whose vote he needed, Maupertuis found that unnamed sources had denounced him to the bishop as a deist and "author of a certain manuscript *Cosmologie,* where he set out to demonstrate the existence of God algebraically." The bishop not only refused his support in the strongest terms, but threatened to veto a result in Maupertuis's favor.[82]

Until the vote, however, no one was sure what would happen. "[Maupertuis] is seeking it forcefully, he does not believe he will get it, because old Fontenelle is opposing it with all the strength his age allows [*toutes ses vielles forces*]. They are writing anonymous letters against him."[83] The duc de Luynes heard that "[t]he friends of M. de Maupertuis were very much afraid of M. de Fontenelle, doyen of the Academy, who had a personal quarrel with Maupertuis . . .; but finally they brought M. de Fontenelle to a reconciliation, which took place before the meeting."[84] The vote was not an unmitigated triumph; twenty members were absent, and the total number of voters fell just short of the required quorum. Nevertheless, no one challenged the breach of the rules, and Maupertuis took his place among the "immortals," with the approval of the king. Maurepas had proved more powerful than the bishop of Mirepoix in the end.[85] In spite of the struggle, the election confirmed Maupertuis's place in the most prestigious cultural institution of the realm, and in the literary and social world that sustained it. Although he had enemies, his supporters were more numerous and more powerful; in any case, having enemies signaled the vibrancy and novelty of one's "philosophy." Maupertuis relished his conflicts with his enemies, as he relished performing flamboyantly for admiring audiences at dinner parties, however much he complained about being persecuted. As he reported to Bernoulli, "You would not believe the indignities they have sustained against me; including wanting to make people believe I had written books against religion. You know very well that most people call books opposed to religion, books that are opposed to them."[86] But his enemies had been defeated, which made the victory all the sweeter. The whole

---

his inclination (*Vie de Maupertuis,* 77). Maurepas knew that Frederick II was still trying to get Maupertuis to go to Berlin and may have seen election to the Académie française as an added honor that might keep him in France.

82. For the most detailed, but not necessarily the most accurate, account of this exchange, see La Beaumelle, *Vie de Maupertuis,* 77–80; Velluz follows La Beaumelle (*Maupertuis,* 83).

83. Graffigny to Devaux, 24 May, [1743], *Correspondance,* 4:293. Fontenelle was eighty-six years old at this time.

84. Luynes, *Mémoires,* 5:28. Fontenelle was not present at the meeting where the vote was taken, however; his acquiescence was only tacit (*Registres de l'Académie française,* 531).

85. Académie française, *Régistres,* 531–32.

86. "Vous ne sauriés croire les indignités quon avoit suscitées contre moy; jusqu'a voiloir faire

business "made known to me more friends as well as impotent enemies." He knew that the real mark against him was the *Examen désintéressé,* "where indeed several of these gentlemen were insulted. They invented this wonderful pretext of religion, which covered them in shame. It was Fontenelle and Mairan who were the leaders of the cabal."[87] Even though it had caused him this trouble, and though many of his friends saw it as overkill, Maupertuis was proud of his scheming about the *Examen.* He had kept silent about it for several years; by the summer of 1743 he described it as "one of the best anecdotes of my life."[88]

---

croire que j'avois fait des livres contre la Relligion. Vous scavés bien que la plupart des gens appellent des livres contre la relligion, des livres contre eux" (M to JBII, 13 July 1743, BEB).

87. "Toute mon affaire de l'Academie françoise a tourné le plus heureusement du monde et ma fait connoitre plus d'amis encor que dennemis impuissants. La source du mal venoit de l'examen desinteressé où effectivement quelques uns de ces Mssrs. avoiet eté persifflés. Ils inventerent ce beau pretexte de la Relligion qui les a couvert de honte. C'etoient Fontenelle et Mayran qui etoient les Chefs de la Cabale." (M to JBII, 22 August 1743, BEB).

88. Ibid.

# 7

# Toward a Science of Living Things

MAUPERTUIS TOOK CONSIDERABLE SATISFACTION in the decline of the Fontenelle-Mairan "cabal," not only in the Académie française, but in the Academy of Sciences as well. On certain subjects, Cartesian voices could still be heard, but they were increasingly drowned out. In the aftermath of a prize competition on the tides, he noted smugly that "Cartesianism is done for [*foutu*] even in the Academy."[1] In the early 1740s, Maupertuis maintained a presence in the Academy, but wrote fewer papers than in previous years. When he began to pursue the subject of generation and heredity, he did so outside the Academy, for the same readers he had addressed in the *Letter on the Comet.* But if the style of this work did not belong in the Academy, the subject was not completely foreign to it. Questions about the relation of soul to body, or mind to matter, and the definition of life itself, provided fertile ground for cultivating an identity as a savant and as a man of the world.

At first glance, Maupertuis's frankly speculative and eclectic writings about organic reproduction look like a discontinuity in his career and oeuvre, perhaps an interlude between the mathematical and geographical work that made his reputation and the metaphysical reflections of his later years in Berlin. But this compartmentalization implies an artificially rigid demarcation between academic and non-academic worlds, and fails to take account of the chronological overlap in his various projects. Although he never directly mentioned his biological works in the Academy, as best we can tell, he wrote them from the perspective of an insider familiar with decades of experimental results and arguments. He also had firsthand experience with natural history through his own observations and experiments with many species of animals. But he published his controversial books on generation anonymously, formally walling them off from his reputation as academician, only admitting authorship once they had been widely read and discussed. His writings on living organisms and the methods for investigating them

1. M to JBII, 22 August 1743, BEB. The prize on the tides (1740) was shared by Daniel Bernoulli, Leonhard Euler, Colin Maclaurin, and Antoine Cavalleri, a relatively unknown Jesuit. According to Clairaut, the latter was added to the list of winners because Réaumur (one of the commissioners) insisted that the Cartesian position must be represented among the winners (ibid.).

remind us of the overlapping and intersecting venues and audiences for science in this period.

## Natural History

Maupertuis's work on what we would call biological questions was built on the texts and disputes of natural history and anatomy—from the ancient canon to moderns like Leeuwenhoek and William Harvey—and on his own longstanding fascination with animal behavior and physiology. As a boy, Maupertuis had collected and dissected reptiles and other poisonous creatures. He kept dogs, cats, exotic birds, and monkeys as pets throughout his life, sometimes taking them along on his travels and breeding them for his friends.[2] Their common interest in natural history had formed a bond between Maupertuis and Réaumur from the late 1720s, although they subsequently fell out as a consequence of the dispute with Cassini. Réaumur applauded Maupertuis's "spirit of observation," noting that his mathematical skills in no way interfered with "his penchant [*goût*] for insects; no one has more love for them than he." Maupertuis supplied not only "singular" insect specimens for Réaumur's collection, but also "curious remarks and ingenious opinions" about them.[3] In later years, Samuel Formey, secretary of the Berlin Academy of Sciences, recalled walking across the courtyard of Maupertuis's house with some trepidation, because of the resident menagerie. Some of these animals were used for breeding experiments, others were pets whose offspring were given as gifts to acquaintances in high society.[4] "Mme. d'Aiguillon . . . has just shown me another great mark of her thoughtfulness by sending me a cat. You would not believe the multiplication of animals of all species I have at my home, and how comfortable I am with this. When one has lived like this for a time one finds almost as much novelty and stimulation from them as from people."[5]

Two early papers to the Paris Academy attest to this abiding interest in natural history. Shortly after his election, he presented the results of experiments with salamanders, and when he spent several months in Montpellier in 1729 being

2. Moreau de Primerais, "Ecrits biographiques," AS Fonds Maupertuis.

3. Réaumur, *Mémoires pour servir à l'histoire des insectes,* 1:49. See also Geer to Réaumur, 6 December 1744, AS, dossier Geer, for ichneumon fly that Maupertuis brought back from Lapland for Réaumur's collection.

4. Formey, *Souvenirs d'un citoyen,* 1:218. Mme du Deffand discussed trading an Angora cat for a copy of Maupertuis's book (Deffand to M, 18 June [1742], in Hervé, "Les correspondantes de Maupertuis," 763); see also Mme. d'Aiguillon to M, 28 December [1747], BN, n.a.f. 10398. Tressan discussed his pet dogs, some of which were gifts from Maupertuis (Tressan to M, 20 July [1756], 22 May 1756, in LS, 334–35).

5. M. to Mme du Deffand, January 1748, SM, fol. 141v.

treated for syphilis, he collected scorpions for experiments on the toxicity of their venom.[6] In a self-conscious demonstration of the power of rational empiricism, Maupertuis described experiments designed to test common lore or superstitions about the salamander's resistance to flames and the scorpion's compulsion to seek death by fire, These two papers show the reader a "modern" naturalist, skeptical of traditional wisdom but willing to brave the scorpion's sting in the name of science. The particular creatures he examined fall into the category of curious phenomena because of their peculiarities and their mythological or emblematic meanings; Maupertuis transformed these familiar but strange (and perhaps dangerous) creatures into experimental subjects. He threw salamanders into the fire, to show that they burn, and allowed scorpions to bite dogs, to test the effect of their venom on different victims.[7] He enclosed scorpions with spiders to watch their battles; he observed scorpions with their young and located the organ on the tail where the venom is ejected (fig. 18). In these investigations, we see a modification of the traditional practice of collecting and observing, to incorporate the methods of experimental philosophy. Jacques Roger, the preeminent historian of eighteenth-century life sciences, argued that Maupertuis was free to propose a theory of generation precisely because he was not a naturalist.[8] But Maupertuis was actually quite experienced in the ways of naturalists and even participated in their practice, as confirmed by Réaumur's testimonial to his observational skills. Even so, Maupertuis wanted to extend the boundaries of natural history to include experiment and theoretical synthesis, and to draw conclusions from the masses of details reported by colleagues like Réaumur, who viewed the mechanism of generation as an inscrutable "mystery of nature."[9] The evident tensions between Réaumur and Maupertuis over questions of theory, speculation, and even style, were accompanied by a shared interest in the prolific variety of forms and behaviors in the animal kingdom.

In the first decades of the eighteenth century, the *Histoire* and *Mémoires* of the Paris Academy included descriptions of all sorts of strange organisms, from corals to chameleons to two-headed calves.[10] Maupertuis's early investigations fit nicely

6. On early interest in poisonous animals, see Moreau de la Primerais, "Ecrits biographiques," AS Fonds Maupertuis. Beeson, dates the trip to Montpellier to spring 1729 (*Maupertuis,* 74).

7. Maupertuis, "Observations et expériences sur une espèce de salamandre," and "Expériences sur les scorpions." He was following in the tradition of the academician Claude Perrault, whose *Description anatomique* (1669) debunked superstitions about chameleons, among other animals (Harth, *Cartesian Women,* 100–03; Sutton, *Science for a Polite Society,* 123–24).

8. Roger, *Les sciences de la vie,* 473.

9. Réaumur, *Mémoires pour servir à l'histoire des insectes,* 6:lxvi.

10. Daston and Park, *Wonders and the Order of Nature,* 231–40.

Fig. 18. Anatomy of scorpion's stinger, viewed through a microscope. Maupertuis, "Expériences sur les scorpions," *MAS* 1731, plate 16.

into this literature. When he revisited natural history in the 1740s and '50s, he went much further, to propose a theoretical account of the process of generation and the phenomena of heredity, synthesizing evidence from anatomy, natural history, animal breeding, and travel literature. His theoretical model of active matter led to speculation about the possibility of change in organic forms and the role of chance in these changes. Attention to the underlying forces at work suggested a theory of matter endowed with properties responsible for organization and heredity, so that the static structures of anatomy and the collections of natural history gave way to a dynamics of forces and contingencies.

## Theories of Generation

The 1740s saw a resurgence of interest, both inside and outside the Academy, in the question of how to account for the generation of organisms. French naturalists and anatomists generally agreed on the broad outlines of a theory that assumed the preexistence of germs, created by God as the seeds of all individuals. Sexual reproduction amounted to a trigger for the expansion or unfolding of organisms already fully formed in miniature. Since organization took place at the creation, it was not a process open to scientific investigation.[11] There was still plenty of contention, however, about just where the germs resided during their long period of latency. They might be in the male semen, perhaps in the microscopic "animalcules" that swarmed there in profusion, or in the female "egg." All such theories, whether ovist or animalculist, located the incipient organism in one sex or the other, leaving the procreative role of the opposite sex more or less mysterious. Anatomists and naturalists argued about the nature and function of these microscopic or submicroscopic bodies, as well as about the physiology of conception. None of the available evidence could settle the question of the location of the germs definitively.[12]

Just at this time, naturalists observed several startling phenomena that challenged their assumptions about the reproduction of organic forms. The freshwater polyp's ability to regenerate a whole body from severed parts, parthenogenesis in aphids, the production of life in infusions of grain or meat: these phenomena became topics of conversation and speculation in the Academy of Sciences, in genteel society, and in the Republic of Letters more generally. Observations sparked further experiments, correspondence, demonstrations, and publications, but these phenomena never quite lost their aura of strangeness. Carefully documented by reputable observers, these peculiar, and usually tiny, creatures made their way into the discourse of natural facts. They prompted persistent experimentation and investigation of law-like natural processes, while arousing the curiosity of those who witnessed them.[13]

Although new evidence fueled the debate about generation, it certainly did not settle the question. Explanations for generation and organization depended on more than the collection of empirical results or observations. Several people, including Maupertuis, Georges-Louis LeClerc de Buffon, and John Turberville Needham, combined speculation and experiment in their forays onto this contested ground. Their books raised questions about the relation of the seen to the unseen, of general laws to particular observations, of the abnormal to the normal, and of the singular to the familiar. They brought the marvelous and the bizarre within reach of experiment, without denying the wonder experienced by witnesses or readers. No longer spontaneous and unanalyzable sports of nature, as they had been in the collections of cabinets and museums over the previous two centuries, strange phenomena fueled the investigation of principles of life itself, investigations

11. On preexistence theories, see Roger, *Les sciences de la vie,* and Gasking, *Investigations into generation.*

12. Although Jacques Gautier d'Agoty saw the features of a fetus in a drop of (his) semen, even without a microscope. See Spary, "Enlightened Natures, 276–78."

13. On polyps, see Dawson, *Nature's Enigma;* on the Haller-Wolff controversy, see Roe, *Matter, Life and Generation,* on Haller-Wolff controversy; see also idem., "John Turberville Needham."

that often went beyond direct observation to speculation.[14] These authors used the surprising features of their evidence to open up the possibility and the legitimacy of new kinds of explanations. They cultivated a sense of the marvelous in their readers that would lead to an appreciation of theoretical gambits that were often quite speculative, and hence distinct from the descriptive tradition of natural history.

The freshwater polyp, or hydra, was the most salient example of this kind of strange phenomenon. It captured the imaginations of naturalists and their readers alike, since it was able not just to propagate by budding, but to regenerate whole organisms from pieces of itself, or to grow several heads where one had been. The polyp became emblematic of nature's extremes and reminded naturalists how poorly they understood the mystery of generation—even though the polyp was a perfectly ordinary, if minuscule, organism collected from pond water, and hardly exotic at first glance. Abraham Trembley, a Swiss naturalist working in provincial obscurity in the Netherlands, first noticed and recorded its remarkable properties around 1740. Trembley considered these "little organized bodies" deeply enigmatic and initially could not decide whether to class them as animals or plants. Embarking on a meticulous program of experiments and observations to chart their life cycle, he showed that, however unusual the species might appear, it had its own characteristic habits and properties.[15]

Lorraine Daston and Katharine Park conclude their history of wonders with reflections on the Enlightenment's "metaphysics of uniformity," in which universal laws replaced the collation of singularities and skepticism replaced wonder as the philosopher's reaction to strange phenomena. They note that this rejection of the marvelous was accompanied by the elevation of "enlightened" over "vulgar" perceptions of nature, with marvels relegated to the territory of the vulgar. Maupertuis's analysis of the salamander's resistance to burning, and the scorpion's compulsion to kill itself when threatened with fire, would be examples of this; modern methods and skeptical sensibility worked together to debunk common superstitions about surprising qualities. Enlightened philosophy was supposed to be useful rather than pleasurable, and in order to be useful, nature had to be utterly predictable.[16] But what happened when reputable observers reported some-

14. On collections of curiosities, see Pomian, *Collectors and Curiosities;* Daston and Park, *Wonders and the Order of Nature.*

15. Trembley and Réaumur initially called them "little organized bodies"; once they decided to classify "these bodies" as animals, Réaumur named them polyps, by analogy to marine creatures of the same name (Baker, *Abraham Trembley of Geneva;* Dawson, *Nature's Enigma;* Vartanian, "Trembley's Polyp"). Trembley published his own account: *Mémoires pour servir à l'histoire d'un genre de polypes d'eau douce.*

16. Daston and Park, *Wonders and the Order of Nature,* 208–10.

thing as peculiar and unanticipated as regeneration or parthenogenesis? In the case of the polyps, predictability undermined neither the strangeness of the phenomena nor the curiosity of the viewers. The responses of the learned adapted the language of curiosity and wonder to enlightened discourse about these "prodigies," which they coopted for philosophy, even as they displayed them to genteel audiences. Alongside the trend, chronicled by Daston and Park, to castigate the marvelous as vulgar amusement, new kinds of marvels emerged to fascinate genteel and philosophical audiences. These natural wonders were demonstrated to be regular, even predictable, and thus appropriate objects of enlightened attention. What were learned and experienced observers to make of these facts? They stimulated speculation about the nature of organized matter, the soul, and life, the very topics that resisted easy reduction to general laws. The attested fact of the polyp's regeneration occupied a key node in the discourse of natural history by virtue of its unexpected and inexplicable qualities.

Like many singular or exotic objects, the polyps from the Dutch pond arrived in due course in Paris to be presented to fashionable society as well as to the Academy. Réaumur read Trembley's detailed report to the Academy in March 1741 and displayed living specimens as soon as he had some that survived the journey. "[Réaumur] made the Academy a visual witness [*témoin oculaire*] to the discovery, and along with the Academy, the court and the city [*la Cour et la Ville*], which in our enlightened century hardly differ from savants in this regard."[17] As with the shape of the earth, the genteel public took up the questions raised by regenerating polyps. Although the tiny organisms were not easy to see, they were evidently considered the kind of remarkable phenomena appropriate for display to elite audiences, and judging by references in correspondence, they must have been a frequent topic of conversation.[18] Even the secretary of the Academy stressed their marvelous nature.

> The story of the Phoenix reborn from its ashes, as fabulous as it is, is no more marvelous than the discovery we are going to speak about.... Here nature goes farther

17. *HAS* 1741, 33. Réaumur read Trembley's letter to the Academy in installments, 1, 8, and 22 March 1741, AS p-v. Maupertuis was in Berlin at the time.

18. Mme. Geoffrin described visiting Réaumur at his estate in Poitou and being shown his specimen polyps (which she called "worms"). They "were the length of a pin and the thickness of a sewing thread; they moved very little and I could not distinguish the head from the tail." They were, in fact, decidedly unspectacular for the casual observer. She was disappointed that she could not witness regeneration, which would have taken several weeks. She also mentioned a conversation on the subject with Maupertuis (Geoffrin to Martin Folkes, 16 January [1743], in Brown, "Mme. Geoffrin and Martin Folkes.") See also Folkes to Réaumur, 6 August 1743, AS, dossier Folkes, for investigations of polyps in England.

> than our fantasies [*chimères*]. From each piece of one animal cut in 2, 3, 4, 10, 20, 30, 40 similar parts, and, so to speak, minced, there are reborn as many complete animals just like the first. Each of these is ready to undergo the same division, and to be reborn from its debris, and so on. We do not yet know where this astonishing multiplication will stop.[19]

The reader could hardly miss the sense of astonishment engendered in witnesses to the reproduction of complete organisms from an original that had been "minced" beyond recognition. Such phenomena as regeneration of a whole from its parts or the spontaneous generation of organisms from apparently lifeless matter raised questions not easy to answer. Réaumur, for one, steadfastly refused to reason from physics to metaphysics, from direct observations to "truths of another order . . . that have as their object beings [*êtres*] that are neither bodies nor matter."[20] Others, notably Maupertuis and Buffon, were less circumspect and took natural history in new directions, beyond collection and description to investigation of the forces of organization and life.

Curiosity and curiosities linked naturalists and anatomists with readers and spectators. With their detailed accounts of deformed fetuses, two-headed animals, or novel species, naturalists literally brought curiosities and spectacle into the Academy and into the pages of its publications. Men of science claimed the right to analyze curious and unexpected phenomena, but they did not try to reduce the wonder experienced by audiences witnessing the objects of their scientific attention. The insect world supplied many examples of curious structures and behaviors; another category of curiosities collected by natural historians and dissected by anatomists was that of the "monster." This broad category, encompassing all congenitally malformed animals, presented a problem for preexistence theories. Dissections of abnormal fetuses and deformed animals were reported routinely to the Academy, often illustrated with detailed engravings. The same kinds of objects filled the curiosity cabinets of aristocrats across Europe.[21]

19. "L'histoire du Phoenix qui renaît de ses cendres, toute fabuleuse qu'elle est, n'offre rien de plus merveilleux que la découverte dont nous allons parler. . . . Voici la nature qui va plus loin que nos chimères. De chaque morceau d'un meme animal coupé en 2, 3, 4, 10, 20, 30, 40 parties, et, pour ainsi dire, haché, il renaît autant d'animaux complets et semblables au premier. Chacun de ceux-ci est prêt à subir la même division, et à renaître de même dans ses débris, et ainsi de suite, sans qu'on scache encore où s'arrêtera cette étonnante multiplication" ([Mairan], "Animaux coupés & partagés en plusieurs parties" (*HAS* 1741, 33–34).

20. Réaumur, *Histoire des insectes,* 6:lxvii. See also Needham, *An Account of Some New Microscopical Discoveries* (London, 1745). On Needham, see Roe, "John Turberville Needham"; Mazzolini and Roe, *Science Against the Unbelievers.*

21. Hagner, "Enlightened Monsters."

Discussion about generation in the Paris Academy from the 1720s to 1740 took place in the context of a dispute about the origin of monsters. The pages of the *Mémoires* in this period recorded numerous cases examined by the anatomists Louis Lémery and Jacques Winslow.[22] These two agreed about the preexistence of germs, but they disagreed radically about how to explain defective structures. Lémery argued for the accidental rearrangement of parts in the germ due to accidental causes after conception. Two fetuses might be forced together, for example, at a very early stage, resulting in the confusion or mixture of their incipient organs and the development of conjoined twins. Winslow, in deference to the inscrutability of divine wisdom, asserted on the contrary that monstrous organisms must have developed from monstrous germs, created by God for unknown purposes. Lémery saw Winslow's position as an insult to divine intelligence; imperfect products could never be attributed to God. The actual dynamics of abnormal development could not be directly observed, although Lémery sought clues to this process in his anatomical analysis of monsters. In general, the anatomists viewed such congenital monsters as anomalies to be dissected and tested against normal organisms. Were the internal organs displaced? Which parts were missing and which were duplicated? Could the skeleton be out of proportion while the internal organs remained normal? The Academy collected reports and specimens from anyone who cared to send them to Paris.[23]

Less grotesque, but even more provocative, was another sort of monster brought to Paris as a curiosity in 1744. This was an albino boy born to African slaves in colonial South America. The young boy was displayed in aristocratic homes and brought to the Academy to be examined and described. For a time, he was the talk of the town, as he "amused the curious and exercised the philosophers."[24] The *Histoire* reported on him as one of a series of monsters and other odd phenomena witnessed by the academicians, giving a brief description of the boy's features and coloring without speculating on the cause of his curious appearance. The report also noted that travelers had reported seeing whole tribes ["*nations*"] of albino Negroes in Panama and in Madagascar, putting the boy into the context of the strange things chronicled in travel literature.[25]

22. Roger, *Les sciences de la vie,* 409–18; Hoffheimer, "Maupertuis." In *L'ordre et les monstres,* Tort analyzes the monster debate for the whole century, with a bibliography of the primary sources.

23. Six papers on monsters appeared in *MAS* for 1740; Fontenelle summarized the Leméry-Winslow debate ("Sur les monstres," *HAS* 1740, 37–50).

24. Formey, "Eloge de M. de Maupertuis," 491. I was not able to discover who brought the boy to Paris or what happened to him after he left. He visited the Academy on 8 January 1744; Maupertuis was in attendance. The anatomist Morand "was assigned to report on him to the company" (AS, p-v, 8 January 1744).

25. "Diverses observations anatomiques," *HAS* 1744, 12–13.

This was not the Academy's first encounter with an albino Negro. Ten years earlier, they had received a report of such a child based on the eyewitness testimony of a Dutch physician in Surinam. In his account of the "very singular" phenomenon, Fontenelle dwelt on the question of the child's parentage, wondering whether the father of the albino might have been a white Negro like those observed by travelers in Africa.[26] The albino boy brought to Paris in early 1744 sparked the curiosity and speculations of the fashionable world more vividly because he was actually present.[27] We can only imagine what it was like for him, a child "four or five years old," to be put on display for the amusement and edification of endless gatherings of Parisians. In addition to recalling the exotic world of the distant European colonies, the albino boy made palpable some of the theoretical questions raised by monstrous animals and by polyps. Maupertuis, who observed the boy in both academic and salon contexts, used the prodigy as an excuse for writing a short anonymous book on generation. The book's title, *Dissertation physique à l'occasion du nègre blanc* (*Physical Dissertation Occasioned by the Albino Negro*), implied a scientific ("*physique*") account of albinism, although the book actually concentrated on a critique of current theories of generation. Maupertuis used the albino boy as a hook to attract the attention of those witnesses who had seen him and argued about his origins, much as Fontenelle had speculated about the parentage of the boy from Surinam. The book circulated through the milieu where the albino had been viewed, provoking speculation about the identity of its author. It sold out in short order.[28]

Questions about the process of generation must have interested Maupertuis for some time, but he used the public awareness of the albino boy to address them in a speculative mode he could not, or would not, adopt in the Academy.[29] He published the *Nègre blanc* quickly, capitalizing on the topical curiosity of the public, as a trial balloon for the risky hypotheses about active matter that emerged from his critical assessment of preexistence theories. He soon expanded it substantially by adding chapters on racial differences, albinism, and other hereditary attributes

26. "On a encore bien des éclaircissements à souhaiter sur ce Père qu'il seroit si curieux de connoître" ([Fontenelle], "Observations de physique générale," *HAS* 1734, 16–17).

27. Mme de Verteillac had the portrait of an albino Negro in her curiosity cabinet. See Maupertuis, *Vénus physique*, 137.

28. In a review of the second edition, Desfontaines, noted that the *Dissertation physique sur le nègre blanc* was "much sought after and the edition exhausted" (*Jugemens sur quelques ouvrages nouveaux*, 9:217). He said readers were surprised at the absence of a discussion of the albino himself. As Graffigny noted, the albino "was mentioned only in the title." For another essay occasioned by the same boy, see Voltaire, "Relation touchant un Maure blanc amené d'Afrique à Paris," in Beuchot, ed., *Oeuvres*, 38:521–24.

29. See Terrall, "Salon, Academy and Boudoir." Maupertuis later referred to conversations with

like sexidigitism (extra digits on hands or feet). This version, with the new title *Vénus physique,* went into several further editions, still anonymously.[30] The original preface to *Nègre blanc* (omitted from *Vénus physique*) referred to the display of the albino in an aristocratic home, rather than in the Academy, and justified the anonymous publication of a book that might be deemed "dangerous." By publishing anonymously, the author situated his theoretical innovations in the sociable world of the salon, taking on the role of gallant wit and freethinker, much as he had done in the *Letter on the Comet* two years earlier.

The book opened with the disingenuous claim, "Nothing was further from my mind than writing a book, when I started the following work. I had found myself the night before in a house where someone had brought the albino Negro who is currently in Paris. . . . Everyone reasoned endlessly about this prodigy."[31] Implying an easy familiarity with the social scene where such prodigies exercised the company, the author suggested that his book too belonged in this setting, where it would stimulate further conversation. Indeed, he claimed that he had only put his reflections into writing at the insistence of "someone in the party to whom I could refuse nothing." His inability to refuse her request set the gallant tone that cropped up throughout the text, mixed in with accounts of experiments and learned arguments. The unnamed lady, by pushing him not just to speak, but to publish, connected his efforts to the well-established literary tradition of men enlightening women through conversation spiced with sexual double-entendres. Maupertuis took as his epigraph a quotation from Virgil: "*Quae legat ipsa Lycoris,*" or "[verses] worthy of being read by Lycoris herself." (Gallus, in love with Lycoris, begged Virgil to put his love-induced despair into poetry for her to read.)[32] This phrase not only figured the reader as female, but made a connection between love poetry and this book about the physical science of love.

Maupertuis enhanced his gallant pose by referring to the dangers of publishing such a book, taking risks for the lady in the name of freedom of thought. "I did not have enough vanity to refuse. My only weakness was not daring to put my

---

Buffon on these matters, especially the resemblance of children to their parents, which would have taken place in the early 1740s (M to La Condamine, 5 November 1750, SM, fol. 97).

30. The *Nègre blanc* came out in three editions in rapid succession, all anonymous, from Leiden. The first edition of *Vénus physique* appeared in 1745 (n.p.). Subsequent editions appeared in 1746 (the Hague), another in 1746 (n.p.), 1751 (listed as "6ième ed.," n.p.). Buffon mentions a printing in Paris in December 1750 (probably the 1751 edition), Buffon to Formey, 6 December 1750, in Buffon, *Correspondance générale,* 1:74–75. Citations to Maupertuis, *Vénus physique.*

31. Maupertuis, *Dissertation physique sur le nègre blanc,* (n.p., 1744).

32. "Quae legat ipsa Lycoris" (Virgil, *Eclogues* 10.2). I thank Ann Blair for the translation and interpretation of this phrase. Maupertuis borrowed the epigraph from Algarotti, who had used it for his *Newtonianisme per le dame* (1738). Maupertuis explained to Algarotti: "On a pris . . . la liberté de

name to it." He regretted that he was writing "in a time when they wish to forbid us all operation of the mind, and where a powerful party is undertaking to demonstrate that we neither know how to write nor ought to know."[33] Implying that he was in the vanguard of the assault on ignorance, he took on this mission with an eclectic wit suited to an urbane readership. The language of risk, coupled with gallantry, reminded the reader of the author's close acquaintance with the literary tastes of the elite. Anonymity might protect him from the authorities, as he claimed, but it was also a mask that marked him as fluent in a culture that valued veiled references and literary secrets. He refrained from openly admitting his authorship well after many readers had identified him.[34] Speculation about his identity fueled the book's success as much as its arguments about the nature of matter and the origin of monsters. Abbé Desfontaines, for example, noted in his review that Maupertuis was only one of several names linked by literary gossips to the *Nègre blanc* and went on to reject the attribution. Some thought it was by "some young Doctor, a *bel-esprit,* who had struggled to appear to be a gallant Philosopher, like the one in the *Plurality of Worlds,* but with less decency and taste."[35] Desfontaines claimed personal knowledge of the author, "neither Academician nor Doctor," although he refrained from naming him;[36] Mme. de Graffigny reported: "Only half of Paris thinks it is by [Maupertuis]; the other half attributes it to a doctor from Versailles."[37]

---

vous voler le *Quae legat ipsa Lycoris.* On a trouvé le mot si charmant, qu'on n'a pu résister à la tentation de s'en emparer" (M to Algarotti, 15 April [1746], in Algarotti, *Opere,* 16:209).

33. "[D]ans un temps où on veut nous interdire toute opération de l'esprit, et où un puissant parti entreprend de démontrer que nous ne savons écrire, ni ne devons le savoir" (Maupertuis, *Nègre blanc,* "Preface," n.p.). Preface reprinted in Maupertuis, *Vénus physique, suivi de la lettre sur le progrès des sciences,* 73–74.

34. "Maupertuis n'avoue ny ne nie le *Negre,* mais il laisse voir qu'il est de lui" (Graffigny to Devaux, 23 August 1744, *Correspondance,* 5:426). Graffigny and Devaux discussed the *Nègre blanc* often in the spring and summer of 1744; she reported on the rumors about its authorship. She sent him a copy, having bought it for "40 sols" (*Correspondance,* 5:258). Devaux enjoyed the cleverness of the book and agreed that Maupertuis was a likely author (Devaux to Graffigny, 5–7 May 1744, cited in *Correspondance,* 5:264, n.3).

35. "[Q]uelque jeune Médecin bel-esprit, qui s'étoit efforcé de paroître Philosophe galant, comme celui de la *Pluralité des Mondes,* mais avec moins de décence et de goût" (Desfontaines, review of *Vénus physique,* in *Jugemens sur quelques ouvrages nouveaux* 9 (1745): 217). The reference is to Fontenelle's *Conversations on the Plurality of Worlds* (1st ed. 1686).

36. Ibid., 9:218. His candidate for author sounds like Voltaire.

37. Graffigny to Devaux, 11 May 1744, *Correspondance,* 5:258. A few weeks earlier she had told Devaux, "People are starting to say the *Nègre blanc* is not by Maupertuis, but by a doctor from Versailles whom I know and who has enough wit to have written this silliness" (29 April 1744, *Correspondance,* 5:236).

Other commentators attributed authorship to Maupertuis on stylistic grounds. François Devaux, Mme. de Graffigny's prolific correspondent in Lunéville, thought he spotted signs of Maupertuis's authorship right away: "I gave in to the temptation to read the *Nègre blanc*. . . . All the reasoning is rendered with so much clarity and gracefulness that I could not help being pleased. I think it is by the Flea [Maupertuis], although there are several places that seem to tell me that it is not."[38] According to the reviewer for the *Bibliothèque raisonnée*, "some attribute it to M. de M——s and others claim it is by M. de V——e."[39] The anonymous author of *Critique of the Dissertation on the Albino Negro* had heard people saying "that we had nothing in France written like the *Nègre blanc* and that the only person in France capable of writing in such a manner was this academician [Maupertuis]."[40] This pamphlet, reprinted as *Anti-Vénus physique* after 1745, objected to Maupertuis's stylistic choices; it returned to the question of authorship repeatedly, even reporting a rumor that Maupertuis had tried surreptitiously to encourage an attack on his own book just so that people would talk about it more. In this way, the author of the critique, the same Basset des Rosiers who had lambasted the *Letter on the comet* a few years earlier, implicated himself in the process of constructing authorial identity, and of blurring the line between irony and truth. He made it plain that matters of style and authorial posture could be as contentious and controversial as substantive interpretations of evidence.

## The Text

The first chapter of *Vénus physique* set the question of the origin of life and organization in the double context of eroticism and philosophy, of love and physiology. The author addressed the unnamed lady directly: "Do not be annoyed if I tell you that you were once a worm, or an egg, or a kind of mud. But do not think either that all is lost when you lose that form that you have now, when that body that charms everyone will be reduced to dust."[41] He sets out to teach his charming companion about the mysteries of generation, while suggesting that he is not immune to her bodily charms. As a creature herself, she exemplifies the process he

38. Devaux to Graffigny, [May, 1744], cited in Graffigny, *Correspondance*, 5:264, n.3.

39. Review of *Dissertation sur le Nègre blanc*, *Bibliothèque raisonnée* 35 (1745): 302.

40. "J'ai même ouï dire . . . que nous n'avions en France rien d'écrit comme le Nègre Blanc et qu'il n'y avoit en France que cet Académicien [Maupertuis] qui fût capable d'écrire ainsi" ([Gilles Basset des Rosiers], *L'Anti-Vénus physique*, 30). The first version of this pamphlet, *Critique de la dissertation sur le nègre blanc*, was already printed in April 1744: Mme. de Graffigny wrote to Devaux: "Il y a une autre brochure que je t'envoyerai dimanche qui est contre Maupertuis, que je n'ai pas encore pu avoir" (Graffigny to Devaux, 17 April 1744, *Correspondance*, 5:211).

41. "Ne vous fâchez pas si je vous dis que vous avez été un ver, ou un oeuf, ou une espece de boue. Mais ne croyez pas non plus tout perdu, lorsque vous perdrez cette forme que vous avez

is investigating, and he promises her a kind of self-knowledge, the knowledge of "the origin of your body." The implied erotic tension between male author and female reader, transposed into the text from the conversational setting of the salon, suffuses the behavior of people experiencing love and lust as well:

> A man is in a melancholy state that makes everything seem insipid, until the moment when he finds the person who can make his happiness. He sees her: everything is embellished in his eyes: he breathes a sweeter and purer air; . . . all nature serves her whom he loves. The lover feels a new ardor for all that he undertakes: everything promises him happy success. She who charmed him ignites with the same fire that burns him; she gives herself up to its transports; and the happy lover rapidly traverses all the beauties that overpowered him. He has already arrived at the most delicious spot. . . . Oh, unfortunate man, whom a mortal knife has deprived of that state! If the blade had ended your life, it would have been less deadly. In vain you inhabit vast palaces; . . . you possess all the riches of Asia; the lowliest of your slaves who can taste these pleasures is happier than you.[42]

Desire for the beloved manifests itself as a natural force, driving the movements and actions of individuals subject to its power. The dynamics of love replicate the dynamics of life, and the author plays a triple role as philosopher (or *physicien*), as lover, and as writer of provocative prose. He steps in and out of these three roles throughout the book, linking theory, observation, everyday experience, and literary exposition. His objections to preexistence theories unfold in this stylized literary context, itself referring back to the more personal context of conversation and flirtation that prompted the exposition in the first place.

The attentive reader would have noted that Maupertuis's critique of contemporary theories of generation also implied a breach of the rigid demarcation between the man of science and the objects of his attention. In the context of Enlightenment sociability, this drew women into the process of knowing about nature by engaging them in conversation. Reflections on organization and development ("the origin of your body") prompted reflections on the experience of

---

maintenant; et que ce corps qui charme tout le monde, sera réduit en poussière" (Maupertuis, *Vénus physique*, 78).

42. "L'homme est dans une mélancholie qui lui rend tout insipide, jusqu'au moment où il trouve la personne qui doit faire son bonheur. Il la voit: tout s'embellit à ses yeux: il respire un air plus doux et plus pur; la solitude l'entretient dans l'idée de l'objet aimé; il trouve dans la multitude de quoi s'applaudir continuellement de son choix; toute la nature sert ce qu'il aimé. Il sent une nouvelle ardor pour tout ce qu'il entreprend: tout lui promit d'heureux succès. Celle qui l'a charmé s'enflamme du même feu dont il brule: elle se rend, elle se livre à ses transports; et l'amant heureux parcourt avec rapidité toutes les beautés qui l'ont ébloui: il est déjà parvenu à l'endroit le plus délicieux. . . . Ah mal-

love and sex as well. "It is this moment, marked by such delights, which brings into existence a new creature who will be able to understand the most sublime things; and what is more, who will be able to taste the same pleasures."[43] Individuals emerge not from the hand of God but from the physical congress of two profane creatures. Pleasure, intellectual understanding, and physical organization are thus inextricably linked by the very nature of life.

Maupertuis proposed a theory only tentatively, often couched in the form of queries or conjectures, after demonstrating the inadequacy of existing explanations based on the unfolding of germs in either the sperm or the egg. Instead of such "development," he articulated a theory of epigenesis, or sequential formation of organisms from unorganized material particles.[44] His explanation drew on a wide variety of types of evidence, from anatomy, the study of monsters, observations of racial differences and family resemblances, animal breeding, microscopy, the natural history of insects, and chemistry. But the core of its challenge to preexistence theories came not from the laboratory or the dissection table, but from everyday experience: the obvious resemblance of offspring to either or both parents. Since individuals display the traits of the mother as often as those of the father, and hybrids of two species share the attributes of both, the offspring must be "a compound of two seeds."[45] Furthermore, the anatomical investigations of William Harvey, who dissected a series of does at various intervals after mating, had shown the gradual formation of the deer embryo from unorganized parts (although he could not observe the primordial parts themselves). Harvey claimed to have seen a "living point" in the embryo before he saw any organized structures; after this he observed the successive appearance of recognizable parts. His observations, legitimated by the most illustrious of witnesses, the king of England, gave Maupertuis embryological evidence for epigenesis. "Instead of seeing the animal grow by the *intus-susception* of new matter, as it would have to do if it were formed in the egg of the female or if it were the little worm that swims in the semen of the male; here [in Harvey's observations] an animal forms itself by the *juxta-position* of new

heureux! qu'un couteau mortel a privé de la connoissance de cet état: le ciseau qui eut tranché le fil de vos jours, vous eut été moins funeste. En vain vous habitez de vastes Palais; . . .vous possédez toutes les richesses de l'Asie; le dernier de vos esclaves qui peut gouter ces plaisirs, est plus heureux que vous" (Maupertuis, *Vénus physique*, 79–80).

43. "C'est cette instant marqué par tant de délices, qui donne l'être à une nouvelle créature, qui pourra comprendre les choses les plus sublimes: et, ce qui est bien au-dessus, qui pourra gouter les mêmes plaisirs" (ibid., 80).

44. On M's objections to preformation, see Hoffheimer, "Maupertuis." On the definition of epigenesis, see Lopéz-Beltrán, "Forging Heredity," 217–18.

45. Maupertuis, *Vénus physique*, 111.

parts."[46] The crucial point was that the animal "forms itself," once the appropriate parts come into physical contact.

The suggestion about the mixing of male and female fluids appeared more than halfway into the book, after an account of alternative versions of preformation theory and recapitulation of a variety of observations used to support one or the other. The author framed his account didactically: "I am going to explain to you the different systems that have divided philosophers on the way generation works."[47] The rhythm of the text, moving back and forth from ovist to animalculist, from dissections and observations supporting first one then the other, from theories to objections to those theories, reflected the equivocal nature of the anatomical and microscopical evidence. This made current understanding of generation appear tenuous indeed, ripe for an intervention from a new direction. The rhetorical structure of the argument harked back to the strategy Maupertuis used years before to overcome his readers' resistance to the notion of gravitational force. First, he considered the competing theories, with the empirical evidence for each position. Then he argued that the development of a preformed germ was no more comprehensible than the gradual ordering of parts from different sources, just as he had argued that impact is no more comprehensible than action at a distance. Finally, he showed his own solution to be neither absurd nor contradictory, just as he had shown that "attraction" entailed no absurdity. In short, what might have seemed absurd at first glance, was no more absurd than any of the alternatives. On closer inspection, it turned out to be a good deal less absurd, and hence conceivable. Once he made it conceivable, he could show how it would work to solve the conundrum at hand, even if it could not do so with absolute certainty.

If, then, organisms emerge from components originating in both parents, what are these parts, and how do they come together in an orderly fashion? Rather than occupying either egg or sperm, the material components of the future embryo are dispersed throughout the seminal fluids of both sexes. Neither homogeneous nor infinitely divisible, these fluids are composed of particles distinguished by specific properties or forces. Inherent forces drive the disaggregated parts to combine coherently into functioning organisms. Teleology, or design, thus derives from the

46. "Au lieu de voir croître l'animal par l'*Intus-susception* d'une nouvelle matière, comme il devroit arriver s'il étoit formé dans l'oeuf de la femelle, ou si c'étoit le petit ver qui nage dans la semence du mâle; ici c'est un animal qui se forme par la *Juxta-position* de nouvelles parties" (ibid., 97). "Intus-susception" and "juxta-position" are technical terms from chemistry that describe different ways in which particles of substances can join or mix. On Harvey's experiments on generation, see Gasking, *Investigations into Generation.* For background on the preformation-epigenesis debate, see Roe, *Matter, Life, Generation.*

47. "Je vais vous expliquer les différens systemes qui ont partagé les Philosophes sur la maniere dont se fait la génération" (Maupertuis, *Vénus physique,* 79).

properties of matter itself. However the mechanism of these active properties actually works, they allow for many possible outcomes within certain parameters. Maupertuis's explanations remained self-consciously speculative. He suggested that speculation of this sort, never too far from empirical evidence, could be productive for a science of life, where the mathematical formulation of laws seemed unable to do justice to the phenomena.

## Affinities and Organic Forces

Maupertuis used analogies liberally in devising a language appropriate to the actions and properties of matter at this submicroscopic level. Chemical reactions provided useful analogies for these irreducible properties, especially in the formation of complex crystals that mimicked organic structures. The "tree of Diana," a crystal formed by silver and spirit of nitre combined with mercury and water, has a "vegetative" shape not evident in any of its constituents by themselves. "The parts of these substances come *of their own accord* to arrange themselves into this vegetable form so similar to a tree that one cannot refuse it that name."[48] Such crystals seemed to occupy the borderline separating living from brute matter and were typically described in language appropriate to living organisms.[49]

The concept of selective chemical forces came from the work of Etienne Geoffroy, a Paris academician of the previous generation. Geoffroy had devised a table of affinities (*rapports*) to represent schematically the relative strengths of affinity operating between substances.[50] As he explained it, one substance combines with another for which it has an affinity, forming a compound with new properties. A third substance can dissolve this union if it has a stronger affinity for either of the two components than they have for each other. Geoffroy's hierarchically ranked affinities, as a kind of map of reactions observed in the laboratory, give each substance specificity and direct its movements, although the table implies no theory of how affinities inhere in substances. The table of relative strengths displayed the diversity of reactions to be expected of a given substance, since its activity with respect to another substance will depend on the surroundings. The force could not be an absolute quantity associated with a particular kind of matter, but an affinity *for* something else subject to alteration due to the influence or intervention of a third substance. In this sense, affinities express relations, often in complex combinations. The complexity of empirical chemistry, and the inherent directedness of

48. Ibid., 119 (my emphasis).

49. For example, La Condamine, "Sur une nouvelle espèce de végétation métallique," *MAS* 1731, 466–82.

50. Geoffroy, "Table des différens rapports," *MAS* 1718, 202–13. On affinity chemistry, see Sadoun-Goupil, *Du flou au clair;* Kim, *Affinity, That Elusive Dream.* Maupertuis refers to Geoffroy (1672–1731), "whose loss our sciences will regret for a long time" (*Vénus physique,* 120).

*rapports,* made chemistry a valuable resource for rendering organic forces plausible. The language and metaphors of chemistry recur throughout *Vénus physique.* Particles that belong together in the organized animal have "a greater affinity for joining" with each other than with other particles "destined" for other positions and functions in the whole. Not all particles contributed by both parents need to be used in any given generation either. Two bits linked by the force of their affinity for each other, "being once united, a third which might have made the same union, no longer finds its place and remains unused."[51] This is just how chemical substances behave in the laboratory.

The 1745 edition of *Vénus physique* added a paragraph relating affinities to gravity.[52] The notion of an attractive power which, by virtue of the universality of its effects, seemed to be inherent in matter elucidated the activity that drove generation. "I cannot help pointing out that these forces and these *rapports* are nothing other than what other more daring philosophers call attraction. This ancient term, revived in our times, at first shocked those scientists [*physiciens*] who thought they could explain everything without it." But astronomers realized that gravitational force would explain celestial motions, and chemists extended the notion of attraction to short-range forces. "Why, if this force exists in nature, might it not play a role in the formation of the bodies of animals?"[53] But we might also ask how gravity could drive qualitative relations, since it does not work in the selective manner of affinities; its very universality and mathematical uniformity give it conceptual and explanatory power.

Most commentators, referring to Maupertuis's reputation as a Newtonian, have read this passage as an attempt to reduce all forces to gravitational attraction.[54] I would suggest rather that Maupertuis was using the newfound respectability of gravity to generalize from it to other inherent forces. These forces would provide the ground for understanding organic processes, just as gravity made certain aspects of physics comprehensible. That is to say, the existence of gravity might suggest that forces can belong to matter, like other properties. By this time, Maupertuis no longer hesitated to declare gravity inherent in matter. When he revised *The Shapes of Heavenly Bodies* for a new edition in 1742, he added explicit references to Newton's "discovery" that "gravity is nothing other than a phenomenon resulting from a force spread throughout matter [*répandue dans la matière*]." The idea

51. Maupertuis, *Vénus physique,* 121.

52. He repudiated this analogy in *Système de la nature* (1754); see chapter 10 below.

53. Maupertuis, *Vénus physique,* 121.

54. Brunet, for example, took this line. "Or ces rapports et ces forces n'apparaissaient à Maupertuis que comme une sorte de justification de l'idée newtonienne d'attraction" (Brunet, *Maupertuis,* 2:318). This is just backwards, as Maupertuis used the viability of gravitational attraction to justify other forces that act selectively and teleologically.

that forces may be "spread throughout" matter could equally well apply to other types of phenomena, whether chemical or organic, though he anticipated objections to his strategy. Chemical affinities, he noted, "as incomprehensible as they are, seem to have penetrated even into the Académie des Sciences, where they weigh new opinions carefully before admitting them."[55] Here, he recalled quite clearly the resistance of many of his colleagues to Newtonian action at a distance. Both kinds of forces—attraction and affinities—function heuristically in his argument; however, they remain different in kind from organic forces. The reference to gravity reminded the reader that abhorrent ideas can become acceptable and even useful with familiarity. But the selective nature of affinities made them more appropriate as a model for organic forces than gravity could be, since it was a universal property of matter.

Having established the admissibility of these organic forces, Maupertuis could speculate about how they function. So, he suggested, the seminal fluids of both males and females consist of innumerable particles, no more visible than the smallest elements of silver or water. The fetus forms out of a combination of these elemental pieces from each parent, though there are many more particles in the seminal fluids than will actually contribute to the organism. Properties something like chemical affinities determine the position and order of particles in the fetus; when they come into close enough proximity to allow the selective forces to operate, each constituent element joins with those that belong next to it. The strong affinities between those elements "appropriate for forming traits similar to those of the individual parent" account for resemblances between parents and offspring.[56]

This explanation also left room for offspring that do not resemble their parents, because there are always particles floating around that correspond to features not visible in the parents. In the case of albinism, for example, generation proceeds normally in the production of all but one trait. The final outcome of any mating is never completely determined, since factors such as climate, diet, and even chance combinations of elements, contribute to individual variations. "Once two parts that belong next to each other are united, a third part, which could have achieved the same union cannot find its place and remains useless. It is thus, by these repeated operations, that the child is formed from the particles of the father and the mother, and often carries visible marks that he participates in both."[57] Some of Maupertuis's readers misinterpreted his use of words like "particles" to mean that he was describing the conglomeration of fully formed miniature parts. Voltaire, for example, in his satirical caricature, objected that Maupertuis imagined an eye

55. Maupertuis, *Vénus physique,* 120.
56. Ibid., 139.
57. Ibid., 121.

attracting another eye, a hand attracting a finger, and so on. A more careful reading of *Vénus physique* shows that the author suggested that the components of the seminal fluids are proto-parts, or subparts of some kind, that do not combine mechanically, but chemically. Unlike the analysis of masses subject to mechanical forces, for example, where each mass is considered as a collection of infinitesimally small increments of similar matter, and then summed together mathematically to give the action of the whole, organic bodies have attributes distinct from the properties of their parts considered in isolation.

## Erotic Science

An erotic element runs through *Vénus physique* in small asides and examples recounted in a colloquial tone. Semen examined under a microscope becomes "that liquid which is not ordinarily the object of attentive and tranquil eyes."[58] Leeuwenhoek, "a chaste and religious Physicist, made a great many experiments [on semen], none of which, he assures us, were made at the expense of his family."[59] As for the question of whether semen penetrates to the uterus, "The liquid spilled into the vagina, far from appearing destined to penetrate further, soon falls back out, as everyone knows."[60] Looking for evidence of symmetry between male and female, Maupertuis asserts that he himself has looked for spermatic animals in the female fluids. "I have searched several times with an excellent microscope to see whether there aren't similar animals in the fluid [*liqueur*] that women produce."[61] Although he shed no further light on the circumstances of the experiment, this deadpan reference casually placed the microscope in the bedroom in what can only have been an erotic situation.

Halfway through the book, the reader comes to a long interlude on the mating habits of animals, written in euphemistic but prurient language that shares certain conventions with contemporary pornography. Anthropomorphic descriptions of animal behavior attribute passions, intentions, and morality to bulls, fish, turtledoves, and gall insects in a parody of the flowery style of much natural history writing, and the natural theology it spawned. "What variety we observe in the ways in which different species of animals perpetuate themselves!"[62] In contemporary works of natural theology, this kind of hackneyed exclamation ordinarily

58. Ibid., 86.

59. Ibid., 88.

60. Ibid., 84.

61. Ibid., 124. Buffon looked for organic bodies in the seminal fluids of both male and female dogs (Buffon, "Découverte de la liqueur seminale dans les femelles vivipares, et au reservoir qui la contient").

62. Maupertuis, *Vénus physique,* 102.

expressed wonder at the luxuriance of God's creation. But here awe took on a prurient tint, and eroticism undermined the reverence of such compendia of variations, turning the reader into a voyeur. The queen bee maintains a "harem of lovers and satisfies them all . . . [in] the most unrestrained debauch."[63] "The impetuous bull, proud of his strength, doesn't fool around with caresses. He throws himself immediately on the heifer, penetrates deep into her entrails, and spills out in a great flood the liquor that makes her fertile."[64] The male dragonfly pursues the female, "he catches her; they embrace, they attach to each other; and hardly encumbered by what they have become, the two lovers fly off together and let themselves be carried by the wind."[65]

The subversion of natural theology reflected back on natural history as well. Although the behavior and appearance of animals vary widely, the underlying process that drives reproduction does not, the reader learns. Scientific scrutiny of this common ground must go beneath surface description of prolific variety, the meat and potatoes of natural theology, to causes and general laws. The "prodigies" that traditionally left naturalists and spectators awestruck—whether monsters like the albino boy or novelties like Trembley's polyps—inspire the man of science to make sense of them, even as he uses them to engage his readers. The polyp appeared in *Vénus physique* both as "a Hydra more marvelous than that of the fable" and as evidence of the careful observations of naturalists. Rigorous experiments and impeccable witnesses testify to the polyp's otherwise unbelievable attributes, making it more than a random curiosity. But it still cries out for explanation:

> What are we to think of this strange kind of generation; of this principle of life spread throughout each part of the animal? Are these animals anything other than collections of embryos just ready to develop, as soon as they are allowed? Or do they reproduce by unknown means all that the mutilated parts are missing? Might Nature, which in all other animals attached pleasure to the act by which they multiply, cause these [creatures] to feel some kind of sensual delight when they are cut into pieces?[66]

This "principle of life" permeates the polyp in much the same way that affinities inhere in chemical substances and attraction inheres in brute matter.

Maupertuis's interlude on variations set the stage for the subsequent theoretical chapters by demonstrating the need for a principle that might make sense of nature's variety. The common thread unifying the vast range of habits in the

63. Ibid., 104.
64. Ibid., 102.
65. Ibid.
66. Ibid., 107.

animal kingdom is pleasure. "Nature has the same interest in perpetuating all species. She bestowed on each one the same motivation, which is pleasure. In the human species, pleasure makes everything else disappear before it; in spite of a thousand obstacles to the union of two hearts and a thousand torments that are bound to follow, pleasure directs the lovers to the goal nature intended."[67] Like the anticipation of pleasure that initiates it, generation can be investigated with the methods of experiment, observation, and calculation familiar to men of science. Above all, such a new science of life would be physical, synthesizing empirical evidence and physical processes, while going beyond the descriptive methods of natural history.

Maupertuis extended the meaning of "physical," using the word in several ways. It appeared in his first title, *Dissertation physique à l'occasion du nègre blanc* and again in the expanded version, *Vénus physique.* It suggested that the author would account for the process of reproduction in accordance with the standards of physical science, leaving metaphysics out of the picture. Physics must explain, not mystify. *Vénus physique* also carried the secondary connotation of the physical act of love, which initiates reproduction in animals. This theme resurfaced in the text with sexual innuendo and descriptions of mating habits. The term *physique* occurs most frequently in the text as a capitalized noun, referring to the science of physics, and also in the form *physicien* to designate practitioners of that science. Physics in this sense is the empirical study of all natural phenomena, organic and mechanical, in contrast to either mathematics or metaphysics; it is more akin to modern "science" than to the discipline "physics."[68] All these references disparaged the current state of the science of life and criticized the notion of preexistence for its false claims to the status of physical theory. Thus, the author wondered rhetorically "if the system of developments [preexistence of germs] makes Physics clearer than it would be if new productions [epigenesis] were admitted? . . . What will Physics lose if we conclude that animals are formed successively?"[69] If phenomena were to override preconceptions, old systems would have to be thrown out. "I beg the forgiveness of modern *physiciens,* but I cannot admit the systems they have so ingeniously imagined. For I am not one of those who thinks that one advances science [*la Physique*] by attaching oneself to a system in spite of some phenomenon that is evidently incompatible with it."[70] The contemporary reader would have thought immediately of the familiar contrast between "imag-

67. Ibid., 103.

68. For a discussion of the changing meaning of "physics" in the eighteenth century, see Heilbron, *Elements of Early Modern Physics,* 4–8.

69. Maupertuis, *Vénus physique,* 108, 110.

70. Ibid., 118.

ined" systems like those of Descartes and empirical physics based on sensory evidence, some of it as obvious as the resemblance of parent and child. Once again, Maupertuis insisted on going beyond the "modern" (evoking Cartesian philosophy) to an enlightened sensibility that refused to be shocked by forces and active matter. "If what I am going to say revolts you, I beg you to regard it as nothing other than an effort I am making to satisfy you. I do not hope to give you complete explanations of such difficult phenomena."[71] Thus, the first-person voice of the narrator flagged his adventurous willingness to speculate, his commitment to tying speculation to phenomena, and his ability to stimulate and satisfy his reader.

## Race and Varieties

When he expanded his original small book into *Vénus physique,* Maupertuis added several chapters on hereditary variations and "the varieties of the human species." As the new preface to one of the 1746 editions noted, he wanted to jump from consideration of the albino boy to "more difficult and important" questions such as: "Why are the inhabitants of the tropical zone black? Why do the most numerous and most beautiful people live in the temperate zones? Why are the glacial zones only inhabited by deformed peoples?"[72] Such examples resonated with contemporary fascination with geographical differences, and implied the possibility of producing a global cultural geography tied to a theory of epigenesis and active matter. Attributes like hair and skin color became visible markers of a submicroscopic process that produced and reproduced variations. Again, the style of these chapters was eclectic, drawing examples from classical texts and travel literature, as well as from the experience of animal breeders and pet fanciers. Africans, Patagonians, and Laplanders appeared alongside a mythical albino race in Panama, described in some detail; closer to home, blonde, brunette, and redheaded women filled the gardens of the Louvre with "all the marvels of the world."[73] The diverting catalogue of varieties led to reflections on general patterns: "The most constant law concerning the color of the earth's inhabitants is that the large band that encircles the globe from East to West, called the torrid zone, is only inhabited by black (or very dark) peoples."[74] Flirtatious

71. Ibid., 139

72. "Pourquoi les habitans de la Zone torride sont noirs? Pourquoi les peuples les plus nombreux et les plus beaux se trouvent dans les Zones tempérées? Pourquoi les Zones glaciales ne sont habitées que par des nations difformes?" ("Préface," *Vénus physique,* 1746 edition, n. p.). Each edition included minor revisions; this preface appeared only in this edition.

73. Maupertuis, *Vénus physique,* 131.

74. Ibid., 130.

interludes on female beauty and the merits of variety alternate with reflection on the production of new varieties of animals through controlled breeding. All this serves to demonstrate that epigenesis was the only viable explanation for the origin and perpetuation of varieties. Unusual traits can originate by chance, as the result of the combination of those organic particles in the seminal fluids. The choices of breeders, or the effects of external factors like climate, can perpetuate these variations.[75] Given that fortuitous traits can become established and then replicate with increasing frequency, the question becomes why they revert less and less often. Rare features in general tend to be submerged by the more commonly occurring traits unless some other cause is operating. "These are lapses of nature, which can be sustained only through art or through a regimen. Her works always tend to reassert themselves."[76]

How then did the black race establish itself? Basing his remarks on the reputed existence of separate races of exceptionally large and exceptionally small people, Maupertuis suggested a social explanation for the geographical distribution of distinct races.

> When giants or dwarves or blacks were born among other men, pride or fear would have armed the greatest part of humanity against them; and the most numerous species would have relegated these deformed races to the least habitable climates of the earth. The dwarves would have retreated towards the arctic pole; the giants would have inhabited the lands of Magellan; the blacks would have peopled the tropical zone.[77]

This extremely simplistic physical geography incorporated natural history, and anecdote, into arguments about general laws or principles. These arguments referred empirical observations, whether of skin color or height, back to the submicroscopic model for explanation. Patterns of inherited characteristics became the object of scrutiny, so that something as apparently singular as the albino Negro could exemplify the way submicroscopic forces result in visible attributes, which in turn can be passed on to the next generation. The exotic boy, displayed to a curious public as a spectacle, was no more or less human, or natural, than the lovely ladies promenading in the gardens of the Tuileries.[78]

75. Ibid., 138.

76. "Ce sont des écarts de la nature dans lesquels elle ne persevere que par l'art ou par le regime. Ses ouvrages tendent toujours à reprendre le dessus" (ibid., 141).

77. Ibid., 144.

78. "Dans les Jardins des Louvre, un beau jour de l'été, vous verrez tout ce que la terre entiere peut produire des merveilles" (ibid., 131).

## Doubts and Conjectures

Taking Newton's speculative queries in the *Opticks* as his model, Maupertuis formulated his theory of epigenesis most explicitly in a set of questions at the end of the book, where he suggested that the inherent directing forces of organization resemble not just affinities and gravity, but animal instincts.

> Does not instinct in animals, that which makes them seek out what is good for them and flee from that which is dangerous, belong also to the tiniest parts of the animal? Does not that instinct, although dispersed in the particles of seminal fluid and weaker in each particle than in the whole animal, nevertheless suffice to make the necessary unions among these particles? We see in completely formed animals, that instinct makes their parts [*membres*] move."[79]

Instinct, though not controlled by the will, operates as an unconscious but still purposive cause of movements. The "instincts" of the organic elements stimulate their motion and combinations, just as instinct causes animals to move toward pleasurable stimuli and away from pain.

Instincts, affinities, desires—these forces are all equally intangible, but equally real. So *Vénus physique* has come full circle from the suggestive gallantry of its opening lines to the naturalized physical pleasure motivating copulation, and from there on to the hypothetical instincts of the tiniest parts of organic matter. The reader is left reflecting on the animality of human desires and behavior, within the highly stylized and eroticized framework of polite society and fashionable literature. The hybrid genre of the book suited the speculative content, more provocative than definitive, leaving room for a range of readers' reactions, but nevertheless claiming an authenticity for its interpretation of phenomena.

"I do not expect that the sketch of a system I have proposed to explain the formation of animals will please everyone; I am not completely satisfied with it myself."[80] In fact, though not everyone liked *Vénus physique,* it certainly made a splash in the Republic of Letters.[81] It did not provoke the kind of storm of criticism that welcomed books like Voltaire's *Siècle de Louis XIV* or La Mettrie's *L'homme machine,* but the "public" evidently read and discussed *Vénus physique* with a mixture

79. Ibid., 145. The "Doubts and Questions" chapter was a slightly revised version of the conclusion to *Nègre blanc.*

80. Maupertuis, *Vénus physique,* 144.

81. *Vénus physique* was reviewed in *Bibliothèque raisonnée* (1745); Desfontaines, *Jugemens sur quelques ouvrages nouveaux* (1744, 1745); Fréron, *Lettres sur quelques écrits de ce tems* (1753); and again in *Bibliothèque raisonnée* when Maupertuis's *Oeuvres* were reviewed by Voltaire in 1752.

of approval and alarm. In its first version, it was published in Leiden as the *Nègre blanc,* a small volume of about 130 pages. Demand exceeded supply in Paris, and it was reprinted twice that same year, although it is impossible to estimate the size of the print runs. Starting with a small number was one way of getting a book talked about, since readers had a hard time finding it. Du Châtelet recommended it to Father Jacquier, professor of mathematics and astronomy in Rome: "You must read the *Vénus physique* of Maupertuis, or the second part of his *nègre blanc,* but it is very rare; I don't know if you will find it there [in Rome]."[82] Graffigny reported spending all day with her lover, "reading an anatomical dissertation on generation written by Maupertuis, which makes him look somewhat ridiculous."[83] By combining a gallant, occasionally prurient style with assessments of the latest microscopic discoveries and anatomical controversies, Maupertuis guaranteed for himself a mix of reactions, some admiring and some satirical. Although he claimed to be writing "not as a metaphysician . . . but only as an anatomist," reactions centered on the scandalous implications of materialism and on the book's unusual, but elegant style. As Abbé Raynal noted, "The piece is written clearly, vigorously, with verve, elegantly; it is from the hand of a master, and our ladies have abandoned their novels to read it."[84]

The review journal *Bibliothèque raisonnée* printed a lengthy and accurate extract of the text, attributing it to either Maupertuis or Voltaire. The reviewer elaborated on the materialist implications of the epigenetic theory: "I hardly need to point out to the Reader that this System leads naturally to Pantheism. It gives to Matter all the energy [*énergie*], all the power, that Spinozists, Naturalists, and Materialists attribute to it." The theory gives nature unlimited power, and God becomes extraneous. "This true cause of regular movements, of the organization of Animals and Plants, resides in Matter, which does everything by instinct, by a force that does not come from outside, and that it has in and of itself. This is Fatalism."[85] Without defending the preexistence of germs, the reviewer wanted to avoid deriving organization from the properties of matter, a move that erases God from the picture. Even accepting the existence of attractive forces, rapports, and instincts, there remained the problem of how order can emerge "without the direction of an Intelligence infinitely superior to those Powers or that Instinct."[86] Although this review drew out the *Nègre blanc*'s materialist consequences, giving

82. Châtelet to Jacquier, 12 November 1745, *Lettres,* 2:143.

83. Graffigny to Devaux, 17 April 1744, *Correspondance,* 5:211. Graffigny was aware of Maupertuis's authorship very early.

84. Raynal, "Nouvelles littéraires," in Grimm, *Correspondance littéraire* 1:114.

85. Review of Maupertuis, *Dissertation physique, Bibliothèque raisonnée* 35 (1745): 312.

86. Ibid., 315.

credence to the reference to "danger" in the preface, the assessment was by no means wholly negative and shows how such a book entered into the public sphere of criticism and conversation.

*Vénus physique* also provoked more vociferous attacks, as we have seen. Basset des Rosiers, a professor of philosophy at the Jesuit Collège d'Harcourt, objected to the impropriety of the overt appeal to female readers. He charged in *Anti-Vénus physique* that Maupertuis misled readers into thinking the book would not offend their modesty, and quipped that the title should have referred to Priapus rather than Venus. Warming to his theme, he insisted that the book should have come with sexually explicit illustrations to warn readers of its prurient content: "Apparently he was afraid of wounding the timid gaze of his chaste Lycorises; because he wanted to attract them by implying that they could read [his book] without fear of encountering any indecency." Throughout his challenge, he implied that women cannot be expected to read with intelligence, that they need pictures, and that they are the victims of overbearing authors. He reviled not just the "anatomical" project (epigenesis) but the cultural project that would bring women into the pool of valued readers and potential investigators. The critic implied that the author illicitly seduced readers into becoming "accomplices," that the enterprise was somehow subversive of moral order. Referring to Maupertuis's investigations of female seminal fluids, Basset des Rosiers noted, "The chaste accomplices of these modest experiments are apparently those Lycorises whom he invites in Latin to read his work, worthy fruit of their common efforts, of their taxing attentions."[87] This book may well have been motivated by personal hostility, given its ad hominem attacks on Maupertuis's morality and integrity, but it also attacked literary style, authorial persona, use of evidence, and theoretical content in turn. It cites errors of style and taste alongside purported mistakes in anatomy and misconceptions about theory. The satire reinforces my claim that Maupertuis made a strategic move by writing in this hybrid genre, and that his readers recognized the move, whether they approved of it or not. "The fashion is to write in all genres, and to succeed or to excel in none."[88] In addition to mixing genres and styles, Maupertuis's antagonist ridiculed him for seeking fame and reputation, for being fashionable, with what could be interpreted as a measure of jealousy. He put Maupertuis in a class with Voltaire ("a poet who writes geometry") and La Mettrie ("a doctor who writes about the soul")—what the Encyclopedists would identify as the "*parti philosophique*"—guilty of subverting the stylistic rules that defined genres.[89]

87. Basset des Rosiers, *L'Anti-Vénus physique,* 112.

88. "La mode est d'écrire dans tous les genres, et de ne réussir ou de n'exceller dans aucun" (ibid., 201).

89. Ibid., 201.

Not all readers objected to Maupertuis's style. Elie Fréron, for example, found *Vénus physique* more readable than most books on anatomical subjects, noting that the author "has shown that an anatomical dissertation can become pleasant reading, and that it is permitted even for Anatomy to adorn itself with flowers when an able hand dispenses them." [90] Desfontaines also commented on the style, which he characterized as having "a great deal of wit [*esprit*] but a bit of licentiousness." He noted that Maupertuis's description of mating habits drew heavily on the work of Réaumur, although "he did not borrow [Réaumur's] gravity of style," and went on to quote several pages from the chapter, "to give the Reader pleasure."[91] This reviewer appreciated the same lightness of tone scorned by Basset des Rosiers. Desfontaines noted as well that the author, "no enemy of Newtonianism," may have written his book "to show how this doctrine, supported by Harvey's observations and Geoffroy's experiments, is compatible with the explanation of the generation of plants and animals."[92] Although he did not pronounce on the validity of the theory as such, Desfontaines treated the *Nègre blanc* as part of contemporary discourse about force, matter, natural history, and literary style. By invoking Newtonianism, he placed the book in the "philosophical" camp. Maupertuis purposefully wrote a controversial book, obscuring his identity behind a curtain of anonymity, that raised as many questions as it answered.[93] Once more he made a point of playing with the boundary between learned academic discourse and the literary taste of "the court and the town."

## Leaving Paris Behind

The *Nègre blanc* came out around the same time that Maupertuis read his paper on the refraction of light to the Academy. Both represented efforts to open up new intellectual terrain, in quite different formats for different audiences. Although the book on generation had its amusing side, it also voiced a serious critique of mechanistic "systems" and made a plea for the explanatory value of active matter when addressing the otherwise mysterious phenomena of life. The work on refraction, with its emphasis on final causes and divine intentions, proposed an approach to

90. Fréron, Review of Maupertuis, *Oeuvres, Lettres sur quelques écrits de ce tems,* 8:164.

91. Desfontaines, *Jugemens sur quelques ouvrage nouveaux* 2 (1744):251.

92. Ibid., 263.

93. Maupertuis claimed to be disappointed with its reception. When Algarotti proposed an Italian translation, he responded, "[J]e dois vous avertir en conscience que celui-ci a eu assez peu de succes à Paris: quoiqu'on en ait fait deux éditions en fort peu de tems; bien des gens l'ont trouvé fort mauvais; et les journaux en ont parlé assez mal" (Maupertuis to Algarotti, 23 June [1746], in Algarotti, *Opere,* 16:215–16). Maupertuis was rarely satisfied with the reception of his books and often made remarks like this about his other works.

mechanics quite unlike what anyone else in Paris was doing. Indeed, in spite of widespread public interest in the book on "how to make babies," it seems that no one knew quite what to make of either venture.[94] As a result, only a year after his election to the Académie française, he complained to Bernoulli, "Critiques and libels are raining down here against all those whose names are a bit known. Another one has just appeared against our work [on the shape of the earth], or rather against us, in which after having criticized all our observations, they conclude with songs I wrote at the polar circle." These were songs he had circulated himself, privately; they were widely known well before, but this was the first time they had been in print. "All this is full of injustice and meanness, but I am so accustomed to it, and so bored with this question that I no longer take any interest."[95] Maupertuis had been subject to other attacks recently as well; not only for his *Nègre blanc,* but for his reception speech at the Académie française. His complaint to Bernoulli has been taken as evidence of Maupertuis's dissatisfaction with his position in Paris, and of his readiness to leave France for Prussia, where he ended up in the winter of 1745.[96] His reasons for this reversal of his earlier decision to stay in France, however, cannot be attributed simply to his reaction to libels, which, as he said, were nothing new for him.

There is no record of a renewed correspondence with Frederick at this time, nor of negotiations about coming to Berlin. Maupertuis knew that the Berlin Academy had been reconstituted at the beginning of 1744, under the direction of curators, with no one in the leadership spot that had been promised to him. He had sent a copy of his latest book, *Astronomie nautique,* to the Berlin Academy, and

94. Graffigny called it "How to Make Babies" (*La manière de faire des enfants*) (Graffigny to Devaux, 11 May 1744, *Correspondence,* 5:528. The physician Procope-Couteau objected to Maupertuis's theory, but claimed *Vénus physique* as the inspiration of his manual, *L'Art de faire les garçons,* on techniques for choosing the sex of babies.

95. M to JBII 1 July 1744, BEB. Badinter exaggerates his mortification at the publication of these verses; he had circulated them himself among his friends (*Passions intellectuelles,* 267). Mme. de Graffigny saw them in summer 1739; Gastelier had them; the poem was included in Maupertuis's letter to Mme. de Verteillac, which had circulated even before his return from the north. The anonymous libel in question was *Anecdotes physiques et morales* [1743 or 1744]; many contemporaries attributed the book to Maupertuis, though this seems implausible. His verses did appear at the end of the book (see chapter 4 above). On the attribution of *Anecdotes physiques,* see Beeson, *Maupertuis,* 142.

96. Badinter suggests that Maupertuis had hoped for appointment as perpetual secretary of the Paris Academy of Sciences, when Mairan gave up the post in 1743. The evidence for this is thin, relying primarily on a later recollection by Voltaire, which is highly unreliable. See Badinter, *Passions intellectuelles,* 242–45. For reactions to the Académie française speech, see Badinter, *Passions intellectuels,* 241–42; La Beaumelle, *Vie de Maupertuis,* 80. The text of the speech is in Maupertuis, *Oeuvres,* 3: 259–70.

to Euler, still the only foreign luminary Frederick had managed to attract to Berlin.[97] But Frederick was once again at war, this time in alliance with France, and the outcome was anything but settled. All observers agreed that the prospects for the Berlin Academy depended materially on the resolution of that war. Maupertuis left no trace of his private thoughts about returning to Berlin. He did plan to leave Paris during the academic vacation in September and October of 1744, for a visit to Basel. Once there, he could not resist continuing on to Freiburg, where the army of Lorraine, under the command of his friend the maréchal de Coigny, held the city under siege.[98] Writing to Basel from the army camp under the Freiberg city walls, he asked Bernoulli to send his books, since he was staying with the army longer than expected. "The more I stay, the more I wish to stay. The only thing I lack is books; because it is not everything to see cities taken; however amusing it may be, one needs to add a bit of distraction."[99] He promised to let his friend know when it would be convenient for him to come "to see all that is most beautiful here, if you want to come." A few days later, he wrote, "This is beginning to get interesting. We are firing with all our force, one side against the other, and we have a hundred mouths of fire. . . . The king [of France] arrives Sunday, and his presence will embellish our army even more. As for me, whatever pleasure I take in living with you in Basel, I would not know how to leave this country so soon. And my curiosity increases every day."[100]

By November, Maupertuis was hatching a plot to get to Berlin under the most auspicious circumstances. He asked Coigny and the marquis d'Argenson (the new French minister of war, and a personal acquaintance) to allow him to act as emissary from France to the king of Prussia, once they had some good news to report. In fact, Frederick was annoyed that the siege had lasted so long and that the French alliance had been less useful to him than he had hoped; Coigny thought that using Maupertuis as ambassador might serve French interests well. The siege dragged on until the city surrendered in early December, and Maupertuis, with the blessing of d'Argenson, left immediately to take the news to Berlin, where he was wel-

97. Euler to Maupertuis, 4 July 1744, in Euler, *Opera Omnia,* series IVA, 6:49–54.

98. He wrote to JBII from the maréchal de Coigny's estate in Orly, 1 July 1744; he described the general as "un homme qui mérite beaucoup personnellement, qui est regardé comme le meilleur général que nous ayons, et qui joint à cela un tres bon esprit et un tres bon coeur" (M to JBII, 22 August 1743, BEB).

99. M to JBII, 4 October [1744], BEB. He had left books to be bound in Basel; he also asked for a book by Euler that Daniel Bernoulli had discussed with him. This letter makes clear that he had not planned to stay long at Freiberg.

100. M to JBII, 8 October [1744], BEB.

comed with open arms.[101] Not only was he feted by the court, "overwhelmed with visits," and deluged with kindness from all sides, the French could only applaud him for smoothing relations with Frederick. "I am truly a man for adventures," he wrote to Bernoulli. "A visit to you in Basel led me to Freiburg, Freiburg to Berlin, and the king's favors will keep me here until he leaves to go back to war."[102] Maupertuis was never entirely forthcoming with any of his correspondents about his plans or his motives, and he did not tell Bernoulli very much about his negotiations with the king, nor his own thoughts about the Berlin Academy. Given his enthusiastic reaction to witnessing the hostilities (this time from a safe distance) at Freiburg, and his evident pleasure at his reception in Berlin, his penchant for "adventures" must have played a role in his outlook. He had tried a gambit whose outcome he could not have predicted, but there is every indication that he was looking for a change from his routine in Paris, and part of the appeal of Berlin was just the element of risk and adventure. In the course of his visit, he decided to accept the king's offer of the presidency of the Berlin Academy. The full extent of his power over the institution would only become public later, but the flattering attentions of the king and the royal family this time proved impossible to resist. He also took the radical step of becoming engaged to marry the daughter of an aristocratic family from Pomerania.

In April, he was back in Paris to arrange his release from his French obligations and to get formal permission from his father to leave his homeland and to marry a Protestant. His former ally Maurepas insisted that he give up all his pensions, and his place in the Academy of Sciences. The minister's authority did not extend as far as the Académie française, however, and the literary academy remained Maupertuis's one institutional tie to France. "Everyone blames me for leaving France where my situation was very agreeable, for a country very different and where the advantages I find, whatever they may be, do not seem to have the same security. But I have calculated as best I could, and it seemed to me that I should have done what I have in fact done."[103] The displeasure of the minister meant that he could not think of returning, at least not with the same status.[104]

101. Correspondence between Coigny and d'Argenson, and Maupertuis and d'Argenson, in Archives de la Ministère de guerre, as cited in Velluz, *Maupertuis,* 90–92.

102. M to JBII, 19 January [1745], BEB.

103. M to JBII, 28 May 1745, BEB.

104. "On m'a flatté, on m'a menacé, on est venu à me donner tous les degouts possibles. Le Roy m'asseure quil me dedommager a bien, je l'espere, nous verrons. . . . si cela ne subsiste pas, nous irons faire lepictetes dans un grenier, car on m'a oté tout ce que j'avois icy" (M to JBII, 16 June 1745, BEB). As it turned out, he had some of his prerogatives restored to him; he got a French pension of

The calculation of the risk and advantage of such a move included more than money, of course, and more than love. It was certainly more than escaping unpleasant satires or old animosities. The honor derived from such a close association with the king of Prussia had no price, and Maupertuis recognized that the position he was going to occupy would be unique in the Republic of Letters. He was to have unheard-of power over every aspect of the institution, and over all the individuals in it, and he imagined that he would have the respect and admiration of his colleagues. But there was more. He had established cordial relations with Euler, whom he recognized as an invaluable resource for further work on mechanics.[105] And he expected to attract many more illustrious thinkers to Berlin to join them. The prospect was not just one of unbridled power, but of working at the center of a society of congenial men of science and letters of his own choosing. Whatever disappointments he suffered in Paris with respect to his new mathematical work and his speculative theory of generation, he could reasonably expect to be able to pursue them more productively in Prussia. He would also have the freedom to develop his book on cosmology, the text he had worked on sporadically for some years. His ambitions for the Berlin Academy of Sciences were also, characteristically, ambitions for himself; these were intellectual as well as personal and political.

---

4000 *livres* one year later. After Maurepas fell from favor himself in 1749, Maupertuis's position as veteran in the Paris Academy was restored to him through the good graces of d'Argenson (M to FII, 1756, Koser, *Briefwechsel,* 320–21). See also M to Hénault, [1749?] Bibliothèque de l'Institut, MS. 2514. This letter expressed bitterness about the Academy's treatment of Maupertuis, which he attributes to the machinations of his enemies, especially Maurepas.

105. Euler to M, 10 December 1745, *Opera Omnia,* series 4A, 6:56–57. Euler appreciated Maupertuis's extremum mechanics and responded in the most favorable terms to the paper on the law of rest. "I am convinced that nature acts everywhere according to some principle of maximum or minimum. . . . It seems to me also that it is here that we must seek the true principles of metaphysics. I also think your principle even more general than you propose . . ." (ibid.).

# 8

# The Berlin Academy of Sciences

When Maupertuis moved to Berlin in August 1745, Frederick was still at war with Austria. He did not return to the capital until after the Peace of Dresden at Christmastime. In the interim, Maupertuis had married Eleonor de Borck, a young lady in waiting to the Princess Amelia. She came from the old Pomeranian nobility; one of her close relatives was Kaspar Wilhelm de Borck, a government minister and a curator of the Academy of Sciences.[1] Frederick, urged on by the queen mother, had pushed the union during Maupertuis's previous visit, as a way of integrating him into Berlin high society. The king may have thought of it as insurance against a change of heart on the part of his new acquisition. Whatever Frederick's motives, the marriage seems to have been more than a matter of convenience or politics for Maupertuis, although marrying outside the Catholic church caused him considerable anxiety.[2] His domestic arrangements seem to have suited him, though we know almost nothing about the couple's relations and nothing at all about Mme. de Maupertuis personally.[3] The union enhanced his social status, and reflected his optimism about the options open to him

1. The family had a long tradition of serving the Hohenzollerns. Eleonor is not named in genealogical sources, since she had no children; her birth date is unknown. She must have been considerably younger than Maupertuis. Some sources say her father was Kaspar Wilhelm (1704–1747), one of the curators of the Academy, but this is highly unlikely, based on his age, and other circumstantial evidence. Eleonor's father was probably Adrian Bernhard von Borck (1668–1741) or Friedrich Wilhelm von Borck (1678–1743), both generals, and both dead before the marriage arrangements. I thank Jörg Sacher for tireless efforts to find this information. Frederick congratulated Maupertuis on his impending marriage in a letter from his camp in Semonitz (FII to M, 10 September 1745, Koser, *Briefwechsel,* 190). The wedding on 28 October 1745 was announced in *Berlinische Nachrichten von Staats- und Gelehrten Sachen,* no. 130 (30 October 1745).

2. Maupertuis confessed to having "an uneasy conscience" about accepting Frederick's favors and marrying outside the Catholic faith. See M to Pope Benedict XIV, 28 January 1748, SM, fols. 137v–38. He also mentioned his scruples in a letter to Cerati (M to Cerati, 8 October, 1746, excerpted in Thierry Bodin, Auction Catalogue (Paris, February 1996); copy in AS, dossier Maupertuis. He calls Eleonor "une femme heretique, une demoiselle Borck" (ibid.).

3. The only extant letters date from the very end of Maupertuis's life, when he was dying in Basel. La Beaumelle says the marriage was a great passion. The couple spent large portions of their marriage apart, and Mme. de Maupertuis never traveled to France with her husband on his many trips

in his new situation. Settling into his unaccustomed roles as husband, academy president, and court favorite, Maupertuis reflected with evident relish: "The king and the queens pile a thousand favors on me every day; the greatest of all is to have given me a wife who is the happiness of my life."[4] He regarded his advantageous marriage as striking evidence of the king's beneficence, and his friends in France saw it as such as well. His intimate friend the duchesse d'Aiguillon congratulated him on his marriage to "a woman who is given to you by a hand that enhances the value of the gift."[5] The marks of royal favor made the move to Berlin worthwhile. Marriage to an aristocrat of the best pedigree only enhanced the change in identity, as he left behind (at least temporarily) the freewheeling life of a Parisian bachelor. The feeling of being appreciated, the feeling that he was receiving his due at last, did not last forever, but for the time being, it fueled an energetic burst of work, both intellectual and administrative.

Because of the court's distance from the city, where the Academy was housed, life in Berlin meant a continual circulation among different settings and different groups of people. Valued for his conviviality as well as his intellect, Maupertuis stepped easily into a lively round of social engagements. He maintained a residence in town, but he also had quarters in the palace at Potsdam, where, as time went on, he often went without his wife. "We are nevertheless much more often at court than at home, we are commonly eight or ten days without being able to dine tête-à-tête."[6] His social circle was by no means limited to the king and court, however. In his first years at the Academy, he attended most meetings and kept up with academy business assiduously, even though the king frequently called him to Potsdam.[7] When he was away from Berlin, he maintained his control over all aspects of the Academy's operations through correspondence with Leonhard Euler, its most active member, and Samuel Formey, the energetic secretary.[8] When in the city, he gave dinners for guests, ranging from his academic colleagues to men of

---

there. Her extensive correspondence with La Condamine after Maupertuis's death show her devoted to the memory of her husband. (Letters preserved in AS, Fonds La Condamine.)

4. M to JBII, 9 April 1746, BEB.

5. Duchesse d'Aiguillon to M, 7 October 1745, BN, n.a. fr. 10398.

6. M to JBII, 9 April 1746, BEB. The king had an apartment renovated for him at Potsdam (FII to M, 7 April 1746, Koser, *Briefwechsel,* 204).

7. FII to M, 7 April 1746, Koser, *Briefwechsel,* 204. See ibid. for renovations of the apartment in the Potsdam palace. Out of sixteen Academy meetings in 1746 (when he was not in France) Maupertuis attended ten; out of sixty-four meetings in 1747 (including committee meetings), he attended twenty-five (Winter, *Registres*).

8. Maupertuis-Euler correspondence, in Euler, *Opera Omnia,* series 4A, vol. 6; Maupertuis-Formey correspondence, Preussische Kulturbesitz in deutsche Staatsbibliothek, Berlin (hereafter SBB) and Bibliotheka Jagiellonska, Cracow (hereafter BJ).

letters, visiting dignitaries, and German aristocrats. Formey recalled many such occasions, with Euler and his wife, with the Italian man of letters Francesco Algarotti, with Count Redern (later the king's grand marshal), Countess Bentinck, and countless visiting savants and dignitaries.[9] Several of his regular associates, especially Formey and Euler, did not frequent the court; others, like Algarotti, lived in the city but visited often at Potsdam.[10]

At Potsdam, Maupertuis was part of the intimate masculine circle at the king's private suppers, and he developed friendships with Frederick's two brothers, Crown Prince August-Wilhelm (known as Guillaume) and Henri.[11] The inner circle changed over the years, as poets, artists, and renegade philosophers came and went. The court became known across Europe as a refuge for heterodox philosophers and theologians like Julien Offroy de La Mettrie and the Abbé de Prades, some of whom were given pensions and titles. The tenor and intensity of life at Potsdam varied with the king's other preoccupations. At its liveliest, it revolved around conversation, reading, writing, performing, and criticizing.[12] Frederick was concerned about his reputation for wit and wisdom as well as for his military conquests. He wanted to be known as the protector of philosophical and religious freedom, but he also wanted to control his resident philosophers. He made deals with them, bestowed favors and honors, and often treated them as if they were under contract. Some of the deals included specific tasks (Darget was the king's reader; the Abbé de Prades took his place later); some were simply resident wits like La Mettrie. Maupertuis, at least in his early years in Berlin, found the atmosphere at court especially congenial because of the mixture of intelligence and glory that Frederick brought to his establishment. Advertising the advantages of such a situation to Algarotti, who was weighing an invitation to Berlin, Maupertuis described it as an ideal social world: "Think seriously about what it is to be the friend of a King, who after having done such great things in war, will do no less in peace. Think of the glory of bringing to the midst of an army the taste for sciences and letters, and of causing them to be appreciated."[13] Maupertuis saw his personal tie to the king as the most valuable benefit of his position, far beyond material advantages. In the constellation of people at court, he held a singular rank, because he

9. Formey, *Souvenirs d'un citoyen,* 1:177–79, 219.

10. Euler was perhaps the farthest from the sparkling wits with whom Frederick liked to be surrounded. He was not invited to Potsdam until 1749, in connection with a waterworks project; he had lived in Berlin since 1741.

11. Some correspondence with the two princes survives in the Saint-Malo Municipal Archives. On letters to Guillaume, see Beeson, *Maupertuis,* 228–29.

12. Pomeau and Mervaud, *De la cour au jardin.*

13. M to Algarotti, 12 May [1747?], in Algarotti, *Opere,* 16:211–12.

Fig. 19. The Berlin Academy of Sciences had its quarters with the royal stables in this building designed by Johann Friedrich Schlütter, on Unter den Linden in the Dorotheenstadt section of Berlin. For a view of the building from the other side, see fig. 20. By permission of Bildarchiv Preussischer Kulturbesitz, Berlin.

had been given his own institution, displacing its old structure of governance with his personal rule. And at the Academy, the president was known to be one of the king's favorites, which solidified his power there (fig. 19).

Berlin under Frederick displayed a peculiar mix of German politics and French style, just as Frederick tried to integrate in his own person the attributes of a courageous soldier and strategist with the brilliance of an enlightened poet and philosopher. Although Berlin lacked the intensity and the scale of intellectual life in Paris, the king's commitment to francophone philosophy, art, and letters made it congenial to Parisian expatriates, especially those who were no longer welcome in Paris. The Academy of Sciences and Belles-Lettres, although by no means exclusively French, reflected the king's ambivalence toward German philosophy and literature. But the Academy also made manifest notions about power, knowledge, and obligation characteristic of the absolutist Prussian regime. It became one location, among many, for working out the ideology of autocracy.

## Absolutism in Prussia

Frederick's political ideology grafted the principles of an enlightened rationalism onto absolute monarchy and permeated all aspects of Prussian life. Of course, the notion of a centralized autocracy did not originate with Frederick.[14] He inherited the framework of a bureaucracy designed primarily to collect taxes to support the standing army so dear to his father. But Frederick analyzed autocracy in philosophical terms and then used the same analysis to direct reforms of the economic underpinnings that he saw as vital to a strong centralized state. From his perspective, only an absolutist government could work efficiently to stabilize the economy, to acquire new territory and new subjects to fuel that economy, to strengthen the state's defenses against external enemies, and ultimately to provide prosperity and peace by consolidating power at the center. "A well-run government must follow a system as coherent as a philosophical system can be," Frederick insisted. "All measures taken must be well reasoned; finances, politics, and military affairs must converge on a single goal, which is the strengthening of the state and the increase of its power."[15] What particularly distinguished Frederick from his father was this appeal to the methods and rhetoric of the Enlightenment *esprit systématique.* He was convinced that a single brilliant individual must conceive, implement, and continually monitor such a rationalized system. "Just as it would have been impossible for Newton to delineate his system of attraction if he had collaborated with Leibniz and Descartes, so it is impossible for a political system to be made and sustained if it does not emerge from a single head."[16] Centralization made rationalization possible, and Frederick made every effort to control personally all aspects of his administration, planning military campaigns, leading his troops into battle, controlling the treasury, making bureaucratic appointments, and overseeing developments in royally subsidized industries.[17] He maintained that this kind of control facilitated planning and progress; the alternative, which he saw exemplified in the ministerial government of France, was chaos. If the king delegates power to ministers, he argued, there can be no system: "chance governs" (*le hasard gouverne*).[18]

The Prussian Academy of Sciences and Belles-Lettres figured prominently in Frederick's careful promotion of himself as an enlightened philosopher-king. Many of his reforms, whether actually implemented or only proposed as eventual goals,

14. The literature on Frederick and the Prussian state is vast. A classic study is H. Rosenberg, *Bureaucracy, Aristocracy and Autocracy;* for a more recent account, see Schieder, *Friedrich der Grosse,* translated as *Frederick the Great.*

15. Frederick II, "Testament politique" (1752), 303.

16. Ibid., 351.

17."Prussian administration in the eighteenth century was rooted in the fiction that the king knows everything, that he can do everything and does everything that is done" (Dorn, "Prussian Bureaucracy in the Eighteenth Century," 63)

18. Frederick II, "Testament politique," 303. Frederick had several of his essays on these subjects read to the Academy, E.g., "Dissertation sur les raisons d'établir ou d'abroger les lois" (1750); "Discours

impinged directly on the Academy. The institution's ability to recruit foreigners and to equip its facilities depended immediately on the state of the treasury, which in turn depended on the success or failure of Frederick's military campaigns and taxation schemes. Although the Academy was certainly less vital to Frederick than his economic program or his military conquests, he supported it as a fitting accouterment of enlightened monarchy. In return, the Academy lent credibility and expertise to various of the king's projects. The interaction between state and academy operated on ideological, practical, and even philosophical levels. From its organizational form, to the selection of members, day-to-day operations, finances, and the topics addressed by the academicians, the Academy was enmeshed in the web of Frederick's political ideology. On the one hand, Frederick projected onto the Academy his revamped image of Prussia, native autocracy overlaid with Enlightenment ideals; on the other, the Academy fed that image in abstract and concrete ways. Even as it worked to represent Prussian ideology, it also provided the institutional base for specific kinds of knowledge. Importing a Frenchman to promote Prussian ideology might seem a bizarre twist, one that exposed Maupertuis to hostile reactions from Germans scornful of Frederick's intellectual pretensions. But it reflected the king's wish to transform a provincial culture that he distrusted into his own syncretic, and opportunistic, image.

## Origins and Revival of the Academy of Sciences

In the 1690s, Leibniz had attempted to interest the Elector Frederick III of Brandenburg (the future Frederick I of Prussia and the grandfather of Frederick II) in his project for a German academy of sciences. Leibniz envisioned his academy as a center for the development and promotion of a unified German culture, which would include history, literature, and German language as well as mathematics and physics.[19] The Academy was not formally inaugurated until 1710, after much difficulty in getting a rudimentary astronomical observatory built and in financing the institution's long-term existence. Leibniz himself was forced to leave Berlin in 1711, abandoning his academy to court intrigues and the king's neglect.[20] After 1713, with the accession of Frederick William I, who scorned both science and belles-lettres, the Academy entered a period of quiescence when it

---

de l'utilité des sciences et des arts dans un état" (1772); "Essai sur les formes de gouvernerment et sur les devoirs des souverains" (1777).

19. For the early history of the Berlin Academy, see Harnack, *Geschichte der Königlich Preussischen Akademie der Wissenschaften*, vol. 1; Bartholmess, *Histoire philosophique de l'Académie de Prusse*; McClellan, *Science Reorganized*. On Leibniz's academic projects, see Ramati, "Harmony at a Distance."

20. Bartholmess, *Histoire philosophique de l'Académie de Prusse*, 41–43.

was barely visible to the outside world. Frederick II's plans for the simultaneous reform of the state and cultural life became the immediate context for the revival and redefinition of the Academy in the 1740s.

In the interim period between Maupertuis's first visit to Berlin in 1740, shortly after Frederick inherited the throne, and his installation as head of the Academy in 1746, the institution went through a series of structural changes. The stages of this reorganization expose the Academy's ideological underpinnings and provide an overview of the issues at stake for Frederick, Maupertuis, and the rest of the academicians. In spite of Frederick's avowed commitment to the Academy, he did little to foster it in the early 1740s; the Silesian wars monopolized his attention. His one acquisition was Euler, who accepted a generous pension and moved from Saint Petersburg in 1741 to participate in a still-hypothetical "renewed" academy. Euler mistakenly believed that Maupertuis had also committed himself to moving to Berlin, and was disappointed to find very little infrastructure and few active colleagues in place when he arrived.[21] In spite of his comfortable financial circumstances, Euler expressed his frustration to a former colleague, the astronomer Joseph-Nicolas Delisle, back in Saint Petersburg. He complained of the Academy's lack of resources, especially the dearth of good astronomical instruments. Observations of the comet of 1742 taken in Berlin were not accurate enough to calculate its orbit "for want of instruments; because the King has not yet undertaken the re-establishment of the Society; and everything is still in the same very imperfect state."[22]

By 1743, with Frederick still distracted by politics, a number of Berlin intellectuals decided to found their own new scientific and literary society. The group included local notables, including men of letters, members of the old academy, and several of the king's ministers and courtiers. Frustrated by the lack of activity in the existing academy, Euler joined with these men to form the Nouvelle Société Littéraire.[23] This organization lasted for just under a year, but it contributed several innovations to the reconstituted Académie des Sciences in 1744. The short-lived literary society departed from the organizational form of the old Societät der Wissenschaften, which had been divided into "departments" and administered by

21. Euler wrote to Maupertuis, then on his initial visit to Berlin, from St. Petersburg,: "As for the Berlin Academy itself, although I still know absolutely nothing about its foundation, nevertheless, since you will take the leading position there, I think I will be able to do the maximum under your direction. . . ." (Euler to M, 13 January 1741, in Euler, *Opera Omnia,* series 4A, 6:48.

22. "[F]aute d'instruments; car le Roy n'a pas encore entrepris le rétablissement de la Societé; et tout est encore dans le meme état tres imparfait" (Euler to Delisle, 21 July 1742, Obs. MS).

23. Harnack, *Geschichte der Koniglich Preussischen Akademie der Wissenschaften* 2:262–68.

a council of nobles.[24] The statutes of the new society expressed more democratic ideals. Formed spontaneously by "several inhabitants of Berlin who have a taste for science and literature," the society would be "of increasing use to the public," cultivating knowledge to "instruct and perfect the mind." All decisions were to be taken by vote of the full membership, including election of new members and selection of memoirs for publication. Of the twenty working members, half were either French or of French extraction; the society conducted its business in French. Meetings were held in the houses of one or the other of the honorary members. Although several of the founders had ties to Frederick's court, the king himself had not been involved in establishing the new society and had no stake in it. Whether by design or not, the society was inaugurated in an interval between the end of the first Silesian war and the king's return to Berlin. When the marquis d'Argens informed the king of plans for a new society, Frederick asked for a delay until he could return to Berlin and give the Academy his personal attention.[25] D'Argens apparently ignored this request, and the group proceeded without the king's explicit support. Perhaps inevitably, Frederick refused to allow the Société Littéraire to serve as his royal academy; in November of 1743 he called for the unification of the old and new institutions. He intended to place himself at the head of the new academy "in order to give the society a dazzling mark of his grace and royal protection, and to encourage the academicians to real emulation."[26]

In response to Frederick's order, several proposals for a revised institution were disseminated and discussed.[27] In the end, the king established the new academy in time for a ritual occasion on the eve of his thirty-second birthday in January 1744.[28] Henceforth, the Academy would meet in newly refurbished rooms

24. The departments of the old academy were (1) physics, medicine, chemistry; (2) mathematics, astronomy, mechanics; (3) German language and history; and (4) literature ("particularly Oriental literature, which may be usefully employed for the spread of the Gospel among unbelievers") (ibid., 2:193).

25. "[E]tant actuellement occupé à des affaires sérieuses qui demandent toute mon attention, je serais bien aise si vous vouliez prendre patience sur la susdite jusqu'à ce que je serai de retour à Berlin, et que j'aurai assez de loisir pour y penser" (FII to d'Argens, 18 June 1743, in Preuss, ed., *Oeuvres de Frédéric,* 14:10).

26. Cabinettsordre, 13 November 1743, in Formey, *Histoire de l'Académie,* 63–64, and Harnack, *Geschichte der Königlich Preussischen Akademie der Wissenschaften* 2:260–61. The commission included three government ministers, three honorary and two ordinary members of the Société littéraire, and two members of the old academy.

27. Various versions of proposals and draft statutes and records of discussions attesting to the vigor with which the reorganization was pursued are preserved in the archives of the Academy. Euler submitted one himself (AdW, I.I.5, fols. 76–78v, published in Euler, *Opera Omnia,* series 4A, 6:306–08). There is no evidence of reaction to his proposal.

28. The plan went to the king in December 1743 (AdW I.I.5, fols. 119–22 (German) and

above the royal stables, rather than in the observatory (where the old academy met) or in private houses (where the Société Littéraire had met during its brief lifetime) (fig. 20). Its members were divided into four classes: experimental philosophy, mathematics, speculative philosophy, and literature. Directors of the four classes were elected for life and joined four curators, drawn from the honorary membership, in running the Academy. The curators, all closely tied to the king and all members of the defunct Société Littéraire, alternated in presiding over the directorate, so that no single individual controlled the institution.[29] All the members of the Société Littéraire now belonged to the renewed Academy; the additional members had virtually all been members of the old German academy. The reorganization slotted individuals into a francophone hierarchy directly dependent on the king. For Euler, easily the most productive and most prominent member of the Academy, these changes came not a moment too soon. "I was very much mistaken when I thought that they would put the new Academy on the same footing as that of Paris. The thing is done at present. We have joined into one body the old and the new society under the name of an Academy of Sciences and Belles-Lettres."[30] Still, very few members had pensions, and revenues from the privilege

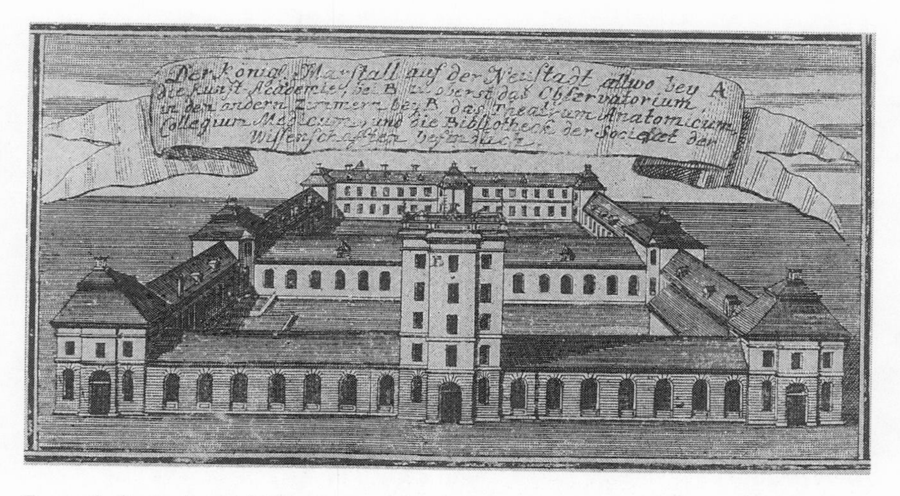

Fig. 20. Berlin Academy of Sciences viewed from Dorotheenstrasse, with observatory tower rising above the stables. The astronomers complained of the warmth and fumes rising up to cloud their instruments from the horses below. By permission of Bildarchiv Preussischer Kulturbesitz, Berlin.

fol. 125–27 (French)). The final version of the statutes is recorded in AdW Registre Général, 24 January 1744 and published in Formey, *Histoire de l'Académie*, 65–74.

29. "Règlement" (1744), in Formey, *Histoire de l'Académie*, 69.

30. "Je me suis bien trompé quand j'ai cru qu'on mettra la nouvelle Acad. sur le meme pied que

on almanacs were insufficient to equip the observatory adequately. At this point, Euler was not optimistic about the prospects for astronomy in Berlin, although he would have liked Delisle to come and set things straight.[31] His frustration was not relieved until Maupertuis finally agreed to take up residence in 1745.

## Rhetoric and Ideology in Maupertuis's Academy

In the event, the 1744 academy was reconstituted yet again only two years later. At the turn of the new year in 1746, anything seemed possible in Prussia. Frederick had returned victorious from the second Silesian war and settled in at Potsdam to pursue his civil reforms and cultivate the arts. Maupertuis sensed the strength of his position as the most illustrious Frenchman to have accepted Frederick's invitation to participate in his projects.[32] He had been named president of the Academy, but just how he was to preside had been left undefined. As a newcomer and a foreigner, his status was awkward and ill-defined with respect to the curators, and he lost no time in asking for a revision of the Academy's statutes to consolidate his control of the institution. "I feel the difficulty of . . . exciting emulation among men of letters governed by ministers of the state and army generals, whose titles alone make them superior to the others." The sorry condition of the Academy, in which "the sciences are in a state of collapse and humility," could not be improved without reorganizing the chain of command.

> I know, Sire, that when I speak to you for the sciences, it seems that I also speak for myself. I shall not hide from you the degree of ambition that I am combining with the favor of serving you. I shall request of you everything that could give me the necessary consideration and credit for the welfare of the Academy and for filling honorably a situation which must be honorable under the reign of Augustus.[33]

Extremely sensitive to the importance of rank and the precarious status of the arts and sciences in Prussian culture, not to mention his own vulnerability as an out-

---

celle de Paris. La chose est faite a present. On a joint en un corps l'ancienne et la nouvelle Société sous le nom d'une Acad des Sciences & Belles-Lettres" (Euler to Delisle: 1 February 1744, MS Obs., Delisle papers).

31. "[O]n n'a d'autres revenues que celles qu'on tire du debit des almanacs, ce qui n'importe gueres grand chose" (ibid.). Delisle was ready to come to Berlin, but only with the promise of adequate support; this never came to pass.

32. There were several other Frenchmen attached to the court, e.g. the marquis d'Argens and Darget, but neither was as well known for his accomplishments as Maupertuis was. Voltaire, Vaucanson, and d'Alembert had all turned Frederick down. Later, other Frenchmen would arrive in Berlin to take advantage of Frederick's policy of toleration, but for the time being, Maupertuis enjoyed special status.

33. M to Frederick, 15 January 1746, Koser, *Briefwechsel,* 202.

sider, Maupertuis argued that authority had to be vested in him by explicit decree of the monarch. Frederick's enthusiastic and generous response gave him the sense that he would be able to mold the Academy as he saw fit.

Frederick instructed Maupertuis to draft yet another constitution for the Academy; once this was done, "His Majesty approved it and annotated it with his own hand."[34] The new statutes, formally presented at the Academy's public meeting in June 1746, named Maupertuis "perpetual president" and gave him precedence over all members, including the curators, regardless of civil or military rank. In Frederick's words, "nothing shall be done except by his orders; just as a general who is only a gentleman commands dukes and princes in an army, without anyone taking offense." The president would control pensions and all other fiscal decisions, appropriating the most important function of the curators.[35] After reading the new statutes to the assembled company, the presiding curator, K. W. de Borck, transferred power by ceding his chair to the new president. The curators were not particularly happy with the change; one of them resigned, and de Borck himself refused to have anything to do with the Academy for some time.[36]

Just a few weeks later, Maupertuis announced the king's acceptance of the title of protector of the Academy. Privately, he interpreted this title in part as homage to himself: Frederick "never wanted to accept this title in the time of the curators."[37] The Academy reiterated its dependence on the king symbolically whenever possible, celebrating the king's birthday every year in its public meeting and designing commemorative medals for Frederick's victories and civil reforms (fig. 21).[38] In the early years of his presidency, Maupertuis showed himself to be adept at manipulating significant ritual moments and rhetorical opportunities for reinforcing the ideological underpinnings of his Academy and his seat at the apex of the institutional hierarchy, his own domain within Frederick's kingdom. The degree of control vested in the president under the new statutes made Maupertuis's position unique and unprecedented in the Republic of Letters. His father, sensitive to the financial element, reported proudly to Daniel Bernoulli, "The king

34. *Histoire de l'Académie royale des sciences et des belles-lettres (Berlin)*, 1746, 3.

35. "He will preside over the curators in all economic matters" (*BAS*, 1746, 6, also in Winter, *Registres*, 97.)

36. Description of installation of president in "Histoire de la renouvellement de l'Académie, *BAS*, 1746, 9. On Viereck's refusal to be listed as a curator in the address calendar, see Formey to M, 9 October 1746, BJ; Euler reported on de Borck's dissatisfaction (Euler to M, 12 July 1746, in Euler, *Opera Omnia*, series 4A, 6:68–69).

37. Winter, *Registres* (23 June 1746), 100, for the announcement; M to JBII, 21 November 1747, BEB.

38. Engravings of medals in *BAS* 1746 and 1747. The revival, at the first public meeting in 1744, had been marked with images of rebirth, on the eve of Frederick II's birthday. Winter, *Registres*, 9.

Fig. 21. Medals designed by Berlin Academy to honor Frederick II. Top: prize medal awarded for the monad prize in 1747, with Frederick II's image as protector of the Academy. Bottom: commemoration, based on concept by Maupertuis, of Frederick's judicial reforms. *Histoire de l'Académie Royale des sciences et belles lettres* (Berlin), vol. 3 (1747).

wished that Maupertuis should preside over Marshall Schmettau and the three minister-curators and that he should manage all the economic business and the whole administration, and that he settle all pensions. [The king] added all that to his *lettres patentes* with his own hand. Such a thing has never been seen before."[39] Maupertuis himself boasted of his new status to all his friends and acquaintances: "In the former state of affairs, the president was president in name only, and it was actually the state ministers and field marshals who governed the Academy. Even the great Leibniz was nothing but a very small boy compared to them. The king has changed all that with a new *Reglement,* in the margins of which he added in his own hand clauses so flattering to me that they make me blush."[40]

The 1746 statutes completed the rejection of the undifferentiated and at least partially democratic form of the Société Littéraire. Some business was conducted by the directors of the four classes; very little of substance was decided by general vote. New members were selected by Maupertuis and approved by Frederick, but

39. Moreau to Daniel Bernoulli, 4 June 1746, BEB, L Ia 708. Moreau handled the investments of Maupertuis and many of his friends, especially the Bernoullis. He fell ill shortly after writing this letter and died several weeks later. (Maupertuis arrived in Paris in early July 1746, just after the death of his father.) See correspondence with Frederick, June and July 1746, Koser, *Briefwechsel.*

40. M to Cerati, 8 October 1746, quoted from Thierry Bodin, Auction Catalogue (Paris, February 1996); copy in AS, dossier Maupertuis.

not by the membership. Although foreign members nominally had to be approved by the whole body, no one proposed by the president was ever rejected. Almost immediately, Maupertuis began selecting eminent foreign men of science and letters to swell the ranks of the Academy with illustrious names.[41] Although the king did not personally attend its meetings, he wrote historical and philosophical memoirs for the Academy as well as occasional eulogies. He liked to joke that he was just another academician when he contributed manuscripts to be read at meetings. In this way, he maintained his credentials as a philosopher-king who took time from his responsibilities as ruler to participate in the intellectual life of his realm. Maupertuis reminded his colleagues that their king and protector, "if he had been born to another station, would have been the ornament of the Academy."[42]

## The Academy in Print

For all his autocratic ways, Maupertuis maintained cordial and respectful working relations with both Euler and Formey, each of whom proved indispensable for the viability of the bilingual academy. Both had lived through the transitional years of Frederick's early reign and were already established at the Academy when Maupertuis took over. There was no love lost between the two, and neither belonged to the sparkling court circle where the Academy president spent time at the behest of the king. But both men were loyal supporters of Maupertuis and often were called upon to mediate between the French president and their German colleagues, as well as to deal with printers, technicians, builders, tax collectors, gardeners, carpenters, and accountants. To the consternation of some long-time academicians, French predominated in all proceedings, especially in publications, even though not all members could use the language fluently. In Formey's recollection, "those who knew a bit of French wrote and read in that language, which sometimes produced lectures peculiar for the gibberish of the speakers."[43] Some members gave papers in German or Latin, but in order to have them printed, they had to arrange for translation into French, often an onerous task.[44]

41. E.g., Montesquieu, Musschenbroeck, all the Bernoullis, Martin Folkes, Henry Pemberton, Linnaeus (Winter, *Registres,* 100).

42. Maupertuis, "Discours prononcé dans l'Académie . . . le jour de la naissance du roi," *Oeuvres,* 3:276.

43. Formey, *Souvenirs d'un Citoyen,* 1:166. Formey attributed the choice of language to Frederick, although Maupertuis's lack of fluency in German was equally significant: "Since the king did not want to read anything except in French, it was clearly necessary to follow his will" (ibid.). See also Bartholmess, *Histoire philosophique de l'Académie de Prusse.*

44. E.g. J. P. Süssmilch, who delivered thirty-two papers in the course of his academic career, only two of which were translated and published in the *Histoire.* Aarsleff, "The Berlin Academy under Frederick the Great," 196–97.

Jean-Henri-Samuel Formey, a Protestant pastor born in Berlin to a French Huguenot family, was professor of philosophy at the Protestant Collège Royale and one of the original members of the renewed academy in 1744. He was a prolific writer of books, essays, reviews, extracts, and especially letters.[45] For his first few years in the Academy he worked under the perpetual secretary, Philippe de Jariges, as "historiographer," with a small pension. Subsequently, he took over the considerable task of managing the Academy's correspondence, publications, and records of meetings. In 1748, he succeeded to the position of perpetual secretary and filled it for nearly fifty years, until his death in 1797. Maupertuis had not been particularly impressed with Formey on his arrival in Berlin; the Protestant cleric-philosopher lacked the brash and dazzling wit of the king's favorites. Maupertuis initially hoped to entice Johann Bernoulli to take on the secretary's position and to give it the kind of visibility Fontenelle had achieved in Paris.[46] This came to nothing, however, and Formey proved an able and reliable secretary, with considerable energy and resilience. He seems to have borne the veiled reprimands of his president with good grace as he learned what was required of him. One of his main tasks was to supervise the translating, editing, printing, and correcting for the production of the annual *Histoire,* almanacs, and collections of prize essays. Especially in the first few years, he worked closely with Maupertuis to produce high-quality volumes that would represent the Academy to the Republic of Letters. Maupertuis shaped the contents of each volume, often adding contributions from foreign members to counter what he saw as the mediocrity of many of his academicians. For his part, Formey had to get everything ready for the printer: obtaining legible manuscripts from authors or copyists, arranging for translations of German and Latin papers (or doing them himself), supervising the printing, and correcting the printer's proof sheets.

When he took the position, Formey planned to tell the printer, Ambroise Haude, to find his own proofreader, "this function not seeming to me to be part of the historiographer's job," but Maupertuis pressured him into doing this too. "I'm sure, Monsieur," the president chided him, "that you will not refuse to do something so useful for the honor of our Academy. M. Haude may have a proofreader, but your overview and final correction will be nonetheless necessary. In the

45. Formey's papers in the Berlin Staatsbibliothek and in the Biblioteka Jagiellonska include some 18,000 letters. Formey contributed to the *Encyclopédie* of Diderot and d'Alembert as well as editing the *Bibliothèque impartiale* and other journals at various times. One volume of correspondence has been published: Fontius et al., eds., *Correspondance passive de Formey.*

46. Maupertuis complained to Bernoulli, "Nous avons un secretaire [Jariges] qui n'en fait aucunes fonctions; nous avons un Historiographe qui les fait mediocrement. . . ." He lamented the waste of pension funds on such positions, but recognized that the situation was unlikely to change (M to JBII, 29 October 1747, BEB).

French Academy [of Sciences], after the first proofs were corrected at the royal printing house, each author corrected his own several times, up to two or three proofs. And M. de Fontenelle corrected the whole historical part himself. I will gladly correct mine and would wish that each of our Academicians were well enough versed in the French language to do the same, but that is not possible.... This work concerns the glory of the Academy and consequently concerns us all equally."[47] Maupertuis took as his model the high standards of the Paris Academy and exhorted Formey to emulate Fontenelle, the ideal secretary. The honor and glory of the Academy had to be fiercely defended, given the Prussian context, where neither the printer nor many of the academicians bothered with the elegant standards familiar to the French.

The volume for 1746, the first product of the new regime, was especially important as a showcase for the renewed Academy's talents. The annual volume, more than anything else, would be the tangible evidence that science was flourishing in Berlin. Maupertuis wanted it to impress readers across Europe, and he was also anxious to impress Frederick, to justify his expectations for the Academy. Maupertuis was quite conscious of those in France who thought he had made a major mistake in giving up his place in Paris for a distant German outpost, almost beyond the bounds of civilization.[48] He took a personal interest in the quality and aesthetic effect of the volume, criticizing the spacing of lines on the page, the layout of titles, and the quality of the paper, all of which were relevant to "giving our book the best possible appearance."[49] His close attention to all details of production, from the selection and organization of its contents to the supervision of printing and correction, made the *Histoire* a monument to his presidency and to the monarch who made possible this enlightened absolutism of letters and science (fig. 22).

47. "Quant à la correction, je crois Monsieur que vous ne vous refuserés pas à quelque chose d'aussy utile pour l'honeur de notre Academie. M. Haude peut avoir un correcteur, mais votre coup d'oeil et votre derniere correction n'en sera pas moins necessaire. Dans l'Academie de France apres que les premieres epreuves avoient eté corrigées à l'imprimerie Royale, chaque autheur corrigeoit les siennes quelquefois jusqu'à 2 et 3 épreuves. Et M. de Fontenelle corrigeoit toute la partie historique luy mesme. Je corrigeray volontaires les miennes; et souhaitterois que chaquun de nos Academiciens fust assez versé dans la langue Francoise pour pouvoir en faire du mesme. Mais cela n'est pas possible: cependant il faudra que MM. Euler et Kies ayont la bonté de revoir les leurs ne fust ce que pour la correction des calculs. Cet ouvrage interesse la gloire de l'Academie, et par conséquence nous interesse tous également" (M to Formey, 12 May 1747, SBB).

48. M to JBII, 28 May 1745, BEB. There was considerable interest in France in the output of the Berlin Academy as attested by the long extract of the second volume of the *Histoire* in the *Mémoires de Trévoux* (December 1748): 2486–94, 2481–2500.

49. M to Formey, 12 May 1747, SBB.

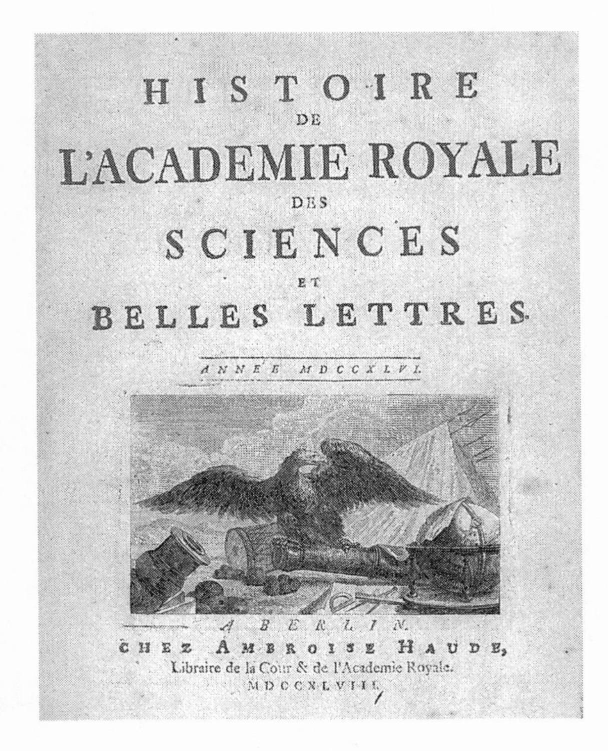


Fig. 22. Title page, vol. 2 of *Histoire de l'Académie Royale des sciences et belles lettres* (Berlin), 1746. Note Prussian eagle, Frederick's symbol, perched on the cannon next to the globe and measuring instruments.

At the head of this volume, readers encountered the complete text of the academic statutes, emphasizing Maupertuis's prominent and "perpetual" position. Underscoring the personal and immediate nature of his mandate, a distinct typeface signaled the king's handwritten emendations to the statutes.[50] Continuing in the ceremonial vein, the next piece was Maupertuis's oration commemorating the king's birthday in 1747. This text and several others were pulled forward to embellish the flagship volume of the renewed academy. "Although it seems to belong to the *Histoire* for the next year, the Academy could not defer publishing its gratitude for the favors of the king."[51] It also put Maupertuis's oratorical skills on display, as he incorporated into his panegyric to Frederick's many virtues a history of the Academy, his own expedition to Lapland, and a justification for the Silesian wars. Even Louis XV came in for praise, as "a prince born for the glory and happiness of his people," who recognized the value of deciding the question of the shape of the earth. "The most reliable means was to measure degrees of the merid-

50. *BAS,* 1746, 3–8; the same text appeared in Formey, *Histoire de l'Académie;* Maupertuis, *Oeuvres* 3:303–11. The official minutes noted: "Ceci est écrit de la main du roy" (Winter, *Registres,* 96–99).
51. *BAS,* 1746, 9.

ian towards the equator and towards the pole. But what an enterprise! What expense! What an array of instruments we had to carry in deserted and wild lands! Louis ordered it, and all difficulties were overcome."[52] This is only one of many examples of the exclamatory style of this formal academic discourse, carefully crafted to link the speaker to his royal patron. "Without my voyage to the pole, my name very likely would never have been known to the [Prussian] king." With further development of the theme of Frederick's heroism and genius, on and off the battlefield, Maupertuis drew, however respectfully, a parallel between his own career and that of the sovereign. The journey to the pole for the progress of science and the improvement of navigation is perhaps a pale reflection of the grand actions of the king, but the comparison is telling. The orator implied that he would reform the Academy just as Frederick had reformed Prussian justice, commerce, transportation, urban design, and social programs. The discipline required by Frederick of his soldiers should remind the academicians of the discipline required for scholarly work, as they "cultivate [their] talents under the eyes of such a master."[53]

A series of medals, designed by the Academy to commemorate the conquest of Saxony and the peace of Dresden, further reinforced the theme of service to the state. The king himself contributed to the *Histoire.* As befitted his status, Frederick was never directly named as an author, but over the years he often submitted historical essays for the literature class, as well as an occasional eulogy. Though his identity was formally masked, the absence of a name signified the royal author. The volume for 1746, for example, contained his "Mémoires pour servir à l'histoire de Brandebourg ," read to the Academy at the public meeting in June 1747. The king enjoyed posing in print as the reluctant author: "I demand and I obtain permission to instruct myself in the Royal Archives; . . . and here I am an Author in spite of myself." Encouraged by a "friend," he "offered" the Academy his essay, in which he reconstructed the military and political exploits of his own antecedents.[54] Frederick was not above correcting his own proofs, and Maupertuis bent over backwards to make sure the printer did justice to the essay. "I have no doubt that you will take all the trouble that the nature of the thing demands," he admonished Formey as he sent back several sheets with the king's corrections for reprinting. "I need not charge you with exactitude, not only for the corrections, but also for the beauty of the composition." He also instructed Formey to pay for

52. Maupertuis, "Discours," *BAS* 1746, 10–15; reprinted in *Oeuvres,* 3:271–82.

53. Ibid., *Oeuvres* 3:282.

54. *BAS,* 1746, 337–78. Darget, privy councilor to the king, read an abridged version to the public session, June 1, 1747 (Winter, *Registres,* 112). "Je demande & j'obtiens la permission de m'instruire dans les Archives Royales; . . . & me voilà Auteur en dépit de moi-même" ([Frederick II], "Mémoires pour servir à l'histoire de Brandebourg," *BAS* 1746, 338).

the reprinting from Academy funds, to retrieve all uncorrected copies from the first printing and to deliver them back to Maupertuis "in a sealed box."[55]

The bulk of the 1746 volume contained the best papers from the four classes. Only the experimental philosophy section of the Academy had a core of productive German members. The chemists Johann Theodor Eller, Johann Heinrich Pott, and Andreas Marggraf, all members of the old Academy, filled this section of the *Mémoires.*[56] The mathematics section was anchored by three pieces by Euler and augmented by a paper by d'Alembert on the integral calculus. (Euler sustained the mathematics class throughout his two decades in Berlin almost singlehandedly.) The metaphysics and literature classes were the most problematic and continued to be so for years, in spite of Maupertuis's efforts to recruit new blood from outside. The president himself contributed the centerpiece of the speculative philosophy section in 1746, "Les loix du mouvement et repos, déduites d'un principe métaphysique" (The laws of motion and rest, deduced from a metaphysical principle). This paper built directly on his earlier work on the principle of least action, introduced to the Paris Academy in 1744, but not yet in print. It served as an example of the kind of metaphysics, closely tied to physics and mathematics, that he hoped the Academy would promote.[57] For the same volume, the literature class included, in addition to Frederick's history and Maupertuis's formal response, a speech by the marquis d'Argenson and a paper by La Condamine on the monuments of Peru. La Condamine, finally back from his geodetical expedition to South America, had obliged his old friend with this first-person account of Incan ruins.[58] His engaging travel report contrasted markedly with the examples of traditional philology submitted by two German academicians, Elsner and Heinius, both survivors from the old Academy. The eclectic nature of this

55. "Je n'ay pas besoin de vous recommander l'exactitude, non seulement dans la correction, mais aussy pour la beauté de la composition" (M to Formey, 14 November 1747, SBB). He went on with exact instructions about how to interpret Frederick's notations: "Tout ce qui est ecrit dans les marges est du corps de l'histoire, excepté ce qui est dit de la ville de Nuis pag. 10; qui est une note marginale. Il y a une feuille entiere celle qui est marquée C qui pourroit subsister, a moins que la maladresse du compositeur ne luy permist pas d'ajuster toutes les corrections precisement dans les feuilles precedent sans serrer les mots et sans hiatus. Car alors il faudra reimprimer aussy cette feuille" (ibid.).

56. The annual publication of the Berlin Academy used the title *Histoire de l'Académie royale des sciences et belles-lettres,* but after the second volume (for 1746), there was no *Histoire* separate from the *Mémoires,* as was the case with the Paris Academy.

57. Maupertuis, "Les loix du mouvement et repos, déduites d'un principe métaphysique," *BAS* 1746, 267–94. In spite of his avowed willingness to do his own proofreading, Francheville (a member of the literature class) corrected the proofs. M to Formey, 24 September 1747, SBB.

58. La Condamine, "Mémoire sur quelques anciens monumens du Perou, du tems des Incas," *BAS* 1746, 435–56. The paper had been read to the Academy in 1747 (Winter, *Régistres,* 114).

collection, and its cosmopolitanism, reflected an ideal Maupertuis strove to perpetuate in future years. The contributions from famous French academicians, both close associates of the president, reminded readers of the productive link between Berlin and Paris.

The Academy's troubled relations with its printers epitomized the problem of reconciling ideals with the realities of life and politics in Berlin. Following the death of the printer Haude in 1748, the directorate of the Academy articulated their conditions for the printers, but they were not able to enforce them consistently. For example, they insisted that "in the future the *Memoires* of the Academy will be printed on a paper that is whiter, larger and stronger than that used ordinarily, and that the printer will deposit a printed sheet for approval before starting the job."[59] Nevertheless, Haude's partner and successor Spener continued to have difficulty obtaining an adequate supply of high-quality paper, often causing publication delays.[60] Even when he was back in France on periodic visits to his family, Maupertuis kept abreast of the progress of the printing of the annual volumes, asking that proof sheets be sent to him by diplomatic pouch so that he could guarantee the quality of the printing. He kept Formey busy cajoling and threatening the printer. "Do not let them print a single sheet before the paper has arrived, and verify that it is the same as the rest, because surely M. Spener will deceive you if he can."[61]

## Enlightened Despotism and Civility

Maupertuis played many roles in his years in Berlin: wise ruler, philosopher, mathematician, man of letters, even orator. One of his official functions was to deliver formal speeches suited to various ritual occasions. As we have seen, he used these occasions to elaborate the relations of king, president, academicians, and

59. "Assemblée du Directoire," 1 July 1748, Winter, *Régistres,* 127. See also M to Formey, 19 May 1748, SBB: "Pensés aussy je vous prie à ce que nous pouvons faire pour profiter de la mort de M. Haude, et pour tacher d'avoir nos Memoires mieux imprismés."

60. See, e.g., Formey to M, 28 November 1747, BJ; printing had stopped on the 1746 volume, due to lack of paper: "Parlez je vous prie a M. Haude, et faittes luy trouver du papier, quand l'Academie devroit luy faire les avances c'est une chose honteuse pour luy et pour nous que ce retardement, et à laquelle il faudra necessairement mettre ordre , et une ordre facheux pour luy" (M to Formey, 2 December 1747, SBB). The next year it was the same story: see Formey to M, 31 December 1748, BJ, and Maupertuis's reply, 30 January 1749 (from Saint-Malo): "Ce qui me fasche beaucoup, c'est l'interruption de notre impression: et il nous faudra absolument prendre d'autres mesures dorenavant." They were never as far behind as the Paris Academy, however, where the lag between delivery and publication of papers was often as long as four or five years.

61. M to Formey, 2 June 1749, SBB. Formey had similar headaches with producing the prize essays, almanacs, address calendars, and commercial calendars. The Academy depended on revenue from these latter items, which had to be updated yearly.

Academy. If Frederick used the Academy to represent himself as a philosopher as well as a king, Maupertuis used it in inverse fashion to represent himself as a king as well as a philosopher. In his academic address "On the Duties of the Academician," delivered to the public assembly in 1750, he portrayed himself as the representative of the king, indeed the analogue of the king, functioning as the enlightened despot of the Academy. Posing as a paternal leader admonishing his subjects, he used political and social metaphors to depict the Academy as a mirror image of the enlightened state. The Academy is to be a society within a society, linked to the outside world through its dependence on the king. Like an organism, it cannot exist independent of its environment, the laws decreed by its governors. Indeed, it depends on those external, wise laws, as the plant depends on the gardener, and as subjects depend on their rulers.

From this idealistic perspective, the Academy enables collaborative, egalitarian, unrestricted science. "The academician acquires in our assemblies that academic spirit [*esprit académique*], that kind of feeling for the truth, which leads him to discover it wherever it is and keeps him from looking for it wherever it cannot be found."[62] Academic spirit complements the more mundane material benefits of appointment to the Academy, allowing members to produce knowledge superior to anything done by individual minds working in isolation. In becoming a member, however, the individual undertook to fulfill certain obligations.

> Why should the philosopher renounce that liberty to which it seems he would have sacrificed everything, to subject himself to obligations? Undoubtedly, he must find some advantage in it; and what is this advantage? It is that which men take from all societies: the mutual aid that all members lend to each other. Each society possesses a common good, from which each individual extracts much more than he contributes.[63]

Drawing on corporate, republican, and autocratic language all at once, Maupertuis articulated an academic ideology that fit perfectly into the Prussian context. Although the Berlin Academy was never constituted as a corporation, as were the universities, Maupertuis emphasized the advantages and obligations of corporate membership, understood metaphorically. But in addition he described his academy, also metaphorically, as a republic, where freedom would promote the growth of knowledge. Addressing his audience as "citizens of the Republic of Letters,"

62. Maupertuis, "Des devoirs de l'académicien," read to Academy at public assembly, 18 June 1750, Winter, *Registres*, 152; published in *BAS* 1753 (1755), 511–21, reprinted in Maupertuis, *Oeuvres*, 3:283–302, quotation on 285–86.

63. Ibid., 284.

the perpetual president asked, "Are the duties which the Academy imposes on you other than what you would do for the love of the sciences alone? Do you feel constrained in the freest Academy in Europe?"[64]

The freedom of the republic was only possible in the larger context of an enlightened monarchy, which combined hierarchical power relations with the *esprit philosophique.* "Everything is permissible for the philosopher, as long as he treats everything with the philosophical spirit."[65] The king dictates the laws of the Academy, guaranteeing intellectual freedom. But the academician must not forget that his freedom derives from the king, who is both legislator and protector of the Academy.[66] Years later, reading Maupertuis's eulogy to the Academy, Formey stressed Maupertuis's place in the corporate/republican/autocratic academy. Although the academicians were all *confrères,* Maupertuis's rank elevated him above his academic "brothers." "He appreciated the genius and the work of each [academician]; and he liked to be fair. But he demanded application and exactitude. . . . This sometimes gave him a sharpness of tone and an air of severity that was uncomfortable for those who disliked being disturbed, or who entertained false notions of the liberty of the man of letters, and in particular that of an academician."[67] Testifying to the validity and viability of such control, Formey went on: "The *esprit d'ordre* cannot be maintained without the vigilance of a ruler who knows how to use his power without abusing it."[68] An apologist for Frederick's own methods might have uttered the same words. The king supported Maupertuis in his absolutist pose. Both agreed that the Frenchman's established reputation as urbane man of letters and accomplished man of science qualified him for the job of running the Academy. Just those attributes that bothered many Germans about him—his French suavity and sociability—endeared him to the king and held out the promise that the Academy could attract other such characters to Berlin.[69]

## The Reality of Academic Life

Academic expertise was available to the king for services ranging from translation of foreign works or preparing the his own works for publication to more

64. Ibid., 287.

65. "Tout est permis au philosophe, pourvu qu'il traite tout avec l'esprit philosophique" (ibid., 295).

66. Ibid., 288. See also Terrall, "Culture of Science in Frederick the Great's Berlin."

67. Formey, "Eloge de Maupertuis," *BAS,* 1760, 511.

68. Ibid.

69. Wolff characterized the Berlin Academy under Maupertuis as a combination of "so-called Newtonian philosophy with the French world of flattery" (Wolff to Schumacher, 6 May 1748, in Harnack, *Geschichte,* 2:310).

technical problems such as designing fountains for the royal palace and canals for commerce, or making calculations for lottery and insurance schemes.[70] In conjunction with Frederick's program for the invigoration of Prussian industry, the Academy managed an extensive mulberry plantation outside Berlin and experimented with methods for cultivating trees and silkworms. While the Academy's botanist, Johann Gottlieb Gleditsch, was supposed to play the major role in this project, the finance committee also had to engage in extended negotiations with tenants of the property, architects and builders, laborers, and royal ministers.[71] In many ways, the Academy functioned as a state agency, subject to the same controls and stipulations as other branches of the burgeoning Prussian bureaucratic apparatus. But the rhetoric of the communal search for truth and the ideology of intellectual honor contributed essentially to the functioning of the Academy. It could not be *only* an arm of the state, since its very reason for existence required integration into the pan-European Republic of Letters. The ideology of cooperation and equality provided more than just a smokescreen to fool academicians into serving the state in exchange for inadequate pensions.

Nevertheless, the fit between ideology and purpose did not ensure smooth day-to-day operations or stellar scientific contributions. Official exhortations did not always have the desired effect on Maupertuis's subjects. He was well aware of the gap between everyday practice and the ideally balanced system of duties and rewards articulated on formal occasions. The dominance of French style, associated in Berlin with aristocrats and the court, along with Maupertuis's ignorance of the German language, aggravated tensions.[72] In private, especially in letters to Johann II Bernoulli, Maupertuis revealed the adversarial quality of his relations with the German academicians. He saw his efforts to uphold the "honor" and public image of the Academy thwarted at every turn, and he characterized many of the pensioners as mediocre and lazy:

> the more authority the king gives me in the Academy, the more I try to make good use of it. Nevertheless, I see every day that a German (or at least a Prussian) Academy, must be administered completely differently from a French Academy, where

70. Euler made extensive calculations for waterworks at Potsdam and for a state lottery. Maupertuis was called upon to correct the king's French style and grammar.

71. Documents concerning the Academy's mulberry plantation in Köpenick in AdW Regître général, I.IV.4. See also Euler to M, 16 September 1749, 21 July 1750, 25 July 1750, in Euler, *Opera Omnia*, series 4A, 6:132–33, 155–56.

72. In 1750 Maupertuis claimed to be learning German, although he never became fluent (M to duchesse de Chaulnes, 21 April 1750, SM). The tensions were not just between universities and academy, or between German and French speakers, but also, as Anne Goldgar has shown, between

each academician has several reasons to motivate his work. And I see by the same token that when the king wanted the pensions to be controlled absolutely by the president, he was more perceptive [*éclairé*] than I who proposed that it be otherwise.[73]

Apparently, German academicians did not have enough innate *esprit académique.* The only way for Maupertuis to get his academicians to work was to insist on strict compliance with the Academy's regulations, which required the regular production of memoirs in order to merit the payment of pensions. He hoped to circumvent some of the German lassitude by importing Swiss members, but many of these recruited members failed just as miserably.[74]

In November of 1747, Maupertuis instructed Formey to inform the directors of the four classes that the statutes were to be strictly observed: not only were sufficient papers to be produced, but they were to be provided to the president for his review two weeks before presentation to the full Academy.[75] His repeated admonitions became increasingly caustic over the next month; he was particularly concerned with the classes of literature and metaphysics.[76] In 1749, he wrote to Frederick, "Our classes of speculative philosophy and of belles-lettres are of the greatest feebleness and would perhaps have been obliterated without the most urgent and most powerful help they have found in Your Majesty yourself."[77] This was still an issue two years later, when Maupertuis chastised the Academy:

> The king, having been informed of the lack of exactitude which several academicians are bringing to the execution of their duties, has ordered me to inform the Academy that he has decided irrevocably that all members of this body [*corps*], pensioners as well as associates, who go for one year without producing a paper will be assigned to the rank of veteran, and that their pensions, if any, will be suppressed and returned to the Academy's funds so that His Majesty can assign them in favor of those who have merited encouragement and recompense for their work.[78]

---

traditional scholars of the old Republic of Letters, on the one hand, and wits and philosophers of the new Enlightenment, on the other (*Impolite Learning*).

73. M to JBII, 21 November 1747, BEB.

74. M to JBII, 6 September 1749, BEB, regarding Battier and Passavant.

75. M to Formey, 15 November 1747, SBB; Formey read the letter at the meeting of 16 November (Winter, *Registres,* 119).

76. Euler took pains to inform him that the mathematics class was in strict compliance, but he also reported resistance to the new order from other classes (Euler to M, 18 November and 25 November 1747, in Euler, *Opera Omnia*, series 4A, 6:90–92).

77. M to Frederick II, [September 1749], in Koser, *Briefwechsel,* 241.

78. Winter, *Registres,* 30 October 1749, 143. Note that the king had actually set two years as the mandated interval, not one (FII to M, [Sept. 1749], in Koser, *Briefwechsel,* 242; also in Harnack, *Geschichte,* 2:277.

This struggle with the membership reflects the difficulties Maupertuis faced in implementing his version of academic culture in a foreign setting. "It's a misfortune," he complained to Formey, "that all the pains I take to put our Academy on a good footing fail so miserably."[79]

In addition to the internal problems of staffing and quality of work, ideals and reality frequently clashed over financial matters. One of the major privileges of the Academy was a monopoly on the production of certain almanacs and calendars as well as an excise tax on other almanacs sold anywhere in Frederick's lands. In 1748, an excise on all maps was added, as were exclusive rights to publish royal edicts.[80] Frederick had his own money problems, aggravated by his penchant for fighting expensive wars. Although he did occasionally offer to pay a particular pension from the royal purse, Maupertuis could not rely on the king's personal largess for the Academy's operations.[81] In 1747, when he was trying to recruit Nicolas Béguelin for the literature class, Maupertuis asked the king to supplement the small sum available in the Academy's coffers for a new pension. "The king is as poor as a church mouse," Frederick replied. "He is founding a great number of colonies of peasants; when they are provided for, we will think of astronomers."[82] A few years later, Maupertuis tried again, asking to have the considerable expenses of operating the anatomy theater reassigned to other funds, since the Academy saw no benefit from dissections.[83] The Prussian military conquests of the early 1740s had had an immediate impact on the academic treasury, expanding the market for its almanacs, but efforts to improve the financial picture for the Academy centered on further privileges that Frederick might assign with no cost to his treasury.[84]

79. M to Formey, 26 July 1749, SBB.

80. Reprinted in Harnack, *Geschichte,* 2:274–75. In 1755, in response to a complaint from Maupertuis, the king ordered the General Directorate to renew the order requiring all provincial officials to buy the annual volume of royal edicts, to benefit the treasury of the Academy ( FII to M, 28 March 1755, in Koser, *Briefwechsel,* 302).

81. Maupertuis's own generous pension was paid by the king. In response to Maupertuis's letter of 23 February 1750, Frederick agreed to pay Merian's pension until the Academy could afford it (Koser, *Briefwechsel,* 251).

82. FII to M, 5 July 1747, in Koser, *Briefwechsel,* 221.

83. The anatomists had generous pensions, predating Maupertuis's tenure, and rarely contributed papers to the Academy's meetings (M to FII, [September 1749], Koser, *Briefwechsel,* 241–43).

84. The acquisition of Silesia was particularly profitable to the Academy, since it nearly doubled the population of Prussia, and hence also the market for almanacs and calendars (AdW, "Régistre général," 85.141, 142). The Academy produced calendars for many different regions with information on markets and fairs; it also sold genealogical and address calendars, and calendars for the Huguenot and Jewish communities. Proposals recorded in an anonymous document from the period 1744–1746 addressed to the king asked for the privilege for book censorship (AdW I.I.7, fols.

Maupertuis was acutely aware of the crippling effect of his limited budget on his ability to attract new members to Berlin, an effort he considered crucial to the quality of the Academy. In the winter of 1747–1748, extensive negotiations about privileges pitted Maupertuis against the ministers of Frederick's general directorate, the council of ministers at the top of the government hierarchy. In November, Maupertuis informed the Academy of a significant new privilege for book censorship, a privilege he had obtained by appealing directly to the king.[85] This stipulated that "the authors of each work published in His Majesty's states must pay to the Academy for the censorship of these books two *groschen* per sheet and six *groschen* for each wedding announcement or funeral oration."[86] Although this gave the Academy unprecedented power over what could be published in Prussia, Maupertuis was more interested in its financial repercussions. "I feel real joy at the thought that the increase in the Academy's revenues will place us in a better position to recompense the work and the assiduity of its members. I do not doubt that the Academy will apply itself more and more to merit such a marked and glorious protection."[87] And to Bernoulli he gloated, "I have just in the last several days obtained for [the Academy] wonderful privileges from His Majesty, that gave me as much pleasure as if they had been give to myself."[88]

The Academy had no set procedure for actually doing the work of reviewing manuscripts. The first assignment went to Formey for "a work which concerns matters of theology and philosophy." Maupertuis instructed him to pay attention to anything relevant to "the state, religion, and good manners."

> The aim of our privilege is not to establish an inquisition. We must allow honest freedom to philosophize, but that liberty must not be abused. For that which concerns the government, I do not believe anyone here would be tempted to speak indiscreetly, but in that case, the censor could not be too severe; as for good manners, you must try to banish that habit too common in Germany of injuriously attacking one's enemies. As for the merit of works, it is not the object of our censorship; if we were rigid about that, our privilege would bring us but little profit.[89]

10–11). Although the old Academy had been granted a limited privilege for censorship of historical works in 1708 (Harnack, *Geschichte,* 2:182–83), little use had been made of it, and by the 1740s it was no longer being applied.

85. Winter, *Registres,* 118. Maupertuis wrote to Bernoulli: "Je viens depuis quelques jours d'obtenir pour [l'Academie] de fort beaux privileges de Sa Majesté, qui m'ont fait autant de plaisir que si c'eust eté à moy quils fussent accordés" (M to JBII, 21 November 1747, BEB).

86. AdW, "Régistre général", 145. Official documents pertaining to the privilege are found at AdW I.IV.44, ff.177–86.

87. M to Formey, 15 November 1747, SBB.

88. M to JBII, 21 November 1747, BEB.

89. M to Formey, 27 November 1747, SBB. This was a brochure by Süssmilch on a theo-

In the event, the Academy had scant opportunity to profit from its new privilege, as the king revoked it four months later.[90] Although he had granted it more or less as a personal favor to Maupertuis, his advisers insisted that the privilege would harm the publishing industry and ultimately bankrupt printers and booksellers.[91] Maupertuis's easy access to the king's attention sparked substantial hostility in the government. His feud with the ministers simmered, and he repeatedly appealed to the king to reconfirm academic autonomy.

Frederick's sudden revocation of the privilege also underscored the Academy's dependence on the royal will. Maupertuis persisted in trying to preserve some measure of autonomy for the Academy, particularly with respect to the general directorate. When the ministers tried to get the Academy to take charge of producing and maintaining a set of printer's type, Maupertuis wrote directly to the king to complain. He objected to the ministerial claim that the Academy was "forced to accept" the assignment, as well as to the inappropriate nature of the task. Academicians are "unsuited to such an undertaking, which is the province of printers; such foreign tasks are an affront to academic occupations."[92] The Academy could only maintain its position by controlling the definition of appropriate services to the state.

In spite of Frederick's availability as adjudicator of such disputes, the Academy remained dependent on the cooperation of officials throughout Prussian domains for the policing of its monopolies. Maupertuis's supervision in turn was mediated by his dependence on people who spoke German. The *General Register* of the Academy records numerous instances of violations of academic privileges. In 1744, an attempt was made to force recalcitrant booksellers and printers, who had taken advantage of the quiescence of the Academy, to comply with the taxes on almanacs.[93] Two years later, the royal edict forbidding the import of foreign almanacs was renewed.[94] The next year, the Academy outlined to the general directorate proce-

logical controversy with Edelmann (Euler to M, 25 November 1747, in Euler, *Opera Omnia*, series 4A, 6:93). When Formey requested the manuscript from Haude, the printer, he discovered that it had already been printed, without being censored. Formey considered the brochure "indiscreet" (Formey to M, 28 November 1747, BJ).

90. The Academy was left with its old privileges for calendars and almanacs, as well as for the printing and import of maps (AdW I.IV.44, fols.223–26).

91. FII to M, 10 March 1748, Koser, *Briefwechsel*, 227. For the Academy's tensions with the general directorate, see Schmettau to M, 22 March 1748, AS Fonds Maupertuis.

92. M to FII, [October 1750,] Koser, *Briefwechsel*, 258. The Economic Commission of the Academy had discussed the directorate's actions on 8 September (Winter, *Registres*, 154). Further recriminations against the ministers appear in M to FII, 9 November 1750, Koser, *Briefwechsel*, 259–60.

93. AdW, "Régistre générale," 409.

94. Ibid., 511.

dures for collecting its revenues on its map privilege; local excise commissioners would have to oversee the distribution of maps in their districts to make sure they displayed the requisite stamps before being sold.[95] In 1752, the Academy began a protracted dispute with a Berlin bookseller over his right to sell certain almanacs without paying the tax. He was repeatedly enjoined from abusing the privilege of the Academy, finally by royal edict.[96] Even with the backing of the king, the Academy devoted substantial effort to enforcing compliance with its privileges and ensuring the collection of revenue, with only limited success.

## Monads in the Academy

As one of its first official actions, the reconstituted academy advertised a prize essay competition on "the doctrine of Monads."[97] As in Paris, prize competitions allowed institutions to act in the public sphere, opening channels between academies and the wider Republic of Letters. They attracted attention to the authority vested in the academic elite to adjudicate unresolved questions, and submissions arrived from quite a variety of authors. Anyone outside the Academy was eligible to send an anonymous submission, and essays typically came in from provincial and foreign savants. The process of blind judging culminated in the opening of the sealed envelope containing the key to the coded identity of the winner (or winners). The author then received a monetary reward, and the Academy published the essay with its seal of approval.

In the Berlin question for 1747, contestants were asked to "set forth in an exact and distinct manner" the doctrine of monads, a key component of Leibnizian-Wolffian metaphysics. Then the essayists were either to refute the doctrine with unanswerable arguments or to "deduce from [monads] an intelligible explanation of the principal phenomena of the Universe and in particular the origin and motion of bodies."[98] The framing of the question, giving contenders the opportunity to write for or against monads, reflected Maupertuis's commitment to developing the Academy's reputation as independent of any particular philosophical dogma. Appropriate to the German context, the question certainly enhanced the Academy's visibility, both at home and abroad. There is no evidence that Maupertuis

95. AdW, I.IX.2, 11 December 1747.

96. AdW, "Régistre générale", 7 July 1753. Additional documents pertaining to this dispute can be found in AdW, I.VIII.247.

97. Harnack, *Geschichte,* 1(1), 402–03. There had been a competition the year before on a question proposed by the mathematics class, on the cause of winds, won by d'Alembert. For list of prize questions by year, see Harnack, *Geschichte,* 2:305–07.

98. Harnack, *Geschichte,* 2:305. The speculative philosophy class proposed three possible questions, and the Academy as a whole chose the monad topic. See Winter, *Die Registres,* 99. On the

tried to use the occasion to challenge Wolff's authority or that he pushed for one side of the question over the other.[99] He did beg Johann II Bernoulli to contribute an essay and "revive the honor of metaphysics."[100] Bernoulli demurred, and his friend in Berlin found that the few contributions he read did not impress him as particularly worthy: "I am wearing out my eyes on the piece on Monads," he told Formey as he struggled with one of them, "and I think that if I could read it I would wear out my mind. It is written like the edicts of Caligula."[101]

As anonymous essays accumulated in the Academy, administering the competition turned out to be more contentious than Maupertuis might have imagined, in large part because of Euler's hostility to Leibnizian metaphysics and Formey's reputation as a Wolffian.[102] The contest pushed this internal division into public view and set the tone for the Academy's relations with philosophers in the universities and in the German-language press. It became a matter of considerable public interest, fueled by a series of anonymous books and journal articles published before the closing date for the contest. According to one retrospective account, "All of this gave rise to great upheavals [*fortes agitations*] in Germany and made a great deal of noise in Berlin, where in the court, in the city, in *clubs des savans,* and in associations of all kinds, everyone spoke of nothing but monads."[103] This may be something of an exaggeration, although Euler recalled it similarly: "There was a time when the dispute about monads was so lively and general that one spoke of them heatedly in all social circles, even in the *corps de garde.* There was almost not a single lady at court who had not declared herself for or against monads. All conversations touched on monads, and no one spoke of anything else."[104] Formey, writing in the midst of the furor, reported that the question aroused more interest outside the Academy than within it: "the members in general are not dis-

---

monad competition, see Harnack, *Geschichte,* vol. 1, passim; Bongie, "Introduction to *Les Monades*"; Clark, "Death of Metaphysics."

99. Maupertuis's role in the choice of question is not documented.

100. M to JBII, 22 October 1746, BEB. Several commentators have assumed that Maupertuis took an anti-Leibnizian position; there is no evidence for this. See, e.g., Calinger, "The Newtonian-Wolffian Controversy."

101. "Je perds les yeux sur la piece des Monades, et crois que si je la pouvois lire, j'y perdrois l'esprit. Elle est ecritte comme les edits de Caligula" (M to Formey, 23 March 1747, SBB). Maupertuis did not read German; this must have been one of the French essays.

102. Formey had published an exposition of Wolffian philosophy as a rather clumsy dialogue, *La belle Wolffienne.* For the antagonism between Formey and Euler on Leibnizian philosophy, see Winter, Introduction to *Registres,* 29–32.

103. Merian, "Eloge de Formey," *BAS* 1797, 67. Merian was not in Berlin until 1750, so this is not a first-person recollection.

104. Euler, *Lettres à une princesse d'Allemagne* (St. Petersburg, 1768), Letter 125. Euler recalled the date of the competition incorrectly as 1748; the prize was awarded in June 1747.

posed to excite themselves about such subjects."[105] Several academicians, however, including Formey himself, became heavily embroiled in the controversy.

Academy regulations prohibited members from competing for the prize, but Euler evaded this restriction by publishing, shortly after the announcement of the competition, an anonymous tract denouncing monadology, *Gedancken von den Elementen der Cörper.* He developed an argument that had been used against Leibniz more than once already, claiming the absurdity of immaterial elements as the basis of material bodies. How could metaphysical entities fill space and cause effects on observable bodies?[106] As a polemical piece on exactly the topic set for the Academy's prize, this was a blatant breach of etiquette, only marginally mitigated by anonymity. No one mistook the author. When the pamphlet appeared, Maupertuis was conveniently absent in France, and he did not learn of it immediately. In any case, he would not have been able to read it, since it was written in German. Formey, Wolff, and other Germans saw Euler's intervention as an unconscionable disruption of the impartiality of the academic process. The controversy threw into harsh relief the intellectual distance separating the universities and journalists from the new Prussian Academy. It also revealed a deep rift within the Academy itself, with Formey as the most visible opponent to Euler. And it underscored the alienation of the new president from the pressing concerns of German philosophers.

In its early stages, the public controversy transpired in German, so Maupertuis needed translators and informants to keep up with it. He learned of the debate in the German press from none other than Christian Wolff, then professor in Halle, who was still on cordial, if formal, terms with the Academy's president. Although readers had easily identified Euler as the author of the offending work, Wolff politely assumed Maupertuis did not know of it and would not have "approved this impudent act of audacity." He reported that a Leipzig journal had protested against Euler's pamphlet, and that other refutations were appearing as well. "I thought it my duty," Wolff wrote, somewhat sanctimoniously, "to send you the Review of Leipzig, as well as the two refutations, so that you would not remain ignorant of what is happening in the Universities on questions that concern the Royal Academy of Sciences." Wolff tried to undermine Euler's credit with Maupertuis, without entering into a direct confrontation with the mathematician. "Euler, too overpowered by his favorable fortune, affects a certain supremacy in

105. Formey to Gottsched, 11 February 1747, in Danzel, *Gottsched und seine Zeit,* 60.

106. Euler, *Gedancken von den Elementen der Cörper.* Euler used the Academy's regular printer, not bothering to cover his traces very assiduously. Similar arguments had been used by Voltaire, "Métaphysique de Newton," in *Eléments de la philosophie de Newton* (1738), and by Müller, *Die Ungegründete und Idealistische Monadologie,* among others.

the Republic of letters," and Wolff professed no desire to engage in that sort of game. "In his brochure, he above all wished to refute Leibniz and me. . . . As for myself, I prefer the insight of Leibniz in metaphysics and philosophy to the profundity of Euler."[107]

Maupertuis relied on Formey for a translation of the Leipzig review and for details on other "refutations" of Euler's *Gedancken.* Formey identified the reviewer as Johann Christof Gottsched, a Wolffian professor at Leipzig.[108] He transcribed a long passage from his translation of Gottsched's article that referred to "a distinguished member of the Academy" who had not only denied the viability of monads, but had asserted "that his work should serve as a directive to all those who would undertake to work on the question proposed by the Academy." Gottsched went on,

> One can see that all the trouble that the partisans of monads could take would be a complete waste, that it would be useless to seek to shed light on the doctrine of these elementary substances [monads] and to deliver them from the difficulties that have kept them, up to the present, from gaining universal approbation. . . . But what becomes of the impartiality of the Academy of Berlin? And why does it take away from philosophers with one hand that which it had given them with the other?[109]

Formey himself was disturbed by Euler's breach of faith and told Maupertuis, "I thought that this passage was essential to show you the impression that the anonymous Berlin pamphlet has produced on the *esprit* of German savants."[110]

Surviving letters reflect only a small portion of the rumors, letters, and conversations circulating around Germany on this subject. Apparently responding to questions about some such rumors, Formey disingenuously assured Maupertuis that he himself had nothing to do with the matter: "There is nothing in all this, Monsieur, that involves me, and it seems that the Leipzig gentlemen thought they

107. Wolff to M, 15 November 1746, in LS, 428–29. Wolff wrote to Maupertuis in Latin; he knew Maupertuis would not be familiar with texts published in German. Euler and Maupertuis did not discuss Wolff's accusations, at least in their letters. See also Pons, "Les années berlinoises de Maupertuis."

108. The review was Gottsched, "Gedanken von den Elementen der Körper . . . ," *Neuer Büchersaal der schönen Wissenschaften und freyen Künste,* vols. 3/4 (Leipzig 1746), 355–66 (Wolff to Maupertuis, 15 November 1746, in LS, 429). Formey did not give the German name of the review; he identified the other authors who had written against Euler as Professor Stibnitz (of Halle) and Professor Körber (no university specified) (Formey to M, 25 November 1746, BJ). On Gottsched, see Danzel, *Gottsched und seine Zeit.*

109. [Gottsched], review of Euler, quoted from Formey's translation in Formey to M, 25 November 1746, BJ.

110. Formey to M, 25 November, 1746, BJ.

ought to submit to your judgment what they have written in defense of a question that you judged worthy of the Academy's prize."[111] In point of fact, Formey had already written his own refutation of Euler, in French, and was soon to publish it with translations of Euler's pamphlet and the other German texts.[112] He had been corresponding with Gottsched as early as September 1746 and may well have been responsible for provoking him to publish his irritated response to Euler. Formey explained to his co-conspirator that Euler "however able he might be in other areas, assuredly has not even a superficial knowledge of Metaphysics. Thus it was easy to respond to him, and there exists in effect a Response that I dare to assure you is perfectly solid and includes some things that are completely interesting and even new." This was Formey's own *Recherches sur les élemens de la matière,* though he refrained from admitting his authorship explicitly. "It is time," he declared, "to make it known once and for all to the gentlemen of mathematics [*Messieurs les Géometres*] that they are not competent judges in the matter of metaphysics, and that their imaginary notions would not be able to lead us to ultimate notions of things." He enlisted Gottsched's support in finding a publisher and a translator for his book, "to avenge Leibniz from all these imputations brought against him by Philosophers who follow no other guide than the senses and the imagination."[113] Formey also corresponded directly with Wolff, who insisted that he proceed with plans to publish, in both French and German.[114] Formey's book was out by early 1747, well ahead of the Academy's decision about the competition, so that the whole controversy had been thoroughly aired in print before the essays had even been collected by the prize commissioners.[115]

111. Ibid.

112. Formey, *Recherches sur les élémens de la matière.* In the same volume, Formey printed a French translation of Euler's text, interspersed with commentary. Unfortunately, Maupertuis's letters to Formey from this period do not survive; there is also a gap in Euler's letters to Maupertuis between July and the end of December 1746. Formey told Maupertuis, without admitting that he had anything to do with the project, that the German texts would soon be published in French translation (Formey to M, 20 November 1746, BJ).

113. Formey to Gottsched, 23 September 1746, in Danzel, *Gottsched und seine Zeit,* 59–60. He added a postscript: "Je vous prie, Monsieur, de tenir tout cela secret."

114. Formey to Gottsched, 14 October 1746, excerpted in Danzel, *Gottsched und seine Zeit,* 60. Gottsched acted as Formey's agent in finding a translator and publisher for the book in Leipzig (presumably); no place or publisher appears on the title page. The text was reprinted in Formey, *Mélanges philosophiques.* Merian says that Wolff reviewed Formey's text before publication. ("Eloge de Formey," 66).

115. The question received an unusual number of entries; according to Euler there were thirty altogether. He left notes on twenty-five of them; most of the manuscripts do not survive. See Bongie, "Introduction to *Les Monades,*" 27; Clark, "Death of Metaphysics."

With the publication of Formey's book, Maupertuis could finally read Euler's metaphysics, which derived the properties of matter from the law of inertia. He had little sympathy with either of the antagonists: "As for M. Euler and M. Wolff, it will never be possible to reconcile them: but I do not think the reasonings of the first on the essence of bodies are more solid than all that [the Wolffians] say about Monads."[116] But he needed Euler's good will and mathematical expertise, as well as Formey's industrious application to the many tasks required of the secretary, and therefore remained aloof from the controversy. To dilute the hostility between Euler and Formey, he appointed extra members to the prize commission, with academicians from all classes and an honorary member, Count Dohna, as chair. Even so, accusations of manipulations and unfairness flew back and forth, behind the scenes.[117] Maupertuis chastised Formey for his role in the dispute. "I am pained to see the dissensions that are dividing the Academy and the animosity characterizing an affair where no one should take any other interest that that of the truth."[118] Several members wondered whether it might not be better to avoid the recriminations that would follow on any decision, by splitting the prize between essays taking opposite positions. (This was standard procedure in judging controversial competitions in Paris.)[119] Maupertuis, however, decided that the commission should name a single winner, although there is no evidence that he pushed for any one contender in particular. In the end, the prize was given to a German piece by an avid opponent of monads, Johann Heinrich Justi, a relatively unknown jurist and administrator from Thuringia.[120] The overwhelming interest in the outcome was reflected at the public meeting where the prize was to be an-

116. M to JBII, 18 September 1747, BEB.

117. Maupertuis named seven prize commissioners in April 1747: Comte de Dohna (honorary member of the Academy), Eller, Heinius, Euler, Sack, Kies, and Formey (Winter, *Registres,* 110); see also Harnack, *Geschichte,* 1(1):402–03. Formey was accused of irregularities of various sorts, probably by Euler, and he protested in letters to Maupertuis: 27 May, 29 May and 7 July 1747, BJ; see also M to Formey, 31 May 1747, SBB. For last minute negotiations and accusations, see Dohna to M, 27 May 1747, AS Fonds Maupertuis.

118. M. to Formey, 31 May 1747, SBB.

119. Euler reported on discussion of this option in Euler to M, 16 May 1747, in Euler, *Opera Omnia,* series 4A, 6:76. The other suggestion was to defer the prize to the following year (as was often done when no suitable entries were received) and change the question. "Ce sont donc des propositions préliminaires, sur lesquelles il faut attendre Votre decision, avant que nous puissions opiner en forme" (ibid.). For Euler's assessment of twenty-five of the submissions, see Euler, "Différentes pièces sur les monades," in *Opera Omnia,* series 3, 2:416–29.

120. Euler claimed that the actual vote of the commission was five to two in favor of Justi. If this is accurate, presumably Kies, Sack, Eller, and Dohna sided with Euler, and Heinius joined Formey in oppostion. Thus, the two members of the metaphysics class were outvoted.

nounced, in the unusual presence of the king's sister and three brothers, as well as numerous other nobles and illustrious visitors.[121]

"Our judges say it is excellent," Maupertuis reported to Bernoulli; he himself did not read it until months later, when it was translated for publication.[122] The decision laid the Academy wide open to charges of collusion with Justi, given Euler's declaration in print of his own position, although in fact Euler did not have any contact with Justi before the competition. The essay simply "seemed the best among all those that were against monads," as Euler later admitted. "I always declared to those [on the commission] opposed to monads that if they found it appropriate to give their votes to some other piece against monads, I would join my vote to theirs."[123] To defuse some of the controversy, Maupertuis decided that the Academy should publish the winning essay, with a French translation, along with six others on both sides of the question.[124] "Some of our judges were very much afraid of having M. Wolff or his gang on their heels: and that could well happen. But we will still have half a dozen of the best pieces that could be written on this subject, for and against." Once he had actually read Justi's piece, he was "a bit surprised at the lavish praise I had heard of it: it contained many rude expressions against M. Wolff which I made him cut out, from both the original and the translation."

The debate continued in print even before the official publication of the essays, with polemical pieces on both sides throughout 1748. Justi broke with academic convention by printing his piece in Leipzig immediately after the judgment, with-

121. Winter, *Registres,* 112. Justi later became known for works on finance, economics, and constitutional history. See Bongie, "Introduction to *Les Monades,*" 31.

122. Euler forwarded the translation to Maupertuis four months after the prize was announced. Sack had volunteered to do the translation; by the time he finished, "il ne s'engageroit plus pour tout le royaume de France" (Euler to M, 9 October 1747, in Euler, *Opera Omnia,* series 4A, 6:85).

123. Euler to M, 21 September 1751, in Euler, *Opera Omnia,* series 4A, 6:187. Wolff referred to Justi as Euler's "tool [*Werkzeug*]" (Wolff to Schumacher, 6 May 1748, excerpted in Harnack, *Geschichte,* 2:310–11).

124. *Dissertation qui a remporté le prix proposé par l'Académie royale des sciences et belles lettres sur le systeme des monades avec les pieces qui ont concouru* (Berlin: Haude and Spener, 1748). This contains Justi's essay in the original German, followed by a French translation. For discussion of the runners up, which include pieces by Samuel König and Condillac, see Bongie, "Introduction to *Les Monades.*" (The identities of these authors were not known in the eighteenth century). Maupertuis commented most favorably on the piece by Condillac, "une [piece] Françoise qui n'est ny pour ny contre qui me paroit belle" (M to JBII, 21 November 1747, BEB). Euler worked on coordinating the volume, overseeing Sack's translation of Justi and working with the printer on corrections. Maupertuis reviewed the sheets as they came from the press (Euler to M, 16 September, 30 September, 9 October, 13 October, 11 November, 18 November, 25 November 1747, in Euler, *Opera Omnia,* series 4A, 6:81–93).

out the Academy's permission and without deleting his offensive comments on Wolff.[125] Hardly anyone admired Justi's essay. Samuel König, mathematics and philosophy professor in Franeker, and an admirer of Leibniz, wrote to Formey bemoaning the poor judgment of the commission. He did not mention that he had also entered the competition, and thus had a personal interest in the decision.[126] The controversy played out as the kind of pamphlet war familiar to European readers, with anonymous pieces provoking gossip, rumors, refutations, and reviews in journals. "I have just read," Euler complained to Maupertuis in September 1747, "a malicious refutation of this victorious piece printed in German in Leipzig, in which M. Formey seems to me to have had a great part, seeing that it contains several particulars that took place here in this matter."[127] The anonymous pamphlet may well have been written by Gottsched. According to Euler, it accused the Academy of engaging in "intrigues to suppress the truth" and failing in its mandate to judge the entries impartially. The "intrigues" had to do with Euler's own pamphlet, which had had the desired effect of discouraging partisans of monads to write for the competition. "It is up to you," Euler wrote to Maupertuis, "to decide if the sense [of the refutation of Justi] is injurious to the Academy or not? But I fear we will soon see more impudent [*grossieres*] pieces."[128] Impudence originated from both sides, however, since Justi had ignored academic protocol and Euler himself had sparked the accusation of prejudice with his clandestine publication before the prize essays had been submitted.

Throughout the controversy, Maupertuis acted more as a moderator than a combatant. A few years later, in a confrontation with his old acquaintance König, Maupertuis did respond to what he defined as an affront to the Academy's honor. At this point, however, in his first year as president of a Prussian institution, he diplomatically let the pamphlet war run its course.[129] He was concerned to pro-

---

For arguments among the commissioners about which pieces to publish, see Dohna to M, 28 June [1748], AS Fonds Maupertuis. For the prize medal awarded to Justi, see Fig. 21 above.

125. Wolff to M, July 18 1747, LS 433–34. See also Clark, "Death of Metaphysics," 441–42.

126. König to Formey, 6 December 1748, AS Dossier König. On the attribution of entry X to König , see Bongie, "Introduction to *Les Monades*," 28–29. It was included in the volume of prize essays published in 1748.

127. Euler to M, 30 September 1747, in Euler, *Opera Omnia*, series 4A, 6:84. I have not found the pamphlet he describes.

128. Euler to M, 9 October 1747, in Euler, *Opera Omnia*, series 4A, 6:85–86.

129. Beeson exaggerates Maupertuis's anti-Leibnizian views (*Maupertuis*, 189). Maupertuis was also called upon to mediate between Euler and d'Alembert over another prize question in 1751, d'Alembert having claimed that Euler's prejudice against him had cost him the prize. Euler once again asserted that "l'honneur de l'Académie est fort interessée dans les crimes qu'il m'impute" (Euler to M, 21 September 1751, in Euler, *Opera Omnia*, series 4A, 6:187–88).

ject an image of the Academy as a forum for civilized exchange, rather than acrimony. As it turned out, the proceedings nourished the hostility of Wolff and his followers, exacerbating the isolation of the institution from its surroundings. The Wolffians looked at Berlin and saw foreign mathematicians overreaching the bounds of their competence to dictate to German metaphysicians. Wolff (and presumably others) felt the monad competition had been hijacked by Euler. "You yourself must fully understand that, contrary to your praiseworthy intention, the question you proposed was derailed, with the goal of giving the prize to the essay that would combat me, by throwing sand into the eyes with its ignorant subtleties."[130] From Maupertuis's point of view, the university philosophers formed a defensive "sect" around their master and his dogmatic system. He attempted to mollify Wolff by asking him to send something for the Academy to print in its *Mémoires,* but Wolff would have none of it.[131]

## Metaphysics in Berlin

Although it had much in common with the Paris Academy of Sciences, only half the members of the Berlin Academy worked in areas defined as "sciences" in France. The literature class corresponded roughly to the Paris Academy of Inscriptions and Belles-Lettres, but the speculative philosophy class had no analogue in French academies. Before Maupertuis took over, the Berlin philosophers had proclaimed metaphysics "the mother of the other sciences"—not at all incompatible with the enlightened use of reason—that "demands minds unfettered by the shackles of a certain superstitious respect, which reigns in many countries, where the limits of reason and faith have not been adequately defined."[132] Maupertuis subsequently gave a broader definition of the subjects appropriate to the philosophical class: "The Supreme Being, the human mind, and all that pertains to mind and spirit. . . . The nature of bodies also falls within its competence."[133] He redrew traditional institutional and disciplinary boundaries; the nature of bodies and the nature of spirit and God all fall into the same category.[134] This

130. Wolff to M, 1 July 1747, LS 432. Wolff had a slightly higher opinion of Maupertuis than of Euler, but regretted that the Frenchman did not understand the concerns of Germans. See Wolff to Schumacher, 6 May 1748, in Harnack, *Geschichte,* 2:310–11.

131. Maupertuis's letter to Wolff does not survive; Wolff's response refers to this offer (Wolff to M, 1 July 1747, in LS, 431). Not all university philosophers were sympathetic to Wolff, especially in Göttingen. See Clark, "Death of Metaphysics."

132. "Préface," *BAS* 1745, n.p.

133. Maupertuis, "Des devoirs de l'académicien" (1750), reprinted in *Oeuvres,* 3:293.

134. The commission proposal of 1743, codified in the 1744 statutes, had reserved for the universities all matters pertaining to theology, limiting academic philosophy to "metaphysics, morals, natural law, critical history of philosophy" (AdW I.I.5, fol. 125).

challenge to the universities' old monopoly on theology was possible in Berlin, partly because there was no entrenched religious institution comparable to the Sorbonne in Paris and partly because of Frederick's state policy of tolerance. Maupertuis himself turned increasingly to writing for the Academy on an eclectic mix of metaphysical topics—God, the nature of matter, epistemology—in an effort to demonstrate the viability of academic metaphysics.

The problem of turning this general program for an enlightened metaphysics into reality plagued Maupertuis throughout his presidency. If the Academy were going to pursue metaphysics at all, he was concerned that it make a respectable showing to the rest of Europe: "[I] would just like [the Academy] to present a good image to the world."[135] In the early years, he tried repeatedly to get Johann II Bernoulli to come to Berlin to direct the speculative philosophy class and later encouraged both Johann and his brother Daniel to write for the Academy: "It's a service both of you owe to a man who loves you and who deserves it. . . . It is above all for our speculative philosophy [class] that I implore you. The two of you are the only ones capable of putting our Academy above all the academies of Europe."[136] When Bernoulli declined, Maupertuis continued to rely on his recommendations of Swiss protégés. In recruiting potential metaphysicians, Maupertuis tried to dispel the notion that a German academy must necessarily promote "German" metaphysics. He instructed Bernoulli to inform one such prospect that "he shall find in our Academy no constraint to be a Wolffian and that there is no academy in the world where there is such freedom to think, to speak, and to write."[137]

In the aftermath of the monad competition, Maupertuis was quite conscious of the middle ground he wanted his academy to occupy, between the German dogmatism of Wolffian philosophy and the French neglect of the subject. "It is true," he wrote to Bernoulli, "that German metaphysics is a strange science, but that is not the fault of metaphysics, but rather of the Germans. . . . The French are too disgusted with metaphysics; the Germans are too mired down in the mud. Perhaps the Swiss can find a viable middle ground."[138] Some mathematical training and a good measure of *esprit* would be required of any potential metaphysician; beyond that, Maupertuis did not know what to look for. And while he did import several Swiss philosophers, he was never completely satisfied with them. The first of

135. "[Je] voudrois seulement quelle jouast dans le monde un bon personage" (M to JBII, 18 September 1747, BEB).

136. "C'est surtout pour notre philosophie speculative que je vous implore. Vous etes tous deux seuls capables de mettre notre Académie au dessus de toutes les académies de l'Europe" (M to JBII, [summer 1749,] BEB).

137. M to JBII, 6 September 1749, BEB.

138. M to JBII, 18 September 1747, BEB.

Bernoulli's recommendations, Daniel Passavant, turned out to be "the laziest and the most dissipated of men." Lacking the initiative to work on his own, he was assigned various translation tasks, which he undertook with great reluctance.[139] Reinhard Battier, also Swiss, left the Academy for a post as tutor to the son of the duke of Gotha, after an undistinguished period in Berlin. "Will we never have any but mediocre people?" the affronted president complained. "Will we never manage to have a single great academician [*sujet*]?"[140] Finances continued to restrict his ability to offer pensions to worthy subjects, in spite of his "quite despotic power" to give out positions.[141] Most of the pension funds were already committed to appointments predating Maupertuis's reign, and though he could have reorganized these awards, he was reluctant to do so. His greatest frustration came from those who collected their pensions without doing anything for the Academy: "Experience teaches me that there is nothing so prejudicial to the progress of the sciences in Germany as giving them too much to eat and drink. There is no one who will not abandon Homer when he hears the dinner bell."[142] For the future, he decided to offer small amounts with the understanding that these could increase if merited: "I have been so entrapped by the savants of Germany that I wish henceforth to be more cautious with the pensions and to give gratifications in proportion to work done."[143]

He did have some success in building up the Academy, although only one of Bernoulli's metaphysicians fulfilled his promise to any significant degree. This was Jean Bertrand Merian, who proved assiduous and loyal, even if not a highly original thinker. In other fields, after Albrecht von Haller turned down a generous offer, one of his students, Johann Friedrich Meckel, came to Berlin to revive the practice of anatomy.[144] Astronomy was improved by cultivating connections to Paris. Grischow was sent off for two years for training at the Paris Observatory;

139. "Je l'ay chargé de traduire en Alleman le traitté du Nivellement de M. Picart, que le Roy a souhaitt qui fust traduit. J'ay cru rendre service à Passavant que de le charger de cet ordre, mais je ne scay pa quand le livre paroitra, et sil tarde, il se fera oter la commission" (M to JBII, [1749], BEB).

140. Ibid.

141. M to JBII, 29 October 1747, BEB.

142. "[L]'experience m'apprend quil n'y a rien de si prejudiciable au progrés des sciences en Allemagne que de leur donner trop bien de quoy boire et manger: il ny a personne qui ne quitte Homere lorsquil entend sonner lheure du Marché" (ibid.).

143. "[J]'ai tant eté attrappé par les scavants d'Allemagne, que je veux desormais etre plus reservé pour les pensions, et donner des gratifications a proportion du travail" (M to JBII, [summer 1749], BEB).

144. The anatomists were even more recalcitrant about their academic duties than the philosophers. See M to FII, [September 1749], Koser, *Briefwechsel*, 242–43. Haller's refusal was a bitter disappointment to Maupertuis. For their negotiations, see exchange of letters in LS 180–96.

somewhat later Jérôme Lalande spent a period in Berlin making observations at the behest of the Paris Academy. But disappointments abounded as well. German candidates in particular were reluctant to give up positions elsewhere for the "despotic" atmosphere of Berlin. One, Johann Theophil Waltz, agreed to move from Leipzig, but died unexpectedly before he could get to Berlin.[145] Negotiations with Abraham Kästner were also inconclusive. In 1749, Kästner had written in the most flattering terms about the inspiration he had found in Maupertuis's paper on the principle of least action. Here was someone who belonged in Berlin. He was elected foreign member immediately, and Maupertuis lobbied hard to get him to leave Leipzig, where he held a professorship in mathematics, but to no avail.[146] Kästner, almost alone among German philosophers and mathematicians, remained on good terms with Maupertuis, serving as his eyes and ears in Leipzig.[147]

But Maupertuis could not help wishing that his academy could attract men of science, especially his old friends, the way Frederick's court drew fugitives fleeing the wrath of the Sorbonne and the Paris Parlement. Commenting on Buffon's troubles with the religious authorities in Paris, he wrote to the duchesse de Chaulnes, "I wish that the persecution would go so far as to make him come here, and that we would have a colony of French refugees for Reformed Philosophy."[148] This was only partly facetious; when it looked like the publication of Diderot and d'Alembert's *Encyclopédie* would be suspended in 1752, after being denounced as a threat to religion and morality, Maupertuis dreamed of taking advantage of French intolerance: "It is certain that the *Encyclopedia* is going to the devil, or at least is receiving a terrible setback. . . . It would be a great triumph to attract here [to Berlin] all the Encyclopedic group with their booksellers and their printing presses, and to have the work continued here. Such a colony of refugees of reformed philosophy would be more useful that that of reformed religion."[149] In the aftermath of this

145. M to JBII, 18 September 1747; for negotiations with Waltz, see M to FII, 5 July 1747, Koser, *Briefwechsel,* 226; Euler to M, 14 March 1746, in Euler, *Opera Omnia,* series 4A, 6:59.

146. Kästner had used Maupertuis's notion of dynamics, and especially its application to the collisions of hard bodies, to reexamine the law of continuity. This work became the basis for much of his mathematical physics. See Kästner to M., 12 November 1749, LS, 274; Clark, "Death of Metaphysics," 446. He was elected to the Academy in February 1750 (Winter, *Registres,* 147). See also Kästner to M, 17 February 1750, LS, 275–77.

147. Kästner also won the Berlin Academy's prize on moral philosophy in 1751. Winter, *Registres,* 162.

148. "Je voudrois que la persecution allât jusqu'a le faire venir icy, et que nous eussions une colonie de François réfugiés pour la Philosophie Reformée" (M to duchesse de Chaulnes, 9 February 1752, SM). Buffon had aroused the censure of the Sorbonne with his theory of the formation of the earth in his *Histoire naturelle.* See Roger, *Buffon,* 185–89.

149. "Ce qu'il y a de seur c'est que l'Encyclopédie est au diable ou du moins recoit un terrible echec. . . . Ce seroit un beau coup de fil que d'attirer icy toute la societé Encyclopédique avec leurs

show of power by the Sorbonne and the Jesuits, the Abbé de Prades came to Berlin as the king's reader, but the rest of the fantasy did not come to pass, and the *Encyclopédie* eventually resumed production in Paris.[150]

In spite of the brilliance of Frederick's suppers, and his continual, almost desperate, efforts to attract the best French and Italian musicians, actors and poets to his court, an undercurrent of dissatisfaction periodically rose to the surface. The king's circle of wits and artists was not stable or loyal enough, or even brilliant enough to amuse him indefinitely. He was constantly made aware of the difficulty of attracting "the best" to Berlin, and some of his most prized acquisitions did not stay long. Maupertuis was similarly disappointed in the failure of his nominally unlimited power to draw accomplished men of science and letters to his domain. In spite of the reassuring intimacy with royalty, which he truly relished, the limitations of his academicians (with the crucial exception of Euler), his distance from his childhood home and his family, his wavering health, and perhaps even his failure to attract disciples, all conspired to leave an aftertaste of disappointment, of greater or lesser intensity depending on circumstances. Doubtless nothing could have fully satisfied his excessive vanity, and his tendency to melancholy affected him wherever he was. In spite of all this, he worked with great energy to play the various roles he had taken on.

---

libraires et leurs presses: et de faire continuer cet ouvrage icy. Une telle colonie de Refugiés de la philosophie reformée seroit plus utile que celle de la Religion reformée" (M to Algarotti, "Mercredi 15, [1752]," in Algarotti, *Opere,* 16:226–27, original in BJ). The published version is not complete.

150. On the publication history of the *Encyclopédie,* see Darnton, *Business of Enlightenment;* Lough, *Essays on the Encyclopédie.*

# 9

# Teleology, Cosmology, and Least Action

ONE OF THE DISTINCTIVE FEATURES of the Berlin Academy of Sciences was its metaphysics class. Although the occupants of that class in 1745 did not impress him, Maupertuis hoped that it might provide a platform for pursuing his own program of metaphysical mechanics. In 1744, in Paris, he had already laid the groundwork for this mechanics in his paper on the refraction of light. He planned to continue in this direction and in early 1746 was at work on a new manuscript that combined a critique of the argument from design with the solution of simple impact problems, using the principle of least action. He attached a great deal of significance to this paper, his debut performance in Berlin as man of science rather than president or rhetorician. It staked out his views on metaphysics and linked them to what he considered a universal law of mechanics. Brazenly advertising the project with the expansive title, "The laws of motion and rest deduced from the attributes of God," Maupertuis also aspired to set a new standard for the metaphysics class.[1]

In this paper, Maupertuis applied the principle of least action to a simple problem of the motion of two bodies colliding on a plane. Before presenting it publicly, he sent the mathematical section to Euler for his corrections and comments. "A man like yourself discourages me from geometry, and ought to discourage many others; but I would like to make use of this little piece in a dissertation on final causes and on the abuse that some physicists [*physiciens*] have made of them."[2] Euler was to be an essential ally in the years to come, sharing a commitment to

1. Presented on 6 October 1746 (Winter, *Registres,* 43). The published title changed to "Les lois du mouvement et du repos déduites d'un principe métaphysique," *BAS* 1746, 267–94. The last section of the paper was reprinted as "Recherche des lois du mouvement" in Maupertuis, *Oeuvres,* 4:29–42; the first two sections became the basis for *Essai de cosmologie* (1750).

2. "Un home tel que vous me decourage de la geometrie et devroit en decourager bien d'autres; mais je voudrais faire usage de ce petit morceau dans une dissertation sur les causes finales et sur l'abus que quleques phisiciens en ont fait. Je vous avoue que l'idee que je propose m'a plu; cependant je ny auray aucune confiance, jusqu'a ce que vous l'ayés examinée. Dittes moy si elle meritte d'etre admise, et si vous etes content de la forme; ou si elle doit etre rejettée? Vous etes le juge souverain de ces matieres; et celuy a qui je me soumets d'esprit et de coeur" (M to Euler, 22 May [1746], AS Fonds Maupertuis). The editors of the Euler-Maupertuis correspondence did not know this letter,

final causes and the principle of least action. His outstanding mathematical abilities and insights, which Maupertuis acknowledged as far surpassing his own, lent a credibility to the principle of least action that Maupertuis alone could not give it. Euler appreciated his older colleague's rather general insights and generously corrected and improved on his mathematics where possible. He apparently had no interest in competing with Maupertuis.[3] Since Euler, quite independently, had applied an extremum principle to the case of central forces very shortly after Maupertuis had presented the principle of least action to the Academy in Paris, they might have regarded each other with hostility, but Euler always most graciously ceded priority.[4]

Euler's reputation as the best mathematician in Europe made him a vital resource for the promotion of metaphysical mechanics. The letter Maupertuis sent with his manuscript showed a characteristic mixture of deference and pride. He asked his colleague to "correct whatever mistakes there might be, and to write your comments in the margins." He particularly wanted an assessment of his way of representing the change in velocity of colliding bodies by imaginary "immaterial planes" carrying the bodies in opposite directions at a rate equal to the loss (or gain) in speed. This somewhat cumbersome heuristic device made the problem of instantaneous change in motion into a global problem, extending to the time before and after the collision, and this helped in formulating the extremum analysis. "I admit that the idea I am proposing pleased me; nevertheless I would have no confidence in it until you have examined it. Tell me if it deserves to be used, if you are content with the form; or if it should be rejected."[5] In his response, Euler noted

acquired by the AS from a private collection in 1990. See Euler, *Opera Omnia,* series 4A, 6:65. Most of Maupertuis's letters to Euler from this period do not survive. Euler mentioned "giving lessons" to Maupertuis in the summer of 1746 (Euler to Delisle, 16 August 1746, ms. Obs. Delisle papers).

3. Euler read Maupertuis's "Loi du repos" when Maupertuis gave it to him in 1745 (Euler to M, 10 December 1745, in Euler, *Opera Omnia,* series 4A, 6:56). Euler commented, "I believe your principle to be more general than you propose," and went on to show how his own results for elastic curves followed from Maupertuis's "law of rest" (ibid.). At this point, he had not seen Maupertuis's refraction paper with the explicit formulation of the principle of least action.

4. Euler, *Methodus inveniendi.* Maupertuis was concerned to verify his own priority, once he knew that Euler had applied his new variational methods to dynamics. Without telling Euler about his 1744 Paris paper, Maupertuis inquired as to the exact date of the composition of the *Methodus.* Euler replied that it had been completed in 1743 in Saint Petersburg, but that the appendices were only written in autumn 1744 (Euler to M, 14 March 1746, in *Opera Omnia,* series 4A, 6:59). In June 1747, Maupertuis asked Bernoulli to find out the exact date Euler sent the appendices to his Swiss printer; he wanted to be sure that his own 1744 paper had predated it (M to JBII, 12 June 1747, BEB). For Euler's work on central forces and its relevance to the principle of least action, see Costabel, "Introduction," Euler, *Opera Omnia,* series 4A, 6:13–16.

5. M to Euler, 22 May [1746], AS Fonds Maupertuis.

diplomatically that the immaterial planes were not technically wrong, but he considered them unnecessary; in spite of this advice, Maupertuis proved reluctant to give up his heuristic device, and took Euler's remark as approval for its validity. On the other hand, Euler also remarked that there was no need to worry about the force responsible for the change in motion, a correction that Maupertuis accepted (as evidenced by the published paper).[6] He apparently consulted Euler again when preparing the manuscript for publication; in a letter from December 1746, Euler conceded that the idea of using immaterial planes "is perfectly correct, and can still be used without risking anything."[7] Although he had tried to dissuade Maupertuis from this idea, he recognized Maupertuis's attachment to it, and reassured him that it did not detract from the argument. And indeed the immaterial planes appeared in the published version.

Maupertuis formulated the principle of least action as a mathematical version of the metaphysical principle that nature acts as simply as possible: "Whenever there is any change in nature, the quantity of action necessary for that change is the smallest possible."[8] That is, all motions (or equilibrium states) conform to an extremum condition whereby the quantity "action" is minimized. In his Paris paper on refraction (1744), he defined action as proportional to the product of speed and distance traveled, and showed how the extremum approach brought the rules describing the motion of light under one general principle, implying that this principle could provide the key to mechanics as well. Addressing his new colleagues in Berlin, he solved simple problems of elastic and inelastic collisions, using the same technique and the same principle, at the end of a paper otherwise concerned with metaphysics and natural theology. Instead of tacking metaphysical reflections onto mathematical arguments, as he had done previously, he made the metaphysics the core of the paper. He saw this work as the direct outgrowth of his earlier papers on statics and refraction, now more fully integrated into the framework that tied extremum mechanics to a mathematically based proof for the existence of God. This was to be his last mathematical paper; he reprinted it and defended it repeatedly, but at the age of forty-eight, in precarious health, he no longer aspired to original mathematical work. In the search for generality, he had shifted his attention from mathematics to metaphysics, moving further away from the work he had undertaken under the watchful eye of Johann Bernoulli earlier in his career. His alliance with Euler, solidified as soon as he arrived in Berlin, meant that this

6. Euler to M, 24 May 1746, in Euler, *Opera Omnia,* series 4A, 6:63–64.

7. Euler to M, 28 December 1746, in Euler, *Opera Omnia,* series 4A, 6:69–70. Maupertuis's letter does not survive.

8. "Lorsqu'il arrive quelque changement dans la Nature, la Quantité d'action nécessaire pour ce changement, est la plus petite qu'il soit possible" (*BAS* 1746, 290; also in Maupertuis, *Oeuvres,* 4:36).

most accomplished mathematician could take over whenever Maupertuis sensed his own limitations. Euler did in fact write a series of papers developing the principle of least action in ways that Maupertuis could never have done.

Maupertuis tailored the paper on the laws of motion, which he described as "a dissertation on final causes," for the Berlin Academy class of speculative philosophy. It began with a critique of the argument from design, or "the proofs of God's existence drawn from the marvels of nature."[9] This first section openly attacked Newton's use of empirical facts as evidence for the existence of an all-powerful God. Newton had argued that blind chance could not have produced the ordered regularity of the solar system, for example. But Maupertuis noted that this order only seems surprising if we have no explanation for its origin; if we accept an ethereal fluid carrying the planets in their orbits, the fact that they move very nearly in the same plane no longer seems an adequate proof of divine wisdom. He did not advocate this physical explanation; it merely served to show that there is no necessary connection between the planetary orbits and God's wisdom. The Newtonians also argued from the fitness of animals' parts to their needs: how could these perfectly designed organs have come about at random? Here again, Maupertuis countered with an alternative scenario, developed more fully in his later work on life science:

> But could we not say that in the chance combination of Nature's productions, it is no wonder that we find fitness in all the species actually in existence, since only those with certain appropriate proportions [*rapports de convenance*] could survive? Chance, we would say, produced an innumerable multitude of Individuals: a small number found themselves constructed such that the Animal's parts could satisfy its needs; in an infinitely greater number, there were neither fitness nor order. All these latter perished: animals without mouths could not live; others lacking organs of generation could not perpetuate themselves; and the only ones left were those with order and fitness. These kinds that we see today are only the smallest part of what a blind Destiny had produced.[10]

Naturalists had fallen into the same trap as the Newtonians, he argued, seeing divine purpose everywhere without noticing the coexistence of contradictory

9. Ibid. By the 1740s, several works of natural theology had been successful publishing ventures, both in England and on the Continent. Prominent Continental examples include Pluche, *Spectacle de la Nature,* 8 vols. (Paris, 1732–1750); Friedrich Lesser, *Théologie des insectes,* 2 vols., trans. P. Lyonet (Paris, 1745). Réaumur's work on insects belongs in the same tradition, although less openly theological: Réaumur, *Mémoires pour servir à l'histoire des insectes.*

10. Maupertuis, "Les loix du mouvement et du repos, déduites d'un principe de métaphysique," *BAS* 1746, 272.

purposes. "Follow the production of a Fly, or an Ant: they make you admire the cares of Providence for the eggs of the insect, for the nourishment of their offspring; for the animal enclosed in the layers of the chrysalis; for the development of its parts in metamorphosis. All of this leads to the production of an insect, harmful to people, which the first bird devours, or which falls into a spider web."[11] The brunt of this objection to natural theology rested on the weakness of the proof, so far from the demonstrative certainty of a geometry. Susceptible to contradiction by counterexamples, resting on details rather than principles, the argument from design demeaned God's wisdom rather than demonstrating it. Although Maupertuis took the Newtonian version of natural theology as canonical, there was also a highly developed tradition in the same direction among German writers, precisely in this period. He referred to only two of many, Friedrich Lesser and Johann Albert Fabricius; he may not have been aware of the extent of the literature in German in this vein.[12] In any case, he was challenging a very broad spectrum of theological writing.

The second section of the paper argued that convincing proofs for God's existence must come instead from the general laws of nature. Such are the laws of motion "founded on the attributes of a supreme Intelligence."[13] Mathematics rather than observation will serve as the touchstone for this proof, giving it "the advantage of the clarity and obviousness [*évidence*] that characterizes mathematical truths."[14] Universal laws, expressed mathematically, are more comprehensible as well as more profound than empirical observations. "The organization of Animals, the multitude and minuteness of the parts of insects, the immensity of the celestial bodies, their distances and their revolutions, are more appropriate to astonish our minds than to enlighten them. The supreme Being is everywhere, but he is not equally visible everywhere. . . . Let us look for him in the first laws he imposed on Nature."[15] Here Maupertuis reminded his reader that enlightenment must come from reason rather than awe. Though physical phenomena are certainly the effects of God's designs, the human mind lacks the scope to grasp intu-

11. Ibid., 274.

12. For German natural theology, with a plethora of examples, see Clark, "Death of Metaphysics in Prussia."

13. Maupertuis, "Loix du mouvement et du repos," *BAS* 1746, 277.

14. Ibid., 290.

15. "L'organisation des Animaux, la multitude et la petitesse des parties des insectes, l'immensité des corps celestes, leurs distances et leurs révolutions, sont plus propres à étonner notre esprit qu'à l'éclairer. L'Etre suprême est partout, mais il n'est pas partout également visible" (ibid., 278–79). Maupertuis took aim at the Newtonian appeal to the "marvels of nature," and not at the law of universal gravitation, which is mathematically precise.

itively the infinite complexity of the connections among them. But mathematics can provide the link from human reason to the divine, by revealing divine purposes on an abstract, metaphysical level.

Maupertuis sought nothing less than a new physics and a newly rigorous theology. "I could have started with the laws [of motion] given by mathematicians and confirmed by experience and looked there for marks of God's wisdom and power. But . . . I thought it more certain and more useful to deduce these laws from the attributes of an all-powerful and all-wise being. If those that I find in this way are the same as those observed in the universe, is that not the strongest proof that that being exists and that he is the author of those laws?"[16] How then can the laws of motion be deduced from God's attributes? First, Maupertuis enunciated a principle embodying divine wisdom and power by virtue of its "fitness." "It was necessary to know that all the laws of motion and rest were founded on the most fitting, appropriate [*convenable*] principle in order to see that they owed their establishment to an all-powerful and all-wise being."[17] This formulation shifted the property of fitness from particular organs or organisms to the principle itself, transposing the argument from design from the phenomenal to the metaphysical realm.

With meticulous attention to rhetorical framing, Maupertuis situated his mechanical principle historically as the enlightened resolution to years of confusion and controversy over conservation, force, and the nature of matter. Least action mechanics would subsume limited conservation laws under a completely general extremum principle. The new principle transcended the historical debates by operating metaphysically as well as physically. So, for example, Leibniz's commitment to the conservation of living force (*vis viva*) had led him to claim that truly inelastic hard bodies do not exist. Maupertuis considered this absurd and followed Newton in declaring the ultimate particles of matter to be infinitely hard and inelastic. This move entailed rejecting another mainstay of the Leibnizian system, the law of continuity, whereby all changes in motion must take place by infinitesimal degrees. According to Leibniz, absolutely hard bodies cannot exist because, lacking all elasticity, they would have to stop instantaneously on encountering each other. But Maupertuis responded that we do not know enough about how motion starts and stops to make such categorical claims. "As soon as we reflect on the impenetrability of bodies, it seems that it cannot be different from their hardness; or that hardness must be a necessary consequence of it. If, in most bodies, the component parts separate or flex, this happens only because these bodies are composed of other bodies. Simple, primitive bodies, which are the

16. Ibid., 279.
17. Ibid., 282.

elements of all others, must be hard, inflexible, inalterable."[18] The other controversy revolved around conservation—whether any quantity is always conserved and if so, what that quantity is. Descartes argued for conservation of quantity of motion ($mv$), Leibniz for conservation of *vis viva* ($mv^2$). But since each of these is conserved only in certain cases, neither qualifies as a universal law. In Maupertuis's retelling of the history of mechanics, Descartes's erroneous conservation law drove him to formulate incorrect laws of motion, while the Leibnizians had gone to the absurd extreme of denying the existence of hard bodies altogether.

According to this analysis, the debates surrounding conservation and the nature of matter demonstrated the need for a new approach to mechanics. "After all the great men who have worked on this matter," Maupertuis remarked disingenuously, "I almost dare not say that I have discovered the universal principle on which all these laws are founded; which extends equally to hard bodies and elastic bodies; on which the motion and rest of all corporeal substances depends." Pushing aside his feigned modesty, he went on to the climactic revelation of a new universal principle: "In the collisions of bodies, motion distributes itself such that the quantity of action, once the change has taken place, is the smallest possible. In a state of rest, bodies in equilibrium must be situated such that if they are given some small motion, the quantity of action will be a minimum."[19]

In Maupertuis's vision of his accomplishment, the principle of least action replaced the fragmentation of the argument from design with a global principle to tie together all parts of the universe and all the phenomena in it, with maximal efficiency. The universe becomes "more worthy of its Author" once we know that a small number of laws suffice for all phenomena. "It is thus that we can have a true idea of the power and the wisdom of the supreme being; and not when we judge him by some little part of which we know neither the construction, nor the use, nor the connection that it has with other parts."[20] Maupertuis initiated his argument with an attack on Newtonian natural theology, and then enumerated the limitations of Cartesian and Leibnizian conservation laws, as well as the absurdity of Leibniz's rejection of hardness as a fundamental property of matter. He presented his own principle as superceding the work of his predecessors and rendering the disputes about conservation irrelevant. He also ignored the work of his contemporary, and former protégé, Jean Le Rond d'Alembert, who had published his own resolution of the same controversy with an explicitly antimetaphysical mechanics just a few years earlier.[21] D'Alembert had no interest in

18. Ibid., 284.
19. Ibid., 286.
20. Ibid., 287.
21. D'Alembert, *Traité de dynamique* (1743). Maupertuis knew d'Alembert's book well. On behalf

resolving theological disputes; Maupertuis offered, without saying so explicitly, a reconciliation of the incompatible positions defended by Leibniz and Samuel Clarke in their famous exchange of letters.[22]

In the final section of his paper, Maupertuis applied the principle of least action to several very simple physical problems. Every motion, every mechanical interaction, he explained, has an action associated with it, defined as the product of mass, velocity, and distance. The treatment paralleled his earlier paper on refraction. The calculation of the action takes into account the starting point, the end point, and the velocity of the body traveling from one to the other. In every collision, there is a quantity of action "produced" for each body—mass multiplied by the change in velocity and the distance traveled ($mvs$). This quantity is actually the *change* in the "actions" of the two bodies as a result of their impact, and it is this change that is minimized. Hard bodies, being indivisible, are perfectly inflexible and retain their shapes, whereas elastic bodies are deformed by collisions and then recover their original shapes. Because they are completely inelastic, colliding hard bodies either come to rest or move off together with equal velocities. Perfectly elastic bodies, on the other hand, bounce off each other, such that their relative velocity remains the same before and after the collision. For each type of collision, the paper gives a simple problem, minimizes the quantity of action, and shows that the resulting motions satisfy the known laws of motion.

The last problem was "to find the law of rest of bodies." Without referring to his earlier equilibrium law, Maupertuis derived an expression for the fulcrum point of a lever. He analyzed the problem solely in terms of the virtual velocities of weights on a balance, so that he could use his formula ($mvs$, or $mv(ds)$) for the quantity of action. The earlier treatment of equilibrium, which involved forces attracting masses, transmuted into a problem of virtual motions. The two can be shown to be equivalent—indeed, Maupertuis had solved the same simple balance problem in terms of forces in 1740—but significantly, he did not do so here. Within the rhetoric of the principle of least action, he did not need to discuss forces, as Euler had pointed out. The universality of the principle rested on the

---

of the Academy of Sciences, Maupertuis certified the *Traité* as worthy of publication, commenting on its "clarity" and "profundity" (Maupertuis and Nicole, Report on d'Alembert, *Traité de dynamique*, 22 June 1743, BJ, Autografen Sammlung von Maupertuis). On d'Alembert's mechanics, see Fraser, "D'Alembert's Principle"; Hankins, *Jean d'Alembert;* Casini, "D'Alembert, l'économie des principes et la métaphysique des science." For a comparison of the metaphysics of Maupertuis and d'Alembert, see Terrall, "Culture of Science in Frederick the Great's Berlin."

22. Samuel Clarke published a bilingual English-French edition of his exchange with Leibniz in 1717, after Leibniz's death: *A Collection of Papers which Passed between the late Learned Mr. Leibnitz and Dr. Clarke . . .*, London, 1717. The central disagreement was over the relative weight given to God's power and wisdom, each side accusing the other of denigrating one attribute or the other.

identity of the mathematical expression of the quantity of action for all situations, including equilibrium. "The simplest law of nature, that of rest or equilibrium, . . . has not until now seemed to have any connection with the laws of motion."[23] Only through such universality can the principle of least action be elevated above disputes about hard and elastic bodies, conservation of *vis viva,* and the laws of impact and static equilibrium.[24] It worked for hard and elastic bodies, for motion and rest, just as it worked for the refraction and reflection of light: what could be more general or more unifying?

Although it was developed independently, Maupertuis's approach was crucially reinforced by Euler's mathematically sophisticated work on extremum problems. Euler did not use metaphysics to make theological arguments, but his first work on extremum mechanics shows that his commitment to the metaphysical ramifications of this kind of physics predated his contact with Maupertuis. This predilection surely contributed to his willingness to act in his president's defense when he came under attack. Euler had presented a general theory of the variational calculus in 1744 in a treatise on isoperimetric problems.[25] In two appendices to this work, he applied the method to the motion of bodies subject to central forces and to curves formed by elastic strings. In the opening paragraphs of each of these appendices, Euler took the existence of extrema in nature as axiomatic. "Since all natural effects follow some law of maximum or minimum, there is no doubt that, in the curves described by projectiles attracted by forces, there exists some property of maximum or minimum. . . . [N]othing whatsoever takes place in the universe in which some relation of maximum and minimum does not appear."[26] Whereas questions of causality do not arise in pure mathematics, the application of this mathematics to the physical world leads to both ontological and methodological claims about causality. *Because* maxima and minima are embedded in the natural world, "there is absolutely no doubt that every effect in the universe can be explained as satisfactorily from final causes, by the aid of the method of maxima and minima, as it can from the efficient causes themselves."[27]

In the late 1740s, Euler wrote a series of papers applying his variational method to a variety of problems. These papers were far more sophisticated than Mauper-

23. Maupertuis, "Lois du mouvement et du repos," *BAS* 1746, 282.

24. Euler took the opposite approach, stating the principle of least action in terms of "the action of the forces," and analyzing both statical and dynamical problems from that perspective.

25. Euler, *Methodus inveniendi.* Maupertuis had discussed this book with Daniel Bernoulli in 1744; he asked Johann II to send it to him in Freiberg (M to JBII, 4 October 1744, BEB).

26. Quotations from Euler, "De motu," cited in French translation by Brunet, *Etude historique,* 49; and Euler, "De curvis elasticis," 76.

27. Euler, "De curvis elasticis," 77.

tuis's sketchy treatment of simple collisions and showed how fruitful an extremum approach could be. Instead of analyzing a mechanics problem reductively in terms of forces acting at each instant, Euler looked for an expression that is a maximum or minimum for a whole path or a whole system of bodies and then found solutions to satisfy this extremum condition. The condition applies to the whole path, or the whole configuration, rather than to each instant or position, so it is global and synthetic rather than analytic. Euler showed that any problem can be viewed from either a reductionist or a global perspective, using different mathematical methods. This "double method" reflected the two kinds of causes operating in physics. The direct method gave the effects of forces (efficient causes), and the indirect method used the calculus of variations to solve extremum problems (final causes).[28] The mechanics of least action entailed claims about how the world works and about how the philosopher should understand the world. Maupertuis and Euler agreed that mechanics should make use of the double method. In principle, according to Maupertuis, we should be able to understand God's designs by investigating mechanics, and vice versa. In practice, given the limitations of human understanding, physics must follow both paths. Mechanics alone doesn't take us far enough, while teleology alone can be misleading, since "we can be mistaken about the quantity that we should regard as nature's expenditure in the production of her effects. . . . Let us calculate the motions of bodies, but let us also consult the designs of the Intelligence who causes them to move."[29] Euler put this metaphysical commitment into practice.

## Essay on Cosmology

Maupertuis regarded the principle of least action, and its application to rationalist theology, as the capstone of his scientific career. In the decade after he presented it to the Berlin Academy, he devoted a great deal of effort to promoting it, defending it from criticism, and reworking the text of his paper for a series of published versions. In this same period, he was working to build up the reputation of the Academy. These two projects—the principle of least action and the Academy—went hand in hand. Maupertuis imagined the Academy, with its space for both mathematics and metaphysics, as an appropriate institutional base for his version of metaphysically grounded mechanics. Euler's willingness to lend his

28. Euler, "Recherches sur les plus grands et les plus petits"; "Réflexions sur quelques loix générales de la nature"; "Harmonie entre les principes généraux de repos et de mouvement de M. de Maupertuis"; "Sur le principe de la moindre action." All these papers are reprinted in Euler, *Omnia Opera*, series 2, vol 5. On Euler's work on least action, see Radelet-de-Grave, "La 'Diatribe du Doctor Akakia,'" 239–45; Brunet, *Etude historique*.

29. Maupertuis, "Accord de différentes lois," in *Oeuvres*, 4:22.

prodigious talents and energies to both of Maupertuis's priorities made them seem viable, at least for a time. In the long run, though, their ambitious program had only limited success, sustaining a series of attacks directed against the institution as well as its president. The fate of the principle of least action led to Maupertuis's ultimate disillusionment with the Prussian academy and with the quest for fame and glory.

By the late 1740s, Maupertuis's discovery (as he described it) was still only know to the specialized audience in and around the academies. In taking it beyond this audience, he followed a different route from the one he had followed in earlier works like *Letter on the Comet* and *Vénus physique,* when he published anonymously and played with literary style. In the spring of 1749, he asked Johann Bernoulli to supervise the printing of "a little geometrico-metaphysical work" in Basel, because the printers in Berlin could not produce the quality desired. He was to spare no expense, drawing whatever was required from Maupertuis's banker in Paris.[30] This book was the first edition of the *Essai de cosmologie,* an expanded version of the Berlin Academy paper on least action, the laws of motion, and God. The level of attention to aesthetic and stylistic details in his instructions to Bernoulli confirm Maupertuis's attachment not just to the text, but to his plan for producing a certain kind of book. Their correspondence over the next six months referred over and over to the design, correction, revision, and production of the book. Thanks to these letters, we can see that he wanted this particular edition of this text for a particular, and restricted, set of readers.

> I want only 100 copies on very beautiful paper in octavo, with very wide margins on both sides, so that it should make a kind of book. I would like three different typefaces. One for the preface, the second for page 1 to page 68 inclusive; the third from page 69 to the end, with the letters about the size of those I have marked here, and more beautiful if possible. For the algebraic calculations, choose a type such that each equation will fit on one line. The woodcut [diagram] should be repeated and visible throughout the relevant section. . . . The running title should be enclosed between lines as it is here, in slightly slanted capital letters. . . . None of those ugly gray letters, but a nice plain simplicity [*un bel uni et une belle simplicité*]. . . . You will be able to fill more space by separating the lines until it makes a volume of 130, 140 or 150 pages. Sixty of the 100 copies can be on smaller, but still beautiful, paper.[31]

30. M to JBII, [June 1749], BEB.

31. M to JBII, [summer 1749], BEB. He had asked for assistance from Bernoulli in the printing of *Réflexions sur les origines des langues* in 1740, when he had likewise insisted on attention to typographic details. See Beeson, *Maupertuis,* 153–54; M to JBII, 2 September, 20 September, and 20 October 1740, BEB.

All of these considerations would contribute to making a book with none of the attributes of arcane scholarly treatises, which often used densely printed type on thin paper with narrow margins. He asked Bernoulli to make sure the italic typeface chosen for the preface would not "disfigure" the book by being out of proportion with the rest of the type. If the first version looked wrong, the printer must do it again. They also discussed details of the text itself, especially the opening section, and Maupertuis asked his friend to correct the manner of exposition as well as typographical errors. The edition, printed at the author's expense, was destined for his friends and patrons: "I do not want it to circulate, nor to provoke criticisms or worries."[32] He continued to send small additions and revisions from Berlin for several months. By December, he was becoming anxious to see the results; only in January 1750 did he finally receive his copies. He instructed Bernoulli to send twenty-five copies (ten in large format, fifteen in small) to La Condamine in Paris for distribution there.[33] Several went to acquaintances in Switzerland, and Maupertuis took delivery of the rest in Berlin. The list of recipients included Maupertuis's brother and sister in Saint-Malo, the duchesse de Chaulnes, Abbé Trublet, Musschenbroeck, d'Alembert, Daniel Bernoulli, the Jesuit Louis-Bertrand Castel, Etienne de Condillac, Abbé Sallier (the king's librarian in Paris), and the marquis d'Argenson.[34] He undoubtedly presented the deluxe edition, on large paper with gilded edges and an elegant binding, to Frederick II and the crown princes. This select group represented Maupertuis's ideal readership, a mix of knowledgeable mathematicians, literate aristocrats, and men of letters. This was the social and literary upper crust whose approbation he sought.

Why was the look and feel of the book so important to the author? Maupertuis wanted his *Cosmology* to appear in a form that could be offered to his patrons (Frederick II and the crown princes) and former patrons (d'Argenson); it would show the colleagues and friends he left behind in France that he was flourishing in Berlin, and it would show his circle in Berlin that he was producing books worthy of his status there. For this readership, he needed a book, pleasing to the eye and hand, rather than a scattered selection of academic papers. He was more concerned to showcase what he considered precious than to popularize it. Producing it for

32. M. to JBII, 6 September 1749, BEB.

33. Bernoulli informed La Condamine of the shipment (JBII to La Condamine, January 1750, BEB).

34. All these are mentioned in various letters as recipients, or themselves mentioned the book in letters to Maupertuis. Condillac to M, 10 June 1750, in LS, 388. The copy sent to Sallier, the royal librarian, very likely ended up in the Bibliothèque Nationale, where there is one copy in the rare quarto format. There are two copies in the former library of d'Argenson, the Bibliothèque de l'Arsenal; one came from the library of the Jesuit Collège de Paris.

an exclusive recipient list, rather than the market, enhanced its value as a gift. Its elegance reflected on its author, but also on its content, marking it as both elite and sophisticated. He carefully designed the book, both content and form, to suit this exclusive but eclectic audience. To call the *Essay on Cosmology* a work of popularization would miss the significance of the mixed readership. He meant it for d'Alembert and Musschenbroeck as well as for ministers and great ladies.[35] He addressed the metaphysical argument to the philosophically sophisticated; he did not intend it to be "accessible to everyone" [*à la portée de tout le monde*], a label carried by many vulgarizations of scientific works. He made this clear in a letter to one of his readers, the duchesse de Chaulnes: "I am very flattered that my little Essay pleased you. . . . It is true that of all I have done, it is what I esteem most highly, and I have taken almost half a century to complete it."[36] The book represented his life's work, and he wanted to be known as its author even before it became more widely available. Once it had circulated in this form, he had no objections to a commercial printing. The Paris printer Durand was interested as early as May 1750, although it took more than a year to complete printing of the second edition; the Dutch publisher Elie Luzac also printed it in Leiden.

The first edition's "*Avertissement*" oriented the reader to the text that followed and signaled the contentious points it addressed. The author revised this foreword in each subsequent edition to incorporate more and more explicit responses to his critics.[37] The essay itself consisted of three parts: a critique of natural theology, the proof of God's existence from the principle of least action (both taken with minor changes directly from the 1746 paper), and a descriptive "system of the universe." The mathematical problems from the Berlin Academy paper, along with the Paris Academy paper on refraction, appeared in smaller type at the end of the volume, as a kind of appendix.[38] Displaying the mathematics was essential to the presentation, as it had been in his earlier books, although the essay was constructed so that it could be read independently of these sections. The book was, then, a

35. Beeson says he separated out the mathematics to "make the text more accessible to a lay audience" (Beeson, *Maupertuis,* 224). This ignores the question of the makeup of this audience, and the significance of reprinting the mathematical portion of the argument at all.

36. M to duchesse de Chaulnes, 21 April 1750, SM, fol 90.

37. Tracing revisions through the editions is confusing: the "Avertissement" of the 1750 and 1751 editions became the "Préface" in the 1752 *Oeuvres* and the "Avant-propos" in the 1756 *Oeuvres,* while the "Avant-propos" of the first edition became Part I of the *Cosmology* in subsequent editions. The first section, regardless of its heading, was always printed in italics, though its content changed slightly with different editions.

38. Maupertuis, *Essai de cosmologie,* 139–73. This section included "Recherche mathématique des lois du mouvement et du repos," from "Lois du mouvement," *BAS* 1746, and "Accord de différents lois," *MAS* 1744.

carefully rearranged and expanded version of the text originally printed in the academy's journal.

Let us look more closely at the portions added to the earlier papers to make up the book. The foreword promised an investigation of the general laws followed by God in forming the universe. "These are the laws I set out to discover, to draw from the infinite well of wisdom from which they emanated: I would be happier to have succeeded, than if I had arrived by the most difficult calculations at following their effects in complete detail."[39] He noted scornfully that "some people" had already objected to his rejection of the argument from design, and he invited his critics "to examine [my reflections] with more attention and more justice than they have done."[40] This comment referred to an extract of the 1746 Berlin Academy volume by Castel in the Jesuit *Mémoires de Trévoux*. Not surprisingly, Castel focused on theological issues, accusing Maupertuis of "copy[ing] the trivial objections [to natural theology] made by the impious." Castel reminded his readers of the "triumphant responses that have been made a hundred times [to such objections]." He had nothing against using an economy principle as evidence of God's wisdom and power, but insisted that the memoir would have been more praiseworthy "if it had not attacked the other proofs." The application of the principle to mechanics did not interest him.[41] Maupertuis insisted in his response that he had not tried to "stifle" proofs from the marvels of nature, but only to show that "people have multiplied these proofs too much," without judging them rigorously enough. "I have taken pity on these criticisms and on the reader, in not citing the indecent or puerile reasonings their authors have filled them with. . . . That is all I have to say. If we find in the criticisms that have appeared, or will appear, some ill-considered expressions that have escaped from superstitious zeal or partisan spirit, they do not merit my attention."[42]

The foreword was followed by a preface examining proofs for the existence of God, taken from the published paper. Part of the argument revolved around the limitations of human understanding, especially of the causes of phenomena. Motion itself, the basis of all phenomena in nature, is fundamentally mysterious. "Without [motion] all would be plunged into an eternal death, or into a uniformity

39. *Cosmologie* (1750 ed.), A2–A3; also reprinted in *Oeuvres,* 1:xii.

40. "Avertissement," A3 (1750 ed.). "Je me suis contenté d'en dire un mot sans désigner personne dans la preface de ma Cosmologie" (M to La Condamine, 17 May 1750, SM, fol. 91).

41. *Mémoires de Trévoux,* December 1748, art. CXXI, quotations on 2487, 2493. The application of the principle of least action is barely mentioned in the review.

42. "Avertissement," *Cosmologie* (1750), A3, A4. To La Condamine, he called the review "as full of injustices as of ignorance" (M to La Condamine, 17 May 1750, SM, fol. 91); Trublet commiserated with Maupertuis about Castel's criticisms (Trublet to M, 7 May 1750, AS Fonds Maupertuis).

even worse than Chaos: it carries action and life everywhere."[43] Motion is, indeed, "the most marvelous" phenomenon, in spite of being entirely mundane. But what is the cause of motion? There we are at a loss; we must look for its laws without inquiring into causes. The earlier paper included a very brief treatment of this question: "Motive force, the power a body in motion has to move others, are words invented to substitute for our knowledge, and only signify the results of Phenomena. Only habit keeps us from recognizing what is marvelous about the communication of Motion."[44] The marvels of motion, upon reflection, outdo the marvels of natural theology, but we require different tools to appreciate them. For the *Cosmology,* he added a more extensive discussion of the nature of force, to counter the Leibnizian treatment of "living force [*vis viva*]." The Leibnizians made force an attribute of bodies and even of "beings they have never seen" (i.e., monads). Maupertuis substituted a more sensationalist definition: the word "force" denotes the sensory feeling [*sentiment*] of effort needed to move a body. "The perception we experience is so consistently accompanied by a change in the rest or motion of the body, that we cannot help believing that it is the cause."[45] Force is more a "confused feeling" than a clear idea; the best we can do is to measure it by its apparent effects rather than speculate about its ontological status.

With these reflections on force and motion, Maupertuis situated himself relative to the contending philosophical positions of Descartes, Leibniz (and implicitly Wolff), Malebranche, and Newton. He shifted priority from force to action, from conservation to economy, from an ontological category to a metaphysical one. He articulated a hybrid position, to supercede the longstanding argument about *vis viva* (which was also about conservation principles), but also to undermine the Leibnizian metaphysics promoted by Wolff and his followers—while transmuting final cause into a useful mechanical principle with explicitly metaphysical overtones.[46] Notwithstanding his attack on Newton's natural theology, Maupertuis retained from Newton the immediate dependence of the universe on God's intentions and actions. He made this explicit in a revision to the *Cosmology,*

43. *Cosmologie,* in Maupertuis, *Oeuvres* 1:26.

44. "La force motrice, la puissance qu'a un corps en mouvement, d'en mouvoir d'autres, sont des mots inventés pour suppléer à nos connoissances, & qui ne signifient que des résultats de Phénomènes. La seule habitude nous empeche de sentir tout ce qu'il y a de merveilleux dans la communication du Mouvement" (*BAS,* 1746, 280).

45. *Cosmologie,* in Maupertuis, *Oeuvres,* 1:29–30.

46. D'Alembert had also attempted to declare the *vis viva* debate an illusory conundrum, reducing it to nothing more than a "dispute de mots" (D'Alembert, *Traité de dynamique,* 1743). It is not clear whether Maupertuis appreciated how close his position was to that of Leibniz. On the use of metaphysics for mechanics, see Leibniz, "Specimen dynamicum," in *Selections,* 119–37 (though Leibniz conceived force to be both physical *and* metaphysical).

published in 1752. Reviewing the limited application of competing conservation principles, both of which pretend to universal status, he noted that Newton had recognized that conservation of motion could not be fundamental. "Newton felt that such an unchanging Force not being found in Nature; and there being more cases where the Quantity of motion diminished than where it increased, that all motion would finally be destroyed, the whole machine of the Universe reduced to Rest, if its Author did not renew it from time to time, and impart to it new forces."[47] The citation of Newton legitimated Maupertuis's iconoclastic position relative to conservation laws, although this position could only be called "Newtonian" in a rather limited sense.

The proof for God's existence, leaving the argument from design far behind, relied on the derivation of the laws of motion from a metaphysical principle appropriate to infinite wisdom and power. Nevertheless, the section entitled "Epitome of the System of the World" looks suspiciously like the kind of natural theology lambasted in the beginning of the book.[48] Starting with empirical details about the planets (including the shape of the earth), sun, comets, and stars, it is a compendium of the most recent knowledge of the heavens. He cautions that his system of the world is "only a sketch, not an explanation," and goes on to admit, "There is no doubt a large gap between these two parts of my work; . . . I have only tried to make each as perfect as possible: in the one I tried to penetrate to the ultimate reasons on which the laws of nature were founded: in the other I tried to accurately depict the phenomena of the Universe."[49] One part is profound, the other superficial but pleasing, like a good painting.

Although he did include formulaic expressions of wonder ("Who can view without admiration that marvelous arc that appears opposite the Sun, when after rainy weather the diffused drops in the air separate the colors of light for our eyes!"), he also pointed to the disorder that characterizes terrestrial species, speculating that it might result from an encounter with a comet. The gap between what we know and what we would like to know of the world, he suggests, could be the consequence of such a confusion in the natural order, not at all the image of God's design found in works of natural theology. "Each separate species can no longer

47. "Préface," *Cosmologie,* in *Oeuvres de Maupertuis* (1752); also in an undated autograph manuscript headed "Préface," AS Fonds Maupertuis. It did not appear in editions published after 1752.

48. The section is retitled "Spectacle de l'Univers" in the 1756 *Oeuvres.*

49. "Avertissement," *Cosmologie* (1750 edition), A3. The descriptive section may well have been based on the original "Cosmologie" mentioned to Algarotti some years before (M to Algarotti, 18 Feb. 1741, in Algarotti, *Opere,* 16:183). This justification of the contrast between the two parts disappeared in subsequent editions. The passage about the tableau was a late addition to the text, sent to JBII while the book was in press (M to JBII, 15 November 1749, BEB).

embellish or reveal the others: most beings appear to us only as monsters, and we find nothing but obscurity in our knowledge. Thus the most regular edifice, after lightning has hit it, no longer looks like anything but ruins, in which we cannot recognize either the symmetry the parts had among themselves, or the design of the Architect."[50] Particular phenomena, no matter how fascinating or engaging, cannot lead to knowledge of God's design, because the current imperfect state of the world differs from the original creation in its details, if not in its guiding principles. The eclectic nature of Maupertuis's book—which also included mathematics problems, as we have seen—marked it as appropriate to the mixed but elite readership the author had in mind. The descriptive "painting" of the universe fit into the genre of literary science for which he was already famous. In any case, he needed a substantial body of accessible prose to counterbalance the more subtle and contentious metaphysical and mechanical arguments of the rest of the work. And finally, these pages, amounting to one-third of the text, made it long enough to be printed as the kind of small, elegant book he imagined as the vehicle for announcing his discovery.

## Reactions to the Essay on Cosmology

Although the Berlin Academy's speculative philosophy class gave Maupertuis an appropriate outlet for his metaphysical mechanics, his letters bear traces of an embattled mentality. Many of his correspondents responded in flattering terms, but these could never outweigh, in the author's mind, what he considered the misguided lack of appreciation from his neighbors in Germany and from theologians in France. His overwhelming sense of the significance of his "discovery" led him to exaggerate the negative response in a letter to the duchesse de Chaulnes, an old friend in France: "In good faith, I don't know why I produce books: in truth it is not for the small advantage of reputation, the price and hope of which I am seriously disillusioned. I think I only publish my works as a sort of testament. . . . to some [they] have seemed to be filled with impiety, to others [nothing but] sermons; [they] have pleased almost nobody." These remarks betray an awareness of his own mortality, certainly, but also imply the association of "reputation" and "success" with a public too ignorant to appreciate his works. He aspired rather to the acclaim of readers like his noble correspondent, whose approval "comforts me for the poor success [the book] has had elsewhere."[51] He insisted that he was not

50. *Cosmologie,* in Maupertuis, *Oeuvres,* 1:72, 73–74.

51. "Je suis bien flatté que mon petit Essai vous ait plu. Il n'en falloit pas tant pour me conforter contre tous les mauvais succès qu'il a eus ailleurs. Il est vrai que c'est de tout ce que j'ai fait, ce que j'estime le plus, et que j'y ai bientot employé un demi siecle. . . . Je ne sais en bonne foi pourquoi je fais des livres: ce n'est point en verité pour le petit avantage de la reputation, du prix et de l'espoir

poaching on theological territory: "Theologians can throw in pell-mell all proofs, the good as well as the bad; it is for philosophers to make a choice among them and to establish their strength. Happy only in that research where I had absolutely nothing but the soundness of the reasoning for my goal, I became more and more confirmed in the knowledge of a truth that I never doubted."[52] He did not presume to offer a proof that would convince everyone, just those who knew how to use their reason. His goal, sound reasoning, is that of the philosopher immune to the "superstitious zeal" of the priests. But he was also showing how a philosopher could also be a believer, without giving up a claim to rationality. As a well-known literary and scientific figure with a reputation for freethinking and libertine living, Maupertuis may have seemed an unlikely defender of the faith.[53] At least one of his ecclesiastic readers appreciated this, however. His friend Nicolas Trublet, a man of letters and clergyman, approved wholeheartedly of the *Cosmology*, especially "the lovely section on the existence of God," and commiserated about the book's "maltreatment by the journalists of Trevoux."[54]

Criticism of Maupertuis in German periodicals only escalated with the publication of the *Cosmology*. His most vociferous critic was the Leipzig professor Gottsched, who had crossed swords with Euler at the time of the monad competition. Gottsched edited a journal of criticism, *Neuer Büchersaal der schönen Wissenschaften und freyen Künste*. As a follower of Christian Wolff, he found the takeover of the Berlin

---

duquel je suis fort desabusé: je ne donne je crois mes ouvrages que comme des especes de testamens, et j'en ai encore un tout pret. Les deux precedens ont paru remplis d'impieté aux uns, aux autres des sermons, n'ont plu presqu'a personne: si vous en faites cas, ils sont ce qu'ils doivent etre" (M to Duchesse de Chaulnes, 21 April 1750, SM, fol. 91).

52. "C'est aux theologiens à entasser pele-mele toutes les preuves tant bonnes que mauvaises, c'est aux philosophes d'en faire le choix et d'en constater la force. Heureux que dans cette recherche ou je n'ai eu absolument que la justesse du raisonnement pour but, je me sois confirmé de plus en plus dans la connoissance d'une verité dont je n'ai jamais douté" (M to LC, 17 May 1750, SM, fol. 91v).

53. For Maupertuis's reputation as *esprit fort*, see Boswell's comment in *Life of Johnson*: "Who could have imagined that the High Church of Englandman [Johnson] would be so prompt in quoting Maupertuis, who, I am sorry to think, stands in the list of those unfortunate mistaken men, who call themselves *esprits forts*. . . . There was in Maupertuis a vigour and yet a tenderness of sentiment, united with strong intellectual powers, and uncommon ardour of soul. Would he had been a Christian!" (Boswell, *Life of Johnson*, A.D. 1768, Aetat. 59).

54. Trublet to M, 7 May 1750, AS Fonds Maupertuis. Commenting on Maupertuis's *Essai sur la philosophie morale*, Trublet had noted that "vous auriez donc beaucoup fait pour le christianisme, en disant simplement que vous etes chretien" (Trublet to M, 26 December 1749, AS Fonds Maupertuis). Castel published a new review of the *Essay on Cosmology* in *Journal de Trévoux*, December 1751, reiterating his arguments. Maupertuis developed religious scruples toward the end of his life; there was much discussion of his turn to devotion in Berlin. (E.g., d'Argens, *Histoire de l'esprit humain*, 11:200–02.)

Academy by a Frenchman who fancied himself a philosopher too much to ignore. He objected to the reorganization of the Academy under a despotic president, to the use of French in the Academy's publications, and most of all to the endorsement of philosophical positions opposed to those of Wolff.[55] He had gone so far as to make light of Maupertuis's Lapland expedition and did not spare the principle of least action either.[56] Other German journals also took note of the first edition of the *Essay on Cosmology.* The *Acta eruditorum* reviewed it in December 1750; Maupertuis's ally Abraham Kästner, professor of mathematics at Leipzig, kept him informed and sent him copies of the journal.

Kästner was unusual among German intellectuals in his defense of Maupertuis's metaphysics. He wrote an extract of the *Cosmology,* but he had to approach it with considerable delicacy in order to satisfy Maupertuis in Berlin without alienating his own colleagues in Leipzig. "I did not omit to insert into my extract some reflections that I thought most appropriate to confound those who have undertaken to judge your works without understanding them. Nevertheless I had to keep within certain bounds, to which my present situation obliged me."[57] Kästner advised Maupertuis not to worry about what was said about him in Germany, knowing that most of it was highly unflattering. This advice did not keep Maupertuis from avidly collecting all references to his work and having them translated by his bilingual academicians.[58]

Although he had tried to build support for his new work by distributing it privately to those most likely to read it sympathetically, he must have expected criticism from theologians and journalists. He could hardly have been oblivious to the resonance of his title with Wolff's own *Cosmologia generalis* (1731), which subtly reinforced the impression that he was offering an alternative to Wolffian meta-

55. On Gottsched, see Danzel, *Gottsched und seine Zeit.* The primary bone of contention was the monad essay competition; see Wolff-Maupertuis correspondence, LS, 424–39. Gottsched had been a foreign member of the academy since 1729. He had other enemies in the Academy, notably the chemist Eller. See Kästner to M, 15 April 1750, LS, 281.

56. Kästner translated Gottsched's comments for Maupertuis: "ce meme M. G. a appellé votre expédition: 'une de ces bagatelles, dont la vanité françoise tiroit gloire pour avoir découvert une chose que Newton et Huygens avoient seu longtems auparavant'" (ibid.).

57. Kästner to M, 17 January 1751, LS, 292. Kästner and Gottsched had had their own disagreement over the evaluation of German poetry, a subject dear to both their hearts. Kästner's extract of *Essay on Cosmology, Hamburgisches Magazin* 6 (1750): 321–26.

58. "Je le [Passavant, a new Swiss member] chargeay il y a un mois de me traduire certains endroits d'un journal Alleman qui paroit tous les mois or tous les ans ou toutes les semaines à Leypsik . . . il lui a fallu un mois pour m'envoyer quelques feuillets de ce journal quon attribue a un Mr. or une Me Godchet" (M to JBII, [summer 1749,] BEB). Mme. Gottsched was a scholar of French literature and worked with her husband on the *Neuer Büchersaal,* as well as translating books from French into German.

physics.[59] His letters from 1750, before the publication of the second edition, refer repeatedly to his critics. To Christlob Mylius, who wanted to translate the book into German, he responded bluntly that not enough people would be interested, no doubt recognizing that hostility was likely to be greater among German readers. In his original letter, Mylius must have raised the question of whether the principle of least action was anything other than the principle that nature acts in the simplest way possible. Maupertuis defended his originality with some exasperation: "[Other philosophers] have certainly established that nature must act by the simplest means; but none of them has really determined what these simplest means are, nor the fund that nature saves in the production of her phenomena." And he went on to complain that anyone who had taken the trouble to read his 1744 paper on the behavior of light would have seen that the principle of least action differs from Leibniz's metaphysical principles. Maupertuis argued that his work fulfilled the promise made by Leibniz to deduce the laws of motion from a metaphysical principle, a promise the German philosopher could not fulfill because of the limited generality of the conservation of *vis viva.* "As I showed in my *Cosmology,* if one wants to regard this conservation as a principle, one must assert that there are no hard bodies in nature, which I see as the greatest of absurdities." Ironically, Maupertuis's use of metaphysical principles, and especially of final causes, betrayed a certain sympathy with Leibniz. But he also worked tirelessly to expose the limitations of Leibniz's view of dynamics. Gottsched and "some others" were misguided in claiming that the principle of least action was nothing new: "If they had been more enlightened or more just, they would have recognized that no one yet had discovered this universal principle to which hard bodies and elastic bodies alike are subject."[60] Evidently German philosophers and journalists were raising questions about originality and the relation of Maupertuis's work to Leibnizian-Wolffian philosophy almost as soon as the *Cosmology* was published.

Although he was disappointed in the response to his magnum opus, he also found sympathetic readers, especially among the scientific and literary elite in France. Condillac, for one, appreciated the *Cosmology:* "The soundness, the clarity and the precision that prevail in it are advantages of which I am infinitely appreciative. . . . All that you say on the existence of God is delicately conceived and explained in a simple and elegant manner. It seems to me indisputable that it is not in the little details that the supreme being must be sought. . . . All that you say after

59. Formey pointed out this connection to d'Alembert, who discussed the question in his article "Cosmologie" for the *Encyclopédie.* Wolff also explored natural theology in his *Theologia naturalis* (1736–1741). Maupertuis never named Wolff in this connection.

60. M to Mylius, 18 July 1750, SM. (The letter from Mylius does not survive.) Mylius did translate the *Cosmology,* in spite of Maupertuis's reaction (*Versuch einer Cosmologie,* Berlin, 1751).

that on the laws of motion and the principles that have been dreamed up to explain them seem to me of the greatest exactitude."[61] Condillac claimed, probably disingenuously, that he could not follow the mathematical details, but he seems to have taken Maupertuis's book seriously as philosophy. He pointed to the values—precision, clarity, elegance—that he espoused in his own books. This is just how Maupertuis wished to be read. He had no desire to tangle with theologians, as he pointed out to La Condamine: "I pretend neither to abolish all other [proofs], nor to give mine even as a demonstration, I just apply myself to reasoning soundly. . . . I repeat that [the book] is absolutely nothing but a philosophical work that goes as far as it can on the existence of God considered in the marvels of nature and that examines the validity of the proofs that are taken from them, without fear and without taking liberties."[62]

D'Alembert also found much to admire in Maupertuis's mechanics. In his *Treatise on Dynamics* (1743), d'Alembert had insisted that inherent forces are no more than "obscure and metaphysical entities, which only serve to spread shadows over a science which is itself full of clarity." Forces can only be known *a posteriori* by their effects, on this view, so the science of dynamics should consider motion as geometrical paths through space, without speculating about causes.[63] Only scattered items of the correspondence between d'Alembert and Maupertuis survive, among them Maupertuis's response to a friendly letter from d'Alembert soon after the first copies of the *Essay on Cosmology* had been distributed in Paris. "I am extremely flattered if my latest work has your approbation: what I said about the word 'force' I took from a paper I had done long ago and that I had read to our Academy."[64]

61. "La justesse, la netteté et la précision qui y règnent sont des avantages auxquels je suis infiniment sensible. . . . Tout ce que vous dites sur l'existence de Dieu est vu finement et exposé d'une manière simple et élégante. Il me paroit hors de doute que ce n'est pas dans les petits détails quil faut chercher l'etre supreme. . . . Tout ce que vous dites ensuite sur les lois du mouvement et sur les principes qu'on a imaginés pour les expliquer me paroit de la dernière exactitude" (Condillac to M, LS, 388–89). Maupertuis had recently read Condillac's *Traité des systemes,* with approval: "j'ai vu avec plaisir en lisant le Traité des Systemes de l'Abbé de Condillac que je m'etois à peu près rencontré avec lui sur cela, il n'y a personne avec qui je sois plus flatté de me rencontrer" (M to d'Alembert, 23 May 1750, SM), fol. 92).

62. "Je ne pretens ni abolir toutes les autres, ni donner la mienne meme comme une demonstration, je m'applique seulement à raisonner juste: . . . Ce n'est absolument je le repete qu'un ouvrage philosophique qui va jusqu'ou il peut sur l'Existence de Dieu considerée dans les merveilles de la Nature, et qui examine sans crainte et sans licence la validité des preuves qu'on en tire" (M to La Condamine, 17 May 1750, SM, fol. 91).

63. "J'ai entierement proscrit les forces inhérentes au corps en mouvement, êtres obscurs et métaphysiques qui ne sont capables que de répandre les ténèbres sur une science claire par elle-même" (D'Alembert, *Traité de dynamique,* 1st ed., 1743, xvi, xxi).

64. "Je suis fort flatté si mon dernier ouvrage avoit votre approbation: ce que j'ai dit sur le mot

We can infer that d'Alembert's favorable assessment of the *Cosmology* derived in part from Maupertuis's rejection of forces as real attributes of matter.[65] When he wrote his article "Action" for the first volume of the *Encyclopédie* (1751), d'Alembert called the principle of least action "one of the most general [principles] of mechanics. . . . The author was able to wed the metaphysics of final causes with fundamental truths of mechanics; to make the impact of elastic bodies and hard bodies depend on a single law, which up until now had had separate laws; to reduce to a single principle the laws of motion and those of equilibrium."[66] The general applicability of the principle to widely differing phenomena weighed in its favor. D'Alembert himself energetically avoided final causes, but he chose not to attack his friend and former *confrère* on these grounds.[67] Because of its relevance to refraction, collisions, and planetary orbits (in Euler's work), d'Alembert nonchalantly read the metaphysics out of the principle of least action. He labeled the principle of least action a "geometrical truth," pointedly downplaying Maupertuis's insistence on the value of metaphysics for mechanics, and he ignored completely the argument for God's existence derived from the economy principle. From d'Alembert's point of view, Maupertuis had discovered something profound, though he gave it a rather different emphasis, to fit with his own approach to mechanics. The principle of least action has the same status as many other so-called laws, such as the conservation of *vis viva* in elastic collisions or the uniform motion of a center of gravity: these are nothing more than "mathematical theorems, more or less general, and not philosophical principles." But d'Alembert's recognition of Maupertuis's originality made him a valued supporter, even if he stressed the mathematical over the metaphysical.[68]

---

de force, je l'ai tiré d'un memoire que j'avois fait depuis longtems et lu a notre academie" (M to d'Alembert, 23 May 1750, SM).

65. D'Alembert, *Traité de dynamique* (Paris, 1743), xxi. D'Alembert developed a dynamics based on the geometry of the motions of bodies, considered simply in terms of distances and times. See Hankins, *Jean d'Alembert,* chapter 8.

66. "L'Auteur a sû allier la métaphysique des causes finales avec les verités fondamentales de la méchanique; faire dépendre d'une même loi le choc des corps élastiques & celui des corps durs, qui jusqu'ici avoient eu des lois séparées; réduire à un même principe les lois du mouvement & celles de l'équilibre" ([D'Alembert], *Encyclopédie* s.v. "Action," 1:119–20).

67. D'Alembert's objections to metaphysics solidified with time; by the second edition of his *Traité de dynamique* (1758), he commented negatively about Maupertuis's use of metaphysics (albeit without naming him): guessing at God's aims "runs the risk . . . of taking as a fundamental law of nature that which may be only a purely mathematical consequence of some [more basic] formulae" (*Traité,* 2nd ed., 404).

68. D'Alembert, *Encyclopédie* s.v. "Cosmologie," 4:295–97. Maupertuis left no record of objection to d'Alembert's remarks about the principle of least action. He did not view d'Alembert as a critic; their correspondence continued on a friendly basis for years. "Je vous remercie d'avance de

The Dutch physicist and mathematician Musschenbroeck, probably another designated recipient of the first edition of the *Cosmology*, also commented favorably. In thanking him, Maupertuis reiterated his claim to novelty and universality: "It is only a sketch if you stick to the general title of Cosmology; it is something more if one pays attention to the principle from which I deduce the laws of motion and of rest. Up until now no philosopher has been able to deduce all these laws for hard and elastic bodies from a unique principle, nor even to find a unique principle that was not contradicted by one or the other of these laws." He also reminded Musschenbroeck that Euler had applied the principle of least action to planetary orbits. "I was more than a little pleased when I saw the use that M. Euler made of it [the principle of least action] in his treatise on Curves with Maxima and Minima, which he published some time after I presented my principle to the Royal Academy of Sciences of France."[69]

## The Public Quarrel with König and Voltaire

The objections of Jesuits and German professors anticipated a more pointed attack by the Wolffian mathematician Johann Samuel König. Maupertuis's dispute with König, and then with Voltaire, echoed through the pages of literary gazettes, pamphlets, letters, learned journals and books for several years, becoming one of the most notorious "literary quarrels" of the eighteenth century.[70] It might have remained an esoteric local affair, if Voltaire had not put his satirical talents to work on König's behalf. The story is one of unlikely alliances and personal enmities. As we have seen, Voltaire had cultivated Maupertuis as a useful ally in the campaign for Newtonianism in the 1730s. The two men moved in the same social circles in France, and stayed on nominally friendly terms over the years, although their acquaintances saw signs of strain in their relations. Voltaire's jealousy

ce que vous avez dit de ma Cosmologie, il est vray que je crois cet opuscule moins imparfait que l'autre [*Figures des astres*]: et quoi qu'en disent des devots à Dieu et à Wolff, je le crois bon, et le croit peut etre d'autant meilleur qu'il leur deplait davantage" (M to d'Alembert, 10 January 1752, SM, fols. 99v–100). D'Alembert had his own reasons for maintaining his alliance with Maupertuis, who served as a go-between in d'Alembert's dispute with Euler (see d'Alembert to M, 4 August 1752, AS Fonds Maupertuis).

69. "Ce n'est qu'une ebauche si l'on s'attache au titre general de Cosmologie, c'est quelque chose de plus si l'on a egard au principe dont je deduis les loix du mouvement et du Repos. Jusqu'ici aucun Philosophe n'avoit pu deduire toutes ces loix pour les corps durs et les corps elastiques d'un principe unique, ni meme trouver un principe unique qui ne fut pas dementi par les unes ou les autres de ces loix." (M to Musschenbroeck, 17 April 1751, SM, fol. 97). Musschenbroeck's letter is lost; he may have been one of the original recipients, or he may have read the Dutch edition of 1751. His letter predates the publication of the Paris edition of 1751.

70. A pro-Voltaire account appeared in a compendium of literary and philosophical controversies: S. Iraïlh, "M. de Voltaire et M. de Maupertuis," in *Querelles littéraires,* 2:72–93.

of Emilie du Châtelet's obvious affection for Maupertuis, and Maupertuis's resentment of Voltaire's sharp wit, contributed to the tension. Frederick II exacerbated this tension by wooing both men, each of whom had a tendency to exaggerate favors shown to rivals. Even before he came to power in 1740, Frederick had showered Voltaire with mediocre verses, lavish gifts, and flattering invitations. The *philosophe* visited the king several times, but it was only after the death of Mme. du Châtelet, his long-time companion if no longer his lover, that Voltaire came to Berlin as poet in residence in 1750. By this time, Maupertuis was well established in Berlin, and in the circle around Frederick in Potsdam. In the rarefied atmosphere of Frederick's supper parties, Maupertuis and Voltaire, both in the prime of their egotism, became rivals for the king's attention.[71] It was in this atmosphere that Maupertuis's conflict with König unfolded.

In the spring of 1751, as Durand was printing a new edition of the *Essay on Cosmology* in Paris, König published a piece in the *Nova acta eruditorum* denying the universality of the principle of least action. König tried to undermine the principle of least action by substituting conservation of *vis viva* as a universal principle and applying it to conditions of dynamic equilibrium. He objected to Maupertuis's reinstatement of hard bodies as the ultimate units of matter, because this move cut away at the foundation of Leibnizian physics. König's paper was hardly a model of clear exposition or of convincing mathematical demonstration. To make matters more confusing, not only did he deny the generality and utility of the principle of least action, but he also charged that Leibniz had articulated the same definition of action in a 1705 letter to Jacob Hermann. König thus claimed priority for Leibniz, quoting from this unpublished letter, while simultaneously arguing that least action was inadmissible as a general dynamical principle.[72] The dispute initially centered on the authenticity of the Leibniz letter, but it also concerned the validity of the principle of least action itself. On both counts, Maupertuis perceived an affront to his honor, and pushed the Academy to investigate the provenance of the purported Leibniz letter. A pamphlet war followed the Academy's formal judgment against the authenticity of the letter, as König traded challenges and counter-challenges with Euler and J. B. Merian, the newest member of the metaphysics class and an indefatigable ally of Maupertuis. Then Voltaire jumped in with an unexpected anonymous attack on Maupertuis, which provoked a defense, also in pamphlet form, by the king himself. Voltaire could

71. The literature on Voltaire is immense. For the Berlin period, see Pomeau and Mervaud, *De la cour au jardin.* Voltaire's correspondence is an invaluable source for the intrigues and backbiting that he pretended to abhor.

72. König, "De universali principio aequilibrii et motus," *Nova acta eruditorum* (March 1751): 125–35, 162–76.

not resist printing ever more vicious satires, jeopardizing his position at court. Maupertuis, although he remained involved behind the scenes, refrained from writing his own pamphlets, leaving his defense to his supporters in the Academy and his most powerful patron, the king.[73]

In retrospect, the whole affair looks petty and even ridiculous. To follow it in detail is to drown in a sea of repetitious invective and recrimination. The quarrel was part priority dispute and part personality clash, and the scientific issues were soon obscured by the Academy's treatment of König and by Voltaire's satires. Although trivial in detail, the uproar had an enormous impact on Maupertuis, as he struggled to maintain his equanimity and his reputation. He found it difficult to escape Voltaire's portrayal of him as a ridiculous and arrogant despot, even though Voltaire's own behavior was widely considered dishonest and reprehensible by contemporaries.[74] Without rehearsing every detail of the polemic, I examine this contorted episode for what it reveals about the cultural and political settings of science in the age of absolutism. Apart from personal jealousies, what themes does it illuminate? What tools were used to fight the battle, and what publics did the participants address? The dispute was more about power, patronage, and honor than about physics, or even metaphysics—although the scientific content of the dispute should not be overlooked. Euler in particular took it seriously on this level and reaffirmed his deep commitment to the truth and utility of the principle of least action. Maupertuis certainly recognized his great good fortune in having such a defender. But the struggle between Maupertuis and Voltaire as competing courtiers sidestepped the scientific questions, and Frederick's dual role as royal patron and author made for some strange and novel twists of plot. Then again, Maupertuis himself was a powerful patron, controlling positions and pensions at the Academy, and he exploited this power in his own defense. Voltaire, a client of the king, still styled himself an independent mind, above normal rules of etiquette. The affair also concerned authorship and the authenticity of texts, and the power of print to resist absolutist control. Finally, it illustrates the tight reciprocal relations between scientific and literary circles, as an esoteric technical dispute spun out into a "literary quarrel" that fascinated contemporaries across Europe.

## The Academy's Judgment

König had known Maupertuis since the late 1730s, when they met in Basel through the Bernoullis. They had been on cordial terms, and Maupertuis had several times had the opportunity to patronize the Swiss mathematician, who spent

73. On the König dispute, see Harnack, *Geschichte,* I(1):331–45; Costabel, "L'affaire Maupertuis-König"; Radelet-de Grave, "La 'Diatribe du Docteur Akakia.'"

74. In the historical literature, it is commonplace to say that Euler was Maupertuis's only

years unsuccessfully seeking university posts throughout German-speaking Europe. Maupertuis had introduced him to Emilie du Châtelet in 1739, when she was looking for a mathematics tutor; subsequently, he opened doors for König in the Paris Academy of Sciences.[75] Since then, he had had a rather checkered career, circulating around Europe, from Switzerland, in and out of France, to the Low Countries and various German cities, selling his services as tutor, translator, librarian, and, for a time, professor.[76] He became a foreign member of the Berlin academy in 1749, at the president's instigation.[77] On a visit to Berlin shortly after this, he startled Maupertuis with his vigorous defense of Leibniz's priority over Newton in the matter of the invention of the calculus. Maupertuis recalled the conversation as "a Scene that we had here about Leibniz in the presence of Algarotti; where [König's] zeal for this great man carried him away to the point of saying insulting things to me. If he was just a fanatic, it would be a bad enough, but I think him a rogue."[78] Always conscious of their difference in status, Maupertuis took offense more easily than he might have done with someone he considered an equal. Although they did argue about Newton and Leibniz at this time, König did not reveal his plans to publish a critique of the principle of least action. A few months later, Maupertuis told Bernoulli of his surprise at reading his former

---

supporter, but this is far from true. See Fellman, "Johann Samuel König," *Dictionary of Scientific Biography,* 7:443.

75. König was born in Germany, of Swiss parents, and educated in Switzerland. He was elected a foreign member of the Paris Academy in 1739, through his connection with Maupertuis and Clairaut, and subsequently translated *La figure de la terre* and *Elements de géographie* into German. See the König-Maupertuis correspondence before the quarrel (1735 to 1749) in LS, 106–18; see also M to Kästner, 9 June 1751, SM, fols. 120–120v. On König's relations with Châtelet, see Terrall, "Emilie du Châtelet"; Janik, "Searching for the Metaphysics of Science." In 1740, König had been on decidedly poor terms with Voltaire, who held him responsible for "converting" Châtelet to Leibnizianism, and for violating her trust and her hospitality. See, e.g., Voltaire to M, 10 Aug. 1741, Volt. *Corr.,* 92:95.

76. He had been professor of philosophy and mathematics at Franeker, Holland; in 1749 he moved to the Hague as privy councillor and librarian to Prince William of Orange, and also taught mathematics at the university there.

77. König was elected 4 September 1749 (Winter, *Registres,* 140–41). Formey confirms that König's membership was due to Maupertuis's sponsorship (*Souvenirs,* 1:175). Maupertuis told Bernoulli that he had received König in his home "as if he were my brother" before the critical paper was published (M to JBII 6 July 1751, BEB).

78. "Une Scene que nous eumes ici sur Leibniz en presence d'Algarotti: & ou son zele pour ce grand homme l'emporta jusqu'à me dire des choses injurieuses. Si ce n'étoit qu'un fanatique, ce seroit déja un assez grand mal, mais je le crois un frippon" (M to JBII, 28 August 1751, BEB). Formey heard about this meeting from Algarotti (Formey, *Souvenirs,* 1:177). A fuller account of the original disagreement between the two men, probably based on Maupertuis's recollection, appears in Merian, *Mémoires pour servir,* 148–49.

friend's paper: "Hardly had he returned to Holland when he published in the *Acta Erud.* a piece where he lords it over me and . . . he criticizes what I have written in our *Mémoires* and in my *Essay on Cosmology.* Realizing the hopelessness of convincing everyone that I say nothing worth anything at all, he concluded by citing a fragment of a letter from Leibniz to Hermann by which he would claim for Leibniz the principle of least action, and by which also Leibniz would have discovered before Euler that wonderful property of trajectories described by central forces, in which the element of the curve multiplied by the velocity of the body, that is to say the Action, is always a minimum."[79]

Although it started as a spat between two individuals, the Academy and the king soon entered the fray. König, a client of Maupertuis by virtue of his status as foreign member of the Academy, undermined the president's authority by manufacturing a new priority dispute, oddly parallel to the Newton-Leibniz dispute about the calculus, with himself defending Leibniz against the Newtonian Maupertuis's claims to the principle of least action (hardly a Newtonian principle). Maupertuis represented his opponent as a fanatic whose "zeal for German philosophy" drove him to overstep his position. "He played me this turn on leaving [Prussia], where he had come on a religious pilgrimage to see the great Wolff and the relics of Leibniz. . . . After having spread a lot of manure and trying to erase from my work whatever merit it might have, being unsure of his success, he wants to attribute to Leibniz the principle and its applications to the laws of motion."[80] Since Leibnizian philosophy was a barely camouflaged fault line dividing the Berlin Academy, König's claims were potentially even more incendiary than a personal insult—although such an insult was more than enough to evoke the wrath of the egotistical president. König had submitted an essay for the controversial monad prize competition in 1747 and may well have nursed a grudge against the Academy as a result of his failure on that occasion. The generally hostile treatment of the management of the Berlin Academy in the German press had primed Maupertuis to respond forcefully to König, a more accessible and more vulnerable target than Gottsched or anonymous journalists. Seeing his scientific accomplishments and his personal integrity called into question, Maupertuis took steps to demonstrate his power to silence "scoundrels".

79. M to JBII, 6 July 1751, BEB.

80. "A quelle noirceur le zele pour la Philosophie d'Allemagne a conduit König, il me fit ce tour au sortir d'icy ou it etoit venu faire un pellerinage de devotion pour voir le grand Wolff et les Reliques de Leybnitz . . . apres avoir repandu beaucoup de fumée et voulu oter a mon ouvrage le meritte qu'il peut avoir, n'etant pourtant pas seur d'avoir reussi il veut en attribuer a M. Leybnitz le principe et son application aux lois du mouvement" (M to d'Alembert, 10 January 1752, SM, fol. 100).

Maupertuis made a strategic decision to focus on the authenticity of the letter, suspecting König's vulnerability on this point. As he explained later to d'Alembert, "instead of disputing with him, I limited myself to pressing him to produce the original, regarding it anyway as quite a mark of approval to want to attribute the basis [of my principle] to the great Leibniz."[81] If all went well, the disciplining of König might be turned into positive publicity for the principle of least action. The Academy formally demanded that König produce the letter, which turned out to be a copy, and then initiated a search for the original in Hermann's papers in Switzerland.[82] Maupertuis used his prerogative to make the dispute a matter of honor for his institution and for the king. In asking Frederick for his support, he framed it thus: "In the present matter, it is not only a question of my interest, it is a question of that of Your Majesty's Academy. Mr. König . . . has cited . . . a letter of Leibniz in which one finds some things we [academicians] have produced as original discoveries."[83] Publicly, Maupertuis limited his concern to the existence and authenticity of the letter; privately, he hoped to expose his accuser as a charlatan, "to bury König in the mud as he deserves."[84] Enlisting Bernoulli's support in hunting for the Leibniz letter, he admonished his friend to "keep the thing secret until we are in a position to close off all his escape routes."[85] Euler bent to the task with a will, no doubt harboring his own hostility to the Wolffian professor who had defended monads to the Academy a few years before. When König could not produce the letter, Euler submitted a report, concluding "that [König's] cause is completely untenable and that this fragment was forged, either to wrong M. de Maupertuis, or to exaggerate by a pious fraud the praises of the great Leibniz."[86] The evidence for forgery was circumstantial, but König had muddied his claims with so many improbable and unsubstantiated explanations

81. Ibid.

82. Maupertuis had the full support of the king, who sent a formal request to officials in Berne, in search of Leibniz papers. FII to M, 16 October 1751, in Koser, *Briefwechsel,* 267. The search was made more complicated by the fact that König had gotten his copy from the Swiss Henzy, who was subsequently executed for treason in Berne.

83. M to FII, 13 October 1751, AS Fonds Maupertuis; Euler and Merian developed this line in their defenses later. For the official story, see Euler, "Exposé concernant l'examen de la lettre de M. de Leibnitz, alléguée pas M. le Professeur König dans le mois de mars 1751 des Actes de Leipzig, à l'occasion du principe de la moindre action," *BAS* 1750, 52–62.

84. M to JBII, 6 July 1751, BEB.

85. Ibid.

86. "Que sa cause est des plus mauvaises et que ce fragment a été forgé, ou pour faire tort à M. de Maupertuis, ou pour exagérer, comme par une fraude pieuse, les louanges du grand Leibnitz" (Euler, "Exposé concernant l'examen de la lettre de Leibniz," in Euler, *Opera Omnia,* series 2, 5:64–73). Euler read his "Exposé" in Latin, so it would be understood by the German speakers. See Winter, *Registres,* 178 (13 April 1752). The authenticity of the Leibniz fragment used by König is still

that Euler stated his condemnation in the strongest possible terms. In April 1752, the Academy voted unanimously to approve Euler's report, although there was some private grumbling about the president's tight hold on power, and bitterness about the impossibility of dissent.[87] (Maupertuis stayed away from the meeting himself.) König severed his ties to the Academy, making a show of sending back his membership diploma.[88]

Such was the first stage of the dispute, conducted initially behind the scenes, then in the closed space of the Academy, and finally in the pages of its journal. From the Academy's point of view, König had impugned the honor of its president, not by just disagreeing with him, but by accusing him (and Euler) of lack of originality and perhaps even plagiarism—although this charge was hardly tenable, since neither could have seen the Leibniz letter before König showed it to them. From König's point of view, Maupertuis and Euler exploited their power unfairly in order to stifle his critical voice. From Maupertuis's point of view, König's accusation was particularly offensive because it came from someone of lower status: "this, as you see, from a man who is very much obliged to me, is really German villainy."[89] The difference in rank was evident to all observers. Kästner, for example, told Maupertuis that König was not worth bothering with; to respond to him would be like "a general of an army wanting to fight a duel with some vagabond volunteer."[90] Historical accounts have often regretted Maupertuis's "abuse of authority," but taken in context, the Academy's response was hardly surprising.[91] Unlike his earlier scientific dispute, this one found Maupertuis in a position to defend himself with the full force of the institution behind him, precisely because of the way Frederick had designed the Academy in the image of his absolutist state. Certainly his personal prickliness and egotism contributed to the escalation of the conflict, but Maupertuis had the elevated sense of his own honor appropriate to

debatable. For more on the Leibniz texts, see Costabel, "L'Affaire Maupertuis-König"; Breger, "Über den von Samuel König veröffentlichen Brief."

87. "Weil Maupertuis alle Gewalt in Händen hat, und man nicht sehr laut gegen ihn reden darf, so ist die Verbitterung im Geheim desto stärker, und dises thut der Academie grossen Schaden" (Sulzer to Künzli, April 1752, quoted in Harnack, *Geschichte,* 1(1): 338, n.2 ). Harnack exaggerated the sparseness of attendance at this meeting.

88. "It was judged that this disobliging act did not merit any response" (Winter, *Registres,* 182 [6 July 1752]).

89. "Cecy comme vous voyez d'un homme qui m'a beaucoup d'obligation est bien une villenie d'Allemagne" (M to d'Alembert, 10 January 1752, SM, fol. 100).

90. Kästner to M, 17 June [1751], in LS, 300. The passage goes on: "Si la querelle étoit d'assés d'importance pour être vuidée d'une manière sanglante, il se trouveroit d'autres champions."

91. Beeson, *Maupertuis,* 244, 266. Mervaud describes the judgment as "incontestablement . . . un acte de tyrannie"(*Voltaire et Frédéric II,* 219).

an absolute ruler, and behaved accordingly when affronted. The affront was not trivial, since he regarded the principle of least action as the culmination of his life's work. In defending its significance, he was fortunate enough to have as a colleague a brilliant mathematician who objected passionately to Leibnizian philosophy and who had explored the technical uses of the principle of least action with far more insight than König could muster.

Maupertuis was concerned from the beginning about how to represent the case to "the public" in print. He wrote to Bernoulli, "After König's conviction and the judgment brought by the Academy, it will no doubt be necessary to print the whole history: but what do you think about the form in which it should be presented? My plan was to report the letters themselves, from me to K. and from K. to me, just as they were written."[92] In the months and years following, many collections of the relevant documents appeared in print, in addition to polemical pieces. Many were letters, sometimes private letters published without the approval of the author. The sheer variety of texts in the torrent of published material shows the range of literary means available for exploitation and amplification in such a dispute, which took place almost entirely on the printed page. Observers and participants across Europe filled their letters with references to and comments on the printed texts: what was said, who was responsible for anonymous editions, and how to acquire the latest pamphlets and books. Following the dispute meant following a complicated trail of print, often mediated by journal articles and letters, as authors and publishers printed a bewildering array of old and new texts. (To help his readers make sense of all this, the French journalist Fréron ended his narrative in 1753 with a list of the major works published to date, and the names of booksellers who carried them in Paris.)[93]

From the outset, all sides competed for the authority to judge and to speak out publicly, and as the conflict developed, authors put forward rival versions of the "history" of the affair. Euler told one such history in the pages of the Berlin Academy's *Mémoires;* he and Merian repeated this more or less official account in a series of small books and brochures published in Berlin. Another story came out in König's pamphlets from Leiden; bits and pieces appeared in various German, Dutch, and French journals; several publishers pulled together anthologies

92. "Apres la conviction de K. et le jugement porté par l'Academie, il faudra sans doute imprimer toute l'histoire: Mais que pensés vous de la forme dans laquelle elle doit etre presentée? Mon dessein etoit de rapporter les lettres mesmes tant de moy à K. que de K. à moy telles qu'elles ont eté ecrittes" (M to JBII, 25 January 1752, BEB). Maupertuis had phrased all communications with König politely, anticipating their publication. They appeared in a collection, *Lettres concernant le jugement de l'Académie,* Berlin 1752.

93. Fréron, "Lettre IX," *Lettres sur quelques écrits* 8:197–210; list on 209–10.

of documents.[94] Eventually, Voltaire transformed the story into *Histoire du Docteur Akakia et du natif de St-Malo,* a compilation of the short pamphlets he had released over a span of many months. Maupertuis published several of his own letters (to König, to Euler, to Voltaire) to document his almost aristocratic restraint; he wrote his own narrative only in the 1756 edition of his collected works, and then in measured tones meant to show up the vulgarity of his antagonists.[95]

The status of the combatants was reflected in the media available to them. Euler read his contributions to the Academy and printed them in the *Mémoires.* Thus, three substantial defenses of Maupertuis appeared as officially sanctioned texts, highlighted by their immediate publication in the volume currently in press. Readers could see plainly where the weight of academic authority lay. Euler contrasted the "Illustrious President" with "the Professor," whose grasp of mechanics was worthy only of contempt.[96] He distinguished between expert knowledge and the slanderous and uninformed criticism found in the "gazettes littéraires," periodicals of a different order than the *Mémoires.* He decried "this excessive itching they have to exercise their criticism on all that happens in the Republic of Letters. All their complaints . . . show quite well that they do not even understand the state of the Question, even though it is exhibited in this Judgment with the greatest incisiveness." The matter demanded "a profound knowledge of the relevant sciences," which could only be found within the Academy. The journals were no more than "public quibblers [*chicaneurs publics*]," and König's association with them, from Euler's vantage point, only lowered his status further.[97] Still, whatever their opinion of the journals, the academicians wanted to reach a public beyond the

94. E.g. *Maupertuisiana,* Hamburg [Leiden], 1753, a miscellaneous anthology of the relevant texts, including Voltaire's satires and Maupertuis's letters. Several versions of this collection are extant; for the complicated bibliographical history, see Kaulfuss-Diesch, "Maupertuisiana." See also Voltaire, *Diatribe du docteur Akakia,* cxxxii–cxxxvi, for various editions of satires.

95. Maupertuis told his story in a new preface to *Essai de cosmologie* for this edition (*Oeuvres,* 1:xxi–xxxvi) and in "Lettre XI," *Oeuvres,* 2:275–83.

96. "Lettre de M. Euler à M. Merian," *BAS* 1750, 520–32, constituted the whole contribution of the belles-lettres class for the 1750 volume, published in 1752, just as the pamphlet war was heating up. Euler, "Examen de la dissertation de M. le professeur König. . .", *BAS* 1751, 219–45; idem, "Sur la principe de la moindre action" (originally published *BAS* 1751, 199–218). Euler himself was implicated in König's charges as well, having worked extensively on applications of the principle of least action. Euler's contributions to the polemic are collected in Euler, *Opera Omnia,* series 2, 5.

97. "Cette démangeaison excessive qu'ils ont d'exercer leur critique sur tout ce qui se passe dans la République de Lettres. Car toutes leurs plaintes . . . font assez voir qu'ils n'entendent pas seulement l'etat de la Question bien qu'il soit exposé dans ce Jugement avec la derniere netteté" ("Lettre de M. Euler à M. Merian," *BAS* 1750 [1752], 520–32; reprinted in Euler, *Opera Omnia,* series 2, 5:133, 135).

readers of the *Mémoires* of the Berlin Academy, and they printed their polemical texts in separate editions as well.[98] The texts appeared in both French and Latin, which, as a reviewer for the *Journal des sçavans* noted, "is very appropriate to make them more public."[99] Readers of the periodical press got König's version of the story in anonymous articles railing against the Academy. Flaunting his rupture with that institution, he published his own pamphlet in Leiden, outside Frederick's realm. He appealed to "the public" as a "citizen of the Republic of Scholars" victimized by the tyranny of the Prussian Academy's judges. Writing in French, in the third person, the author claimed that "he recognizes no Superior, no private judge. Only the Public is his natural judge."[100]

König's vociferous objections to being judged by the Academy prompted Merian to articulate and defend the ideology inherent in the institution. He argued that the Academy had approved and adopted as its own "the discovery of M. de Maupertuis and the geometrical operations of M. Euler connected to that discovery."[101] By printing the works of great men, the Academy took on the obligation to defend their honor. Academies are, so to speak, repositories of glory, a powerful force driving new discoveries, and they can only manage this force by preserving the reputations of their great men.[102] "If an academy has the authority to publish the eulogies of its members, even more does it have authority to publicly repulse injuries done to them." In Merian's picture, the reading public played only a distant and dependent role, since outsiders could not legitimately judge merit. But this defender of intellectual elitism was also concerned about the public image of philosophy and philosophers. The quarrel seemed to represent the breakdown of rational and polite order.

98. The first of these, *Lettres concernant le jugement de l'Academie royale des sciences,* included Euler's report, a further defense in the form of a letter to Merian, and a letter from Maupertuis to Euler. Euler's two academic memoirs on the König dispute, originally published in *BAS* for 1751 (see above, n. 98), were reprinted in French and Latin as *Dissertation sur le principe de la moindre action avec l'examen des objections de M. le Prof. Koenig faites contre ce principe* (1753). Merian published his own detailed history in *Mémoires pour servir à l'histoire du jugement de l'Académie* (1752). On the publishing history of these papers and letters, see notes to the Maupertuis-Euler correspondence by Pierre Costabel, in Euler, *Opera Omnia,* series 4A, 6: 208–12.

99. *Journal des sçavans* (December 1752): 818.

100. "Il ne reconnait aucun Superieur, aucun juge particulier. Le Public seul est son juge naturel" (*Appel au Public du jugement de l'Académie Royale de Berlin sur un fragment de lettre de Mr. de Leibnitz, cité par Mr. Koenig* [Leiden: Elie Luzac, 1752], 47). König's original paper had appeared in Latin, in a scholarly journal. As the polemic heated up, König responded to Euler again (König, *Défense de l'Appel au public* (Leiden, 1753)).

101. Merian, *Mémoires pour servir,* 45.

102. The *Journal des sçavans* concurred: "On approuvera sans doute l'intéret que l'Académie de Berlin prend à la gloire de son illustre Président" (*Journal des sçavans* (December 1752): 818).

> Is this the society of cultivated minds, formed by the lessons of wisdom, polished by the commerce of Muses, . . . or an anarchy of barbarians who have given in to all the excesses of an implacable hatred and a fury that knows no limits? A city of philosophers or a hideout of brigands? If men of letters debase themselves to this point, what claims can they have henceforth on the esteem of the public? . . . In reading the story of our dispute with M. König, posterity will be surprised to see in an enlightened century and among civilized nations, that these atrocities have reached their peak.[103]

Leaving aside the hyperbole about König's viciousness and posterity's interest in the case, we see that Merian made his adversary into a wild man who did not understand the rules of civilized discourse. No sooner had the Academy rendered its judgment than "the gazetteers sounded the alarm; once this signal was given, the whole plebeian order of the Republic of Letters ran to take up arms: the world was inundated with anonymous libels filled with ignorance, calumnies and rudeness."[104] Merian asserted the Academy's moral superiority, by equating König with the rabble of the print world, and reserving for the Academy the privileged imprimatur of state sponsorship. In this sense, the issues at stake, as the dispute spread outward from Berlin, became central Enlightenment issues of public access to knowledge, the regulation of print, and the role of criticism in public discourse.

Many of the anonymous libels mentioned by Merian were the work of Voltaire, hardly a "plebeian" in the Republic of Letters, but a self-styled enemy of tyranny and defender of press freedom. He tried to make König into a victim of crass injustice and came to his defense by attacking Maupertuis. When a nasty review of Maupertuis's works and a satire on the Academy's ruling appeared in the September 1752 issue of the *Bibliothèque raisonnée,* they were not immediately identified as Voltaire's work, and he denied responsibility, although his authorship was soon known throughout Europe.[105] His foray into the dispute was unexpected; Merian reported to Bernoulli that Voltaire had taken König's side, "one does not

103. Merian, *Mémoires pour servir,* 146.

104. Ibid.

105. Voltaire published two pieces anonymously in the same issue: a review of Maupertuis's *Oeuvres* and "Réponse d'un académicien de Berlin a un académicien de Paris" (*Bibliothèque raisonnée,* 18 September 1752). Frederick wrote to Maupertuis, "J'ai été indigné d'un nombre d'écrits qui ont parus contre vous, je ne sais qui en sont les auteurs, mais je ne les taxe pas moins de la lacheté et de la plus infâme malice" (18 October 1752, in Koser, *Briefwechsel,* 278). Voltaire denied his involvement: "Je puis vous assurer que je ne me suis melé ni de son affaire ni de son livre, quioique je n'approuve ni l'un ni l'autre. . . . Je l'abandonne à lui meme; mais encore une fois, je n'entre pour rien dans les querelles qu'il se fait, & dans les critiques qu'il essuie" (Voltaire to La Condamine, October 1752, Volt., *Corr.* 97:211–12).

really know why."[106] He was motivated by personal animosity toward his old friend and mentor, stemming from various petty intrigues at court, as well as by hostility toward certain of his more speculative ideas.[107] The principle of least action held little intrinsic interest for Voltaire. He did not understand the mathematical aspects of the question, and he certainly was an unlikely defender of Leibniz or of Wolffian professors. He concentrated on ridiculing Maupertuis as a tyrant and a buffoon, and on making fun of the more speculative parts of his works. According to Merian, "M. de M. is treated [in the satire] like the worst of men, it says that he dishonors and tyrannizes the Academy and that if it weren't for the king all the academicians would desert."[108] With his attack on Maupertuis's personality, literary style, and scientific ideas, Voltaire added another layer to the arguments about König's mechanics and made it into a literary quarrel.

In addition to the review, Voltaire wrote a separate satirical piece in the *Bibliothèque raisonnée,* under the title "Réponse d'un académicien de Berlin à un académicien de Paris," ridiculing the Academy's ruling. When Frederick realized who had written it, he fired off his own pamphlet, "Lettre d'un académicien de Berlin à un académicien de Paris"[109] As he told Maupertuis, "I have waited in silence until now to see what your Academy would do, and if there would be anyone who would reply to the libels printed against you; but as everyone has remained mute, I have raised my voice. . . ."[110] As protector of the Academy and patron of Maupertuis, Frederick interpreted Voltaire's satire as an affront to his own honor. The

106. Merian described Voltaire's review as "la plaisanterie d'un auteur qui ne cherche que de rendre ridicule, sans comprendre meme et sans se soucier de comprendre les ouvrages qu'il attaque" (Merian to JBII, 20 February 1753, BEB).

107. "Platon [Maupertuis] a pensé mourir de douleur de n'avoir point été de certains petits soupers ou j'étais admis, et le roi nous a avoué cent fois que la vanité féroce de ce Platon le rendait insociable" (Voltaire to Mme Denis, 15 November 1752, Volt. *Corr.,* 97:235). Voltaire resented Maupertuis's ties to La Beaumelle, who engaged in a long battle on other matters with Voltaire in print. Maupertuis refers to yet another bone of contention in an unpublished letter to Voltaire (December 1751, SM, fol. 99). There seem to have been innumerable perceived slights and offenses between them.

108. "M. de M. est traité comme le dernier des hommes, il y est dit qu'il déshonore et qu'il tyrannise l'Academie et que sans le roi touts les academiciens déserteroient" (Merian to JBII, 14 November 1752, BEB).

109. Also published in France (1753) and in *Maupertuisiana* (Hamburg, 1753). Maupertuis was in poor health throughout this period: "Je suis très affligé de voir par votre lettre qu'au lieu de penser au voyage de France vous pensez à ce voyage dont on ne revient plus" (FII to M, 28 August 1752, in Koser, *Briefwechsel,* 277).

110. "J'ai attendu jusqu'ici dans le silence pour voir ce que ferait votre Académie, et s'il ne se trouverait personne qui répondrait aux libelles qu'on a fait imprimer contre vous; mais comme tout le monde est demeuré muet, j'ai élévé ma voix" (FII to M, 7 November 1752, in Koser, *Briefwechsel,* 281).

first edition of the king's pamphlet appeared anonymously, but a second soon was circulating with the royal arms on the title page, a tacit admission of authorship recognized by all.[111]

Frederick sought to trump Voltaire with his dual identity as both monarch and author of anonymous libels, and assumed that it would end there. Voltaire professed shock at the unkingly behavior of the king, but proceeded to publish yet another attack on Maupertuis, the *Diatribe du docteur Akakia,* in flagrant violation of Frederick's orders. Instead of sending his manuscript off to Holland, this time Voltaire printed his pamphlet clandestinely, under the very nose of the king, using a faked permission to obtain the services of the royal printing press.[112] The king had the edition seized and burned in his chambers, ordered Voltaire to cease and desist, and again assumed that the case was settled.[113] But Voltaire simply smuggled the manuscript to Leiden for printing and imported a shipment back into Berlin, while sending others on to France. This sort of clandestine publishing was a completely familiar modus operandi for Voltaire, who had longstanding connections with printers, booksellers, and smugglers in Amsterdam, Leiden, and the Hague.

The spectacle of a king scrapping with one of his fallen favorites dazzled Berlin when Frederick ordered the royal executioner to burn confiscated copies of Voltaire's book on the first day of Christmas in the public squares of the city, "in full view of great crowds of people as the churches let out around midday."[114] The king sent the ashes ("a refreshing little powder") to Maupertuis in his sickroom.[115] Especially because it took place at the heart of the Prussian court, the dispute turned into a spectacle that drew the attention of readers and gossips all over Europe. All the journals and gazettes covered it, letters crisscrossed Europe with news of the latest publications and speculations about what would happen to Voltaire, and authors orchestrated the distribution of their pamphlets and books in

111. Frederick II, "Lettre d'un académicien de Berlin. . .", Berlin 1752. Frederick attacked Voltaire as "faiseur de libelle sans génie" and "cet ennemi méprisable d'un homme d'un rare mérite." Mervaud, *Voltaire et Frédéric II,* recounts the exchanges between Voltaire and Frederick in exhaustive detail. At his death, Maupertuis's papers included two copies of this pamphlet, now in AS Fonds Maupertuis.

112. For the details of the publication of *Diatribe du docteur Akakia,* see Tuffet, introduction to Voltaire, *Histoire du docteur Akakia.* Voltaire could not attack the king directly, so he continued his satires of Maupertuis: "M. de V. ne se tut pas. N'osant s'en prendre a l'auteur il s'en prit a M. de M" (Merian to JBII, 20 February 1753, BEB).

113. Frederick informed Maupertuis of this scene, 10 December 1752: "L'affaire des libelles est finie" (Koser, *Briefwechsel,* 285).

114. Merian to Bernoulli, 20 February 1753, BEB.

115. FII to M, 24 December 1752, in Koser, *Briefwechsel,* 286.

France. Voltaire continued to produce short, punchy satires, had them printed as cheap pamphlets, and circulated them as manuscripts as well. He kept the quarrel alive by feeding these works to the scandal-hungry public in small doses, then repackaging them in larger editions.[116]

Francophone readers manifested an insatiable appetite for any bit of news, gossip, or text associated with Voltaire. The authenticity of the Leibniz letter and the principle of least action interested the German public; Voltaire performed his antics primarily for French readers, although he had a local German audience as well. Maupertuis tried to control his public image by overseeing the publication of documents and "histories" that would contrast with the slanderous tone of his enemies. Although he had experience with writing the kind of scurrilous satire now being produced against him (recall his attacks on the Cassinis), this authorial posture would have undermined his status as president of the Academy, a status he was intent on protecting. Of course, the conflict was personal as well as institutional. Formey later remembered that "The tone of equality König adopted in the beginning did not suit M. de Maupertuis, who had a bit of hauteur, and who recalled the distance there had been between them at Cirey."[117] Voltaire also harbored personal animosity, as well as philosophical objections, and he conducted his attacks accordingly. Though they shared a broadly anticlerical and antidogmatic outlook, Voltaire resented Maupertuis's attack on the argument from design as a betrayal of the Newtonian heritage. In spite of their long acquaintance, and even alliance, their interaction in Berlin fueled their antipathy. Well before their animosity became public, friends were aware of the potential for disaster in their proximity. As Buffon commented on hearing of Voltaire's move to Prussia, "Between ourselves, I think that the presence of Voltaire will be less pleasing to Maupertuis than to anyone else; these two men are not made to live together in the same room."[118]

For Maupertuis, the personal, institutional, and intellectual aspects were inextricably bound together, as interlocking components of his identity and his reputation. His position as president was a crucial aspect of his identity in Berlin, and of his image as seen abroad as well, benefiting from the luster cast on him by the

116. See "Introduction" to Voltaire, *Histoire du docteur Akakia,* for an attempt to sort out the complex bibliography of these texts. Many of the pamphlets were ephemeral and do not survive in their original form.

117. Formey, *Souvenirs,* 1:176. Formey also tells a story about the visceral nature of Maupertuis's reaction to König's affrontery: When Formey suggested that Maupertuis might do better to ignore his antagonist, "Ses yeux s'enflammèrent; 'quoi,' dit-il, 'vous voulez donc qu'on me prenne pour un Olybrius, etc.' Son ton fulminant m'effraya, & je compris que j'en avois trop dit" (Formey, *Souvenirs,* 1:179).

118. Buffon to LeBlanc, 22 October 1750, in Buffon, *Correspondance générale,* 1:72.

favor of the king. Voltaire tapped into hostility to Maupertuis among German readers, but also played cruelly on Maupertuis's claims to greatness by virtue of proximity to great political power. By relying on his academicians to defend him, and by refraining from printing invective against Voltaire, Maupertuis attempted to maintain his dignity. His defenders appealed strategically to an elite public of their peers, leaving the wider public, who could not see through König's "paralogisms," to the other side. All participants were quite conscious of the complex nature of the reading public. Merian noted that when König appealed to his public, he addressed only an assembly of individuals, lacking the authority lent by corporate membership. No one can speak for an undifferentiated public, Merian argued. Some parts of the public are worthy audiences, while others are not; ignorant critics and "weekly pettifoggers [*chicaneurs hebdomodaires*]" are "the shameful part" of the public.[119] Reporting the latest from Potsdam in July 1753, the royal engineer Lefebvre, a friend and ally of Maupertuis, reflected on the ignorance and fickleness of public opinion: "Everyone is talking a lot about the conduct of the King towards the poet [Voltaire] and they seem to fear the judgment of the public; but what is the public, in a word, but a hydra with 500,000 heads of which each half dozen makes up a private body that never judges things soundly, that reasons and decides on everything without understanding and without knowing."[120] This was the paradox of turning to the public for approbation, while recognizing the limitations of the same public to make reasonable judgments. The tension between the absolutist control of knowledge and the free movement of texts and ideas through the Republic of Letters characterized much of the intellectual life of the Enlightenment at mid-century. The question was especially fraught in Prussia, where Frederick II himself wanted recognition as a citizen of the Republic of Letters as well as admiration as absolute ruler, seeking public acclaim from cosmopolitan Europeans while scorning German provincialism and ignorance.

Although it has often been said that the bitter feud between Voltaire and Maupertuis ruined the end of the latter's life, this diagnosis is far from definitive. Maupertuis did suffer a series of health crises in this period, but they predated his conflict with Voltaire. His obsession with Voltaire over the next few years did not preclude other work. And the principle of least action did not suffer either. Euler's

119. Merian, *Mémoires pour servir*, 56.

120. "On parle beaucoup de la conduitte du Roy vis-a-vis du poete [Voltaire] et on semble craindre le jugement du public; mais qu'est-ce donc en un mot que le public sinon une hydre à 500,000 têtes dont chaque demi-douzaine fait un corps particulier qui ne juge jamais des choses sainement, qui raisonne et decide de tout sans connoitre et sans scavoir" (Lefebvre to M, 23 July [1753], AS Fonds Maupertuis). Lefebvre kept Maupertuis informed of events in Berlin and Potsdam when he was away; in this case, Maupertuis was in France.

further work on the principle of least action was a direct result of König's challenge. In the winter of 1753, Maupertuis noted with satisfaction that "M. Euler is having a work printed here in which he examines (1) to whom the principle of least action belongs, (2) the extent and the worth of this principle. I believe that when König sees this work he will regret having asked that the Academy examine the thing."[121] In addition to exposing the faults in König's mathematical reasoning, Euler extended his own examination of the implications of the principle of least action in several memoirs in 1753.[122] As we have seen, Euler had already applied the principle of least action to a variety of problems beyond the rudimentary applications worked out by Maupertuis. Now, prompted by his analysis of König's paper, he spelled out the way the principle worked for both statics and dynamics. These papers were designed rhetorically to reinforce Maupertuis's claims, but Euler was also working out his own commitment to the least action approach to mechanics. As the premier mathematician in Europe, his assessment carried a great deal of weight and inspired the early work of Joseph-Louis Lagrange on the application of variational methods to mechanics in the 1760s.[123] Several German mathematicians also followed up on Maupertuis's approach. Kästner, when he wrote a widely used textbook on mechanics, articulated his objections to the Leibnizian law of continuity in terms of the principle of least action.[124] G.-W. Krafft, a mathematics professor at Tübingen, printed a dissertation on the principle of least action in 1753 that recognized Maupertuis as the author of the principle.[125]

Maupertuis was further vindicated by d'Alembert's assessment of the dispute in his *Encyclopédie* article "Cosmologie" (1754).[126] In spite of their quite different approaches to mechanics, d'Alembert gave Maupertuis credit as "the first [to] reduce

121. "M. Euler fait imprimer icy un ouvrage dans lequel il examine (1) a qui apartient le principe de la moindre action, (2) l'etendue et le valeur de ce principe. Je crois que lors que Koenig verra cet ouvrage il se repentira d'avoir demandé que l'academie examinate la chose" (M to La Condamine, 13 January 1753, SM, fol. 129). The book in question was Euler, *Dissertation sur le principe de la moindre action.*

122. These were published in the volumes for earlier years, to avoid the standard delay in printing the *Mémoires.* (The volume for 1751 appeared in 1753.) Euler, "Harmonie entre les principes généraux de repos et de mouvement de M. de Maupertuis," *BAS* 1751, 169–98; idem, "Sur le principe de la moindre action," *BAS* 1751, 199–218; idem, "Essay d'une démonstration métaphysique du principe général de l'équilibre," *BAS* 1751, 246–54; these papers were reprinted in Euler, *Opera Omnia,* series 2, vol. 5.

123. On Lagrange, see Craig Fraser, "J. L. Lagrange's Early Contributions."

124. See Kästner to M, 12 Nov. 1749, in LS, 273–74. Also Clark, "Death of Metaphysics."

125. Krafft to M, 25 March 1753, AS Fonds Maupertuis. Krafft sent Maupertuis a copy of his dissertation, *Theses inaugurales matematico-physico de numero pari, rectis parallelis et principio actionis minimae.*

126. Maupertuis saw d'Alembert as ally in 1752 (M to d'Alembert, January 1752, SM. fols. 99v–101). See also de Gandt, "1744: Maupertuis et d'Alembert entre mécanique et métaphysique."

the impact of hard bodies and elastic bodies to a single law."[127] He found no merit in König's claim for Leibniz's priority and explained the wide utility of the principle, citing also Euler's "elegant and straightforward" applications to a variety of problems. Finally, d'Alembert applauded his colleague for his demeanor in the dispute: "M. de Maupertuis never replied to the insults that were vomited against him on this occasion. . . . This quarrel on action, if it be permitted to say so, resembled certain religious disputes, by the bitterness put into it and by the quantity of people who spoke of it without understanding anything about it."[128] Given the readership of the *Encyclopédie*, such support was particularly valuable.

Maupertuis left Berlin for about a year, from May 1753 to July 1754, for Paris and the healthy "native air" of Saint-Malo. On the way to Paris, he met up with the journalist Fréron in Nancy, showed him a letter he had written to Voltaire, and enlisted his support.[129] He avidly kept abreast of Voltaire's latest moves through correspondence with friends in Berlin and elsewhere, collecting as many copies of pamphlets and books as he could.[130] When a new satire directed at Frederick II appeared in Paris, also by Voltaire, Maupertuis told the king that he was ready to respond to Voltaire at last.[131] "I thought, Sire, that it might be appropriate to publish a short history of the conduct of Voltaire at your court, of your goodness toward him and of the manner in which he responded. I have written something which I have the honor of sending to you; I am perhaps not fully informed of all the facts, it is only a sketch."[132] In the event, Frederick did not encourage this publication, and it never saw the light of day.[133]

127. [D'Alembert], *Encyclopédie*, s.v."Cosmologie," 4:296.

128. "M. de Maupertuis n'a jamais rien repondu aux injures qu'on a vomies contre lui à cette occasion. . . . Cette querelle de l'action, s'il nous est permis de le dire, a ressemblé à certaines disputes de religion, par l'aigreur qu'on y a mise, & par la quantité de gens qui en ont parlé sans y rien entendre" (ibid., 297).

129. The letter (3 April 1753) was soon to be published in *La Nouvelle bigarurre*, vol. 3 (May 1753): 40. For the meeting, see Fréron to d'Hémery, 15 May 1753, in Balcou, *Dossier Fréron*, 55; see also the report on the quarrel in Fréron, "Lettre IX," *Lettres sur quelques ecrits*, 8:197–210.

130. See correspondence with, e.g., Lefebvre (AS Fonds Maupertuis); Knyphausen (AS Fonds Maupertuis); La Beaumelle (LS).Many of the documents Maupertuis collected in this period remained in his papers at the time of his death and can now be found in the AS Fonds Maupertuis.

131. The pamphlet, published anonymously, was *Idée de la personne, de la manière de vivre et de la cour du roi de Prusse*. Frederick had seen it; as he told Maupertuis, "Les libelles qui ont couru à Paris sont de lui. Pour déguiser son style, il les a fait traduire en allemand et de l'allemand retraduire en français" (FII to M, 15 [September 1753], in Koser, *Briefwechsel*, 296). The king affected indifference to Voltaire's calumnies (ibid.).

132. M to FII, 28 August 1753, in Droysen et al., *Nachträge*, 7–8.

133. Fréron reported rumors about this work: "Il a fait une petite histoire de sa querelle avec Voltaire en forme de letter à un ami. Il la ferait imprimer probablement" (Fréron to d'Hémery,

The dispute had certainly added to Maupertuis's notoriety in France, where he was already a controversial figure by virtue of his defection to Prussia: "It has been said rightly that one should not have received either him or Voltaire in France; they are two scorched brains [*cervaux brulés*], who insolently left their homeland."[134] Fréron's letters to the police inspector d'Hémery, filled with literary news of all sorts, show that the "quarrel" provided plenty of material for gossip, as Voltaire and Frederick continued their feud and Maupertuis retailed his story around Paris. But though he undoubtedly burned with hatred for Voltaire, it did not incapacitate him. His other main intellectual interest in this period was a new work on generation, which he was discreetly introducing to literary circles in Paris, and which would be published in the spring of 1754, just before his return to Berlin.

---

12 August 1753, in Balcou, *Dossier Fréron,* 89). Several manuscripts in AS Fonds Maupertuis could be fragments of such an account; some pieces of these were used in revising the preface to the *Essai de cosmologie.*

134. "On a dit avec raison qu'on n'avoit dû recevoir en France ni lui ni Voltaire; ce sont deux cervaux brulés, qui ont insolemment quitté leur patrie" (Fréron to d'Hémery, 20 August 1753, in Balcou, *Dossier Fréron,* 89).

# 10

# Heredity and Materialism

*They will say that all is lost if we admit that thought can exist in matter. But I beg them to listen to me and to respond.*

(MAUPERTUIS, *SYSTÈME DE LA NATURE*)

THE DISPUTE WITH KÖNIG AND VOLTAIRE had embroiled Maupertuis in contorted arguments about the authenticity of texts and the proper interpretation of the principle of least action. The public trajectory of this dispute only makes sense in light of Maupertuis's status as head of an academy and favorite of the Prussian king. In private, while the various books, journals, and pamphlets relative to the König affair rolled off the presses across Europe, Maupertuis was working on quite another problem: how to present his reflections on the interrelated questions of the generation of organisms, the properties of matter, and the possibility of a science of life. In *Vénus physique,* Maupertuis had played with genre, style, and risky ideas about the properties of organic matter. He did not submit his theory for the approbation of the Academy, nor did he write a systematic treatise with rigorously derived conclusions. His queries and speculations suggested routes for readers to follow, without providing authoritative answers. The effect of the queries was to draw readers into a kind of conversation, and he continued the conversation in subsequent revisions of the theoretical parts of the text. The element of play, including the pose of anonymity and the use of gallantry and irony, was integral to the project, as the author manipulated his own identity as man of letters, as naturalist, as wit, and as man of science. But he also wanted his readers to think about methods appropriate for rational understanding of reproduction and about the literary style appropriate for sophisticated and "enlightened" readers. All of this was controversial, as we have seen.

When he moved to Berlin, Maupertuis had to forfeit his seat in the Paris Academy of Sciences, but he continued to play a role in the Francophone Republic of Letters, which valued playful wit, performance, curiosities, and imaginative reflection. In fact, his engagement with this literary culture enhanced his value to the Prussian king and court. While he was occupied with running the Berlin Academy and improving its image in the Republic of Letters, according to Frederick's wishes, he continued his work on mechanics, metaphysics, and generation. His attention to the problem of generation is evident in his revision of *Vénus physique* for a new anonymous edition in 1751 and then in its inclusion in a one-volume collection of his works the following year—the same volume that elicited Voltaire's nasty review in the *Bibliothèque raisonnée.* He also wrote other texts on the subject, published in various editions between 1751 and 1756.[1] In the same period, he was reading new works by Buffon, Needham, Réaumur, and others, as well as conducting his own investigations on the inheritance of traits in dogs, birds, and people.[2] This chapter examines Maupertuis's excursions into the vexed questions of organization, life, and generation in the context of broader debates in the 1750s about natural history, experiment, and the relation of science to theology.

Publication of the first volumes of Buffon's *Histoire naturelle* in 1749 sparked much of this discussion. This work came under attack by naturalists and theologians, who saw it, and the worldly secular culture that lionized it, as a threat to both Christian morality and natural historical practice. In developing the theory he had sketched out in *Vénus physique,* Maupertuis allied himself with Buffon, a personal friend and longtime associate, even though they did not agree on all particulars.[3] They mounted a challenge to the descriptive natural history pursued by Buffon's enemy Réaumur, a challenge that Diderot then folded into the critical project of the Encyclopedists. Although no one questioned Réaumur's skill and dedication, his opponents represented his observational approach as narrow, even trivial, since he insisted on avoiding causes or metaphysics. So, for example, Diderot charged Réaumur with small-mindedness, playing on the naturalist's long years of watching insects: "What would [posterity] think of us if we were able to leave it no more than an incomplete insectology, a vast history of microscopical animals? Great objects to great talents; small objects to small talents."[4] Diderot championed the power of imagination and sweeping vision over the minute observation of particulars. This vision resonated with the program of Buffon's vast publication project and with Maupertuis's theoretical interventions about the submicroscopic realm.

1. *Vénus physique,* [Paris], 1751; reprinted in *Oeuvres* (Dresden: Conrad Walther, 1752; also in 2 vols., Lyons: Bruyset, 1753). This was the first time Maupertuis acknowledged his authorship in print. "Lettre sur la génération des animaux," in *Lettres de M. de Maupertuis,* Dresden: 1752; Berlin, 1753. *Dissertatio inauguralis metaphysica . . . ,* Erlangen [Berlin]: 1751; *Essai sur la formation des corps organisés,* Berlin [Paris]: 1754; reprinted as *Système de la nature,* in *Oeuvres,* 4 vols., Lyon: 1756.

2. For breeding experiments, see below. See also C. Wolff to M, 15 November 1746, LS, 428–29, for discussion of six-fingered families.

3. Roger, *Buffon;* Spary, *Utopia's Garden.* For friendship between Buffon and Maupertuis, see Sallier to M, Sept. 1750, AS Fonds Maupertuis, where he recalls friendly dinners; Duclos to M, 22 September 1753, in LS, 382; and Maupertuis letters to Buffon, SM.

4. Denis Diderot, *Pensées sur l'interprétation de la nature,* para. LIV, 86.

Many threads wove a tapestry of discourse around these questions in books, letters, and conversations. From Prussia, Maupertuis participated in a complex dialogue with his former compatriots. The whole network of exchanges among writers (and readers) on these topics proceeded across international boundaries like a large-scale conversation. Critiques of preexistence, of observational practice, and of natural theology were serious business, but they used the techniques of witty rhetoric and performance familiar to the habitués of salons and coffeehouses.

To set the scene for Maupertuis's own contributions, I look briefly at two works published in 1749: the first three volumes of Buffon's *Histoire naturelle, générale et particulière* and Réaumur's *Art de faire éclorre et d'élever . . . des oiseaux domestiques [On the Hatching and Breeding of Domestic Fowls]*. Both authors aimed their books, quite different in genre and style, at an eclectic audience including genteel women, philosophers, experimentalists, gentleman-farmers, and academicians. These works serve as useful reference points for charting the evolving dynamics of the intertwined issues of generation, natural theology, active matter, and the methods of natural history. Réaumur and Buffon, in many ways at opposite poles of these debates, both commented on *Vénus physique,* assuming that their readers would recognize oblique references to Maupertuis's notorious book, and in turn their books inspired Maupertuis to re-enter the conversation.[5]

## Buffon and the *Histoire naturelle*

Buffon, embarking on the task of describing the royal natural history collection, surrounded these objects with theoretical excursions into everything from the evolution of the solar system and the shape of the earth to generation and the "behavior [*moeurs*]" of animals. He worked self-consciously in an accessible, occasionally even poetic, style that made his lavish books instantly successful with the lettered elite across Europe. The integration of the specimens in the king's cabinet into the grand sweep of nature's profusion downplayed nature as evidence of God's power and wisdom and instead made it analogous to the display of the king's own profuse wealth.[6]

5. I disagree with Roger's assessment that "*Vénus physique* was seen as a pamphlet of little importance." According to this view, Maupertuis's contemporaries did not appreciate the significance of his work "until [his] ideas were taken up and discussed by Buffon and Diderot" (*Sciences de la vie,* 475–76).

6. Loveland has made a close study of Buffon's literary style and rhetoric in *Rhetoric and Natural History.* The first edition of *Histoire naturelle* sold out in six weeks; another quarto edition was published in 1750, followed shortly thereafter by a cheaper duodecimo edition. For the "prodigious" success of the book, see Buffon's correspondence from January and February 1750, *Correspondance générale,* 1:60–72. Buffon to de Brosses, 16 February 1750, mentions the particular success of the chapters on generation (Buffon to de Brosses, 16 February 1750, ibid., 64–65). As Spary notes,

According to John Turberville Needham, Buffon had "long been dissatisfy'd with the Opinion of preexistent Germs in Nature; and he and M. Maupertuis . . . had often discours'd together upon this subject."[7] These conversations must have taken place around the time that the albino boy was brought to Paris in the winter of 1744, before Maupertuis left for Berlin.[8] Buffon endorsed the attack on preexistence theories found in *Vénus physique,* a book he described in the most favorable terms. "This treatise, although very short, brings together more philosophical ideas than are to be found in several large volumes on generation. Since everyone has this book already, I will not analyze it."[9] He admired its "precision" and noted that its conclusions agreed in general with his own. Toward the beginning of his investigation of reproduction, Buffon declared that "the living and animate, instead of being a metaphysical aspect [*un degré métaphysique*] of creatures, is a physical property of matter."[10] By analogy to the composition of inert substances ("a grain of sea salt is a cube composed of an infinity of other cubes visible under the microscope") he reasoned that "there is in Nature an infinity of organic particles actually in existence, alive, whose substance is the same as that of organized beings."[11] Organic molecules are not yet organisms—they cannot reproduce themselves—but they are the material of life, the matter from which organisms build themselves.

Having deduced the existence of "living organic particles," Buffon went looking for them. In collaboration with the English microscopist Needham, who had earlier found teeming animalcules in the milt of squid, Buffon examined fluids taken from the sex organs of dogs and other animals. They found "moving bodies," many with tails, in both male and female fluids, and decided that these apparently ubiquitous entities must be organic elements that had combined to produce "germs" that would in turn generate new organisms. From the modern point of view, these experiments involved considerable confusion about what was actually

"Buffon's [ornithology] incorporated concerns for measurement and precise physical determinations within a narrative of luxury, rarity and value deriving from Nature's boundless creativity" (Spary, "Codes of Passion," 112).

7. Needham, *Observations upon the Generation, Composition, and Decomposition of Animal and Vegetable Substances;* translated in *Nouvelles observations microscopiques.* For Maupertuis's conversations with Buffon, see M to La Condamine, 17 May 1750, SM, fol. 92. For Buffon and Needham collaboration, and their theories of generation, see Roe, "Buffon and Needham."

8. Buffon was not present at the Academy when the albino was presented; he was in Paris by 11 January, and undoubtedly saw the boy on display in a private home (AS p-v, 8 January 1744).

9. "Comme ce livre est entre les mains de tout le monde" (Buffon, *Histoire naturelle,* in *Oeuvres philosophiques,* 285). Buffon preserved Maupertuis's anonymity in this passage.

10. Ibid., 238.

11. Ibid., 239.

seen.[12] But for Buffon and Needham, their observations of seminal fluids led them to conclude first of all that mammalian eggs did not exist, and secondly that male and female functioned symmetrically in reproduction. Following further experiments with infusions of seeds and meat, in which moving particles appeared spontaneously, they concluded that organic particles could be found throughout nature—not just in reproductive organs. They claimed to have actually seen the building blocks of living organisms.

Next they wondered how to understand the organization of these elements into functioning, living organisms. Buffon held that organic molecules come together through the action of "penetrating forces" that guide them into "internal molds" (*moules intérieures*) where they take on the appropriate structure and form of body parts. "In the same way that we can make molds by which we give to the exterior of bodies whatever shape we please, let us suppose that Nature can make molds by which she gives not only the external shape, but also the internal form, would this not be a means by which reproduction could be effected? . . . Nature can have these internal molds, which we will never have, just as she has the qualities of gravity, which in effect penetrate to the interior; the supposition of these molds is therefore founded on good analogies."[13]

The molds may be understood as structures, but they are also immaterial ordering principles; they are "forms" in the sense of a sculptor's forms for casting metal, but they are also forms in a more Aristotelian sense, functioning as guiding principles. If our senses were not limited to the surfaces of things, Buffon tells us, we might be able to conceive of internal molds more immediately. As it is, we have no reason to assume that Nature cannot employ means that are beyond our direct intuition. The internal molds were conceptually slippery because they sometimes operated as forces, by analogy to gravity, and sometimes as constraining structures, by analogy to the sculptor's molds. Buffon wrestled with how to combine a quasi-mechanical explanation of forces, shapes, and motions with a notion of active organic matter that resisted this kind of explanation. The internal mold was his counterintuitive proposal for solving the problem, an attempt to think beyond the limits of the human senses, analogous to Maupertuis's extrapolation from chemical affinities to organic forces. Both of them reasoned from "good analogies," al-

12. On Needham, see Roger, *Sciences de la vie,* 424–520; Roe, "John Turberville Needham and the Generation of Living Organisms"; idem, "Buffon and Needham." On the experiments on seminal fluids, see Roger, *Buffon,* 140–45 and Sloan, "Organic molecules revisited." See also the account of the experiments in Buffon, "Découverte de la liqueur séminale dans les femelles vivipares."

13. Buffon, *Histoire naturelle,* in *Oeuvres philosophiques,* 243–44.

ways recognizing the limits of those analogies. Their critics, in caricaturing these models, missed this crucial point.[14]

This way of drawing conclusions about forces and causes that can only be detected by their effects, incensed naturalists like Réaumur. But Buffon considered it the only way to know about natural processes, invoking the authority of Newton to support his strategy by making the analogy between gravity and internal molds quite explicit.

> I have admitted in my explanation of development and reproduction first the accepted principles of mechanics, then that of the penetrating force of gravity, which we are obliged to accept, and by analogy I thought I could say that there were other penetrating forces that act on organic bodies, as experience assures us. I have proved by facts that matter tends to organize itself, and that there are an infinite number of organic particles; I have thus done nothing but generalize from observations, without having advanced anything contrary to mechanical principles.[15]

Taking the force of gravity as the model for mechanical explanations begged the question for critics like Réaumur, for whom such forces were mysterious and even dangerous, especially when considered as properties of matter.

### *Réaumur*

When Réaumur published his book on the breeding and incubation of domestic fowls, the literary world already knew him well, as the author of the six-volume *History of Insects,* the work that provoked Diderot's pointed comment quoted above. Réaumur avoided grand views or reflections on causal mechanisms in favor of close observations of behaviors and structures. He only allowed himself to reflect on theories of generation at the end of his book on chickens, following a voyeuristic and completely humorless account of the "amours" of a hen and a rabbit. These animals, which Réaumur had kept under observation in his house for a time, had entertained the public, he remembered, "when the whole town of Paris was so desirous to be informed of it, when I met with nobody that did not desire me to tell him the whole truth of their amours, when, in short, the curiosity of seeing them both drew to my house so many people of all conditions and classes."[16]

14. Buffon's most outspoken critic was Joseph-Adrien Lelarge de Lignac, a Jesuit and amateur natural historian; see Lignac, *Lettres à un Amériquain.* For a caricature of Maupertuis's theory, see Voltaire, *Diatribe du docteur Akakia.*

15. Buffon, *Histoire naturelle,* 2:52–53.

16. Réaumur, *L'art de faire éclorre . . . des oiseaux domestiques;* English trans., *The Art of Hatching and Bringing Up Domestick Fowls,* quote on 457. All quotations from this translation.

This interest led, naturally enough, to speculation about the outcome of the interspecies union. A furry chicken or a feathered rabbit would have called into question Réaumur's commitment to preformed germs, so it's perhaps just as well that the hen's eggs never hatched. He ended his book with an attack on contemporary challenges to preformation, followed by suggestive comments about breeding experiments. This amounted to a defense of his style of natural history, along with the theology it supported, in the face of the kind of alternative he had encountered in *Vénus physique.* His remarks about generation articulated the opposition to active matter or directing forces. He found it impossible to believe that the cause of organization "can by any means be supposed to act blindly." Even if we "set loose our imagination" and suppose that "the prolifick liquors of either the male or the female are composed of parts similar to those which all the organs of either the one or the other are made up of," how can we imagine that those parts can organize themselves without an agent to make order out of chaos?[17] He noted that the "fashion" for attraction had led to the suggestion that it might "disentangle all the materials which enter into the prolifick liquors [and] . . . operate the miracle of the formation of the foetus. . . . [A]ll the similar parts fit to make a heart, all those fit to make a stomach or a brain . . . will seek those of their own kind, draw near and unite with them."[18] His reluctance to admit inherent and self-organizing forces in organic matter led him to denigrate Maupertuis's theory as a revival of discredited (and nonmechanical) occult qualities and sympathies. Ultimately, Réaumur fell back on the assertion that knowledge of the causes of the formation of animals is beyond the reach of human intelligence. To speculate, to turn the imagination loose, was tantamount to irreligion, but also undermined the tried and true method of natural history: meticulous observation and description of complex organic phenomena, with the imagination safely under control.[19]

Réaumur did note that chicken breeders might well shed light on the question of generation by crossing normal birds with others exhibiting hereditary peculiarities like extra claws. "For if we suppose, as we have done, that the germ exists before the copulation, and that all we want to know is whether it existed in the male

17. Ibid., 460–61. Buffon's book was not yet in print when Réaumur wrote *L'art de faire éclorre.* Réaumur subsequently encouraged Lignac in his explicit attack on Buffon. See Lignac, *Lettres à un Amériquain;* Roger, *Buffon,* 190–93.

18. Réaumur, *Art of Hatching and Bringing Up,* 461–62. "How will attractions give to such and such a mass the form and structure of a heart, to another that of a stomach, to another that of an eye . . . !" (ibid., 463).

19. Privately, Réaumur accused Buffon of lacking basic observational skills. See correspondence cited by Roger, *Buffon,* 190–92; Torlais, *Réaumur.* On Diderot's objections to Réaumur's natural history, see Spary, "Codes of Passion," 123.

or in the female before their junction, the chickens we are now speaking of, must needs shew us by some parts they shall have, or by the want of some other parts, whether it is to the male or to the female that the germ belonged to originally."[20] Exactly this argument would later provide Maupertuis with some of his strongest evidence against preexisting germs. But here Réaumur most tantalizingly, on the last page of his book, indicated that although he had performed many such crosses, he would save the results for another publication. In fact, he never published these results.

## Maupertuis

In the early 1750s, Maupertuis rejoined the debate about generation, although he continued to keep his work on generation quite distinct from his academic contributions, as he had in Paris. At the peak of his institutional and social power in Berlin, and in spite of recurrent chronic illness, he kept up his position in the Republic of Letters by publishing strategically to maximize exposure and minimize risk of attack. In correspondence and in his books, he entered into exchanges with readers and colleagues that can be traced through the pages of successive versions of his writings. His engagement with this issue was a matter of passionate personal interest, as well as a means of leaving his mark on philosophical questions of crucial concern to his contemporaries. In 1750, a new edition of *Vénus physique* was being printed in Paris; at the same time, Maupertuis "avidly" read Buffon's first two volumes in Berlin. "I admit," he wrote to La Condamine, "that I am not a little flattered to see that without [benefit of] all the experiments he has done, I built a system in my *Vénus* that strongly resembles his. The only difference between us lies in a point I don't understand at all, the internal molds." He also referred to earlier arguments with Buffon: "He is absolutely converted on the resemblances of children to their parents, about which we argued vigorously years ago, and which he would [then] think of as only accidental."[21]

The same summer Maupertuis read another new book, a translation of Needham's work on the generation and composition of living things. Needham reported on his experiments with Buffon on organic particles, and on many other microscopic observations of various organic infusions. His descriptions stunned Maupertuis, who wrote to La Condamine,

20. Réaumur, *Art of Hatching and Bringing Up*, 466.

21. "Je vous avoue que je ne suis pas peu flatté de voir que denué de tout les experiences qu'il a faites j'eusse bati dans ma Venus un systeme qui ressemble fort au sien. Car la seule difference qui est entre nous ne consiste qu'en un point que je n'entens point, qui est les moules interieres. Il est dans cet ouvrage absolument converti sur les resemblances des enfans aux parens, sur lesquelles nous

> Have you read Needham's book? What are we to think? What a new universe! What a shame that a man who observes so well reasons so poorly! After reading his book, my mind was so dizzy [*étourdi*] from all the ideas it presented to me that I had to go to bed like an invalid, and I have not yet completely recovered from the upheaval that this reading put me in. I hope when this tumult calms down a bit to take up again the thread of some meditations that I have begun some time since on this subject, and see if it is possible to pull out something reasonable from it.[22]

Needham's book described experiments with infusions of seeds and meat gravy, all of which produced microscopic "moving Globules" after sitting for a few days. Based on many observations, he decided that "vegetative Powers" belonged to all organic substances, animal and plant alike. Here is one description of the evolution of life in an infusion of macerated wheat:

> To the naked Eye . . . it appear'd a gelatinous Matter, but in the Microscope was seen to consist of innumerable Filaments; and then it was that the Substance was in its highest Point of Exaltation, just breaking, as I may say, into Life. These Filaments would swell from an interior Force so active, and so productive, that even before they resolved into, or shed any moving Globules, they were perfect Zoophytes teeming with Life, and Self-moving.[23]

When he heated a sealed container of mutton gravy, Needham found after some time that "My Phial swarm'd with Life, and microscopical Animals of most Dimensions, from some of the largest I had ever seen, to some of the very least."[24]

Maupertuis was not convinced by Needham's talk of epigenesis through vegetation, but the array of new observations roused him to declare, "Here we can say that the structure of the tiniest insect is more marvelous than that of the whole

---

nous estions fort disputés autrefois, et qu'il ne regarderoit absolument que comme accidentelles. Au reste cet ouvrage me paroit excellent, rempli de choses extremement curieuses et bien ecrit: je ne sais pas d'ou ont pu venu les persecutions qu'on lui a faites en France" (M to La Condamine, 17 May 1750, SM, fol. 92).

22. "Avez vous lu le livre de Needam? Ou en sommes nous? Quel nouvel univers! Quel dommage qu'un homme qui observe si bien raissonne si mal! Il m'est arivé apres la lecture de son livre d'avoir l'esprit si étourdi de toutes les idées qu'il me presentoit qu'il m'a fallu me coucher comme malade et je ne suis pas encor fort bien remis du trouble où cett lecture m'a mis. J'espere quand ce tumulte sera un peu apaisé de reprendre le fil de quelque meditations que j'ai dequis quelque tems commencées sur ce sujet et voir s'il est possible d'en tirer quelque chose de raisonnable." (M to La Condamine, 24 August 1750, SM, fols. 125v–126). The book was Needham, *Nouvelles observations microscopiques*.

23. Needham, "A Summary of some late Observations," para. 26, 647.

24. Needham, *Observations upon the Generation, Composition, and Decomposition*, 24.

planetary system."[25] Overcome with amazement and confusion at the "marvels" revealed by Needham's microscope, he revisited his theory of active matter and thought about how to develop it further. Needham and Buffon had produced a new kind of empirical evidence from the boundary land between life and death. Not only did these observations of infusoria open up novel explanatory possibilities, but the community of natural historians did not adequately appreciate these possibilities. "Is it possible," he wrote to La Condamine, "that in your Academy Réaumur, who truly has so much talent for natural history, regards these experiments . . . as indifferent, and amuses himself with hatching chicken eggs while a new world is being discovered? Assuredly, they are unjust in your Academy not to make as much of these new discoveries as they deserve."[26]

To Buffon, Maupertuis wrote appreciatively of the *Histoire naturelle*, encouraging him to ignore theologically motivated attacks. "I have just been reading Needham's book. What disorder, what a [poor] reasoner! But it also contains plenty of marvels. It's too bad that such a man wants to make systems, and that a maker of bricks wants to be an architect." Maupertuis encouraged Buffon to try further experiments, specifically to observe pure water to see if the same moving particles would appear. "About the uniformity of your organic molecules in all the kinds of matter where you have observed them, I cannot help thinking that they might belong to a fluid that is everywhere the same. This is said with all due respect to the microscope that alone should decide all these questions and which has perhaps already decided them for you."[27] Two weeks later he wrote in a similar vein to La Condamine:

> For myself, if I had the time to spare I would sell all my books to buy a microscope, and would do nothing more than look through it. . . . The similarity of all these animated particles seen in so many different kinds of matter still makes me think that they could belong to some universal and elementary fluid that is the same everywhere. If this fluid were in water, it would soon be found. But it could be also some other fluid distributed and spread everywhere and then the thing would not be easy;

25. M to La Condamine, 24 August 1750, SM, fol. 126.

26. Ibid.

27. "J'ai lû ces jours passés le livre de Needham, quel désordre, quel Raisonneur! Mais il contient aussy bien des merveilles. C'est dommage qu'un tel homme veuille faire des sistemes et qu'un faiseur de Briques veuille etre Architecte. . . [J]e ne scaurois m'empecher sur l'uniformité de vos molecules organiques dans toutes les matieres où vous les avez observés de penser qu'elle pourroient appartenir à une fluide similaire et toujours le meme. Ceci soit dit sauf le respect dû au microscope qui seul doit decider toutes ces questions et qui pour vous l'en a peut etre deja decider" (M to Buffon, 1 September 1750, SM, fol. 126v).

> it would perhaps be up to the chemists to find it. Finally, this is a matter where the microscope ought to precede reasoning. As for Needham, it is certain that he has no common sense.[28]

Apparently Maupertuis did not think he could undertake microscopic observations himself. When he first wrote to La Condamine on the subject, he claimed, "If I were in M. d'Argenson's place [as minister responsible for the Paris Academy of Sciences] I would forbid all the Academy's experimenters [*physiciens*] to occupy themselves with anything else."[29] In principle, he was in such a position in Berlin, but there were no microscopists in the Academy, and he did not succeed in promoting work along these lines. Lacking further experimental evidence, he went back to his speculations. "What keeps me from pursuing my reflections on this matter is that I lose myself in it, and the more I do with it, the more I find myself confounded. Nevertheless, I have pushed the thing pretty far, almost to extravagance." His reluctance to write publicly about it, he claimed, had nothing to do with fear of the consequences. "Do you really believe in good faith that any subject is forbidden in our Academy, or even any way of presenting it? Do you think it is that which bothers me?"[30] Perhaps not, but the fact remains that he never presented this material to the Academy, even in Berlin.

The potential of microscopy to unlock the secrets of generation turned out to be elusive. More than a year after reading Needham for the first time, Maupertuis wrote to Buffon,

> I tell you that I am still bothered by all these deviltries [*diableries*] seen in infusions. I have reread Needham's book once more just now, but it is the work of the Sibyls. Is it true what M. Ulloa told me the other day in passing by here, that in London

28. "Pour moi si j'avois du temps de reste je vendrois tous mes livres pour acheter un microscope, et ne ferois plus d'ici à deux que regarder. . . . La similitude de toutes ces parties animées qu'on voit dans tant de matieres differentes me jette toujours dans la pensée qu'elles pourroient appartenir à quelque fluide universel et elementaire et partout le meme. Si ce fluide etoit en l'eau, il seroit bientot trouvé: mais ce pourroit etre aussi quelqu'autre fluide repandu et délayé partout et alors la chose ne seroit pas aisée; ce seroit peut etre aux chymistes à le trouver. Enfin c'est ici une affaire où le microscope doit preceder et preceder de beaucoup le raisonnement. Pour Needam il est certain qu'il n'a pas le sens commun" (M to La Condamine, 15 September 1750, SM, fol. 95).

29. M to La Condamine, 24 August 1750, SM, fol. 125v.

30. "Ce qui m'empeche de poursuivre mes Reflexions sur cette matiere, c'est que je m'y perds, et que plus j'en fais, et plus je me trouve confondu: quoique pourtant j'aye poussé la chose assés loin, et jusqu'a une espece d'extravagance. Croyés vous point en bonne foy que quelque matiere soit interdite dans notre academie ny meme quelque maniere de la presenter? Croyés vous que ce soit ce qui me gene?" (M to La Condamine, 5 November, 1750, SM, fol. 96v).

they have noticed that one does not see microscopic animals in infusions when they are under a film?[31]

The excitement at the prospect of a "new world" was tempered by frustration at how to make sense of it. These letters from the early 1750s reveal an intense struggle with confusing bits of evidence, about a question with the widest possible ramifications. Not since the years of searching for a way to shed light on the obscurity of Newton's physics had Maupertuis engaged so fiercely with a recalcitrant intellectual problem.

Where Buffon argued that animals formed from "organized and living" particles, Maupertuis presumed that the organic elements were not themselves alive. These invisible elements had to be smaller than the "spermatic animals" in seminal fluids. "Nevertheless I do not deny either the organization or the life of the beings one sees in semen through the microscope. I regard them as real animals, but I do not believe that they are the material of the fetus."[32] However, the ultimate question remained unsolved, regardless of the status of Buffon's microscopic bodies.

> If material organic parts of the bodies of animals had been found, it would not have fundamentally explained generation: because the formation of these organic particles would [also] need to be explained. It is too bad that those who do the experiments hardly attempt these speculations, and those who do the speculating are devoid of experiments.[33]

This combination of speculation and experience/experiment would characterize Maupertuis's subsequent published writings on generation. He complemented the suggestive and tantalizing evidence of Needham's infusions with breeding experiments of his own, and with evidence of inherited abnormalities in human populations. As was his habit, he pursued a variety of tactics for introducing his reflections on this evidence to the public. Different versions of several texts appeared

31. "Je vous avoue que je suis toujours dans la peine sur toutes ces diableries qu'on voit dans les liqueurs. J'ai relu encore ces cours passés le livre de Needham, mais c'est l'ouvrage des Cybilles. Est il vrai ce que m'a dit M. Ulloa l'autre jour en passant par icy, qu'à Londres ils avoient remarqué qu'on ne voit point d'animaux microscopiques dans les liqueurs lorsqu'il s'est formé audessus une pellicule?" (M to Buffon, 14 December 1751, SM, fol. 122).

32. "Cependant je ne nie point l'organisation ny la vie des etres qu'on decouvre au microscope dans la semence. Je les regarde comme de veritables animaux, mais je ne crois point qu'ils soient les materiaux du foetus" (M to La Condamine, 16 October 1751, SM, fol. 98v).

33. "Quand on n'auroit trouvé des parties organiques materiaux des corps des animaux ce ne seroit pas avoir expliqué primordialement la generation: car il faudroit expliquer la formation de ces parties organiques elles memes. C'est domage que ceux qui font des experiences ne s'elevent gueres à ces speculations, et que ceux qui font ces speculations soyent denués des experiences" (ibid.).

over a period of about two years. The most adventurous of these, leaning heavily toward materialism, was published anonymously, its origins strategically shrouded in secrecy.

Maupertuis wrote up a version of his "meditations" on generation and had it printed pseudonymously in Latin to obscure his identity. The edition was sufficiently small that no modern commentator has found a copy, although surviving correspondence indicates that some contemporary readers had seen the Latin text. The history of this little book, which I will designate by its ultimate title, *Système de la nature,* shows how printing could be used to obscure textual origins and stimulate discussion and speculation about a new work. Many authors and printers in this period performed such maneuvers, for a variety of reasons. Access to controversial books printed in limited editions, often through direct distribution by authors, marked readers as members of an elite circle, an exclusive public that only expanded into a faceless "public" once the text was reprinted in a larger edition.[34] These editions circulated much like manuscript news and gossip sheets (*nouvelles à la main*) and surfaced in journals and conversations before being published more widely.[35]

Maupertuis's little Latin book, obscurely titled *Dissertatio inauguralis metaphysica de universali naturae systemate,* purported to be a thesis by one Dr. Baumann of the University of Erlangen.[36] The "dissertation" opened the door to materialism by recasting the inherent forces of organization in psychological terms, giving matter "some principle of intelligence." Although he later claimed that it first appeared in Germany in 1751, there is no corroborating evidence for this date, and it seems more likely that it was not printed until 1752. The first mention of the text occurs in December 1752 when Maupertuis asked La Condamine for his opinion: "You do not say anything . . . about the little brochure on the formation of or-

34. Beeson identifies *Système de la nature* with a manuscript sent to Johann Bernoulli in April 1751 for secret printing in Basel (*Maupertuis,* 207). While this conjecture is intriguing, it is more likely that the "jolie petite brochure" of 1751 was the first version of *Lettre sur le progrès des sciences,* released in a larger edition in 1752. The copy of *Lettre sur le progrès* now held in BN Réserves differs from other editions in length and type face and is probably the copy sent to Sallier, the king's librarian, from the original Basel edition (it has no title page). The list of intended recipients, including patrons and former patrons, made sense for the *Lettre;* none of Maupertuis's correspondents mention *Système de la nature* until late 1752, when he sent it to Buffon and La Condamine, among others (M to JBII, 10 April 1751, BEB).

35. Funck-Brentano, *Les nouvellistes.* Grimm's *Correspondance littéraire* was an example of this genre, as were Fréron's reports to the police inspector d'Hemery, BN f.fr. 22158. See Balcou, *Le dossier Fréron.*

36. The book was small; when published in French, it came to only 67 duodecimo pages. Diderot quoted from the Latin edition; the Latin text was also the basis for a German translation, now exceedingly rare itself (*Versuch von der Bildung der Körpers,* Leipzig, 1761, translator's preface). The anonymous translator, writing in Potsdam, says there were only ten copies printed of the Latin text.

ganized beings. I am curious to know your opinion of it. I beg you also to let me know if Buffon is pleased with what I say of him, and if Réaumur holds against me what I do not say."[37] Thus, it seems that he had recently sent copies to La Condamine, Buffon, and possibly Réaumur as well. The use of the first person made his authorship clear to La Condamine, although Maupertuis also referred to the book as the work of Baumann. La Condamine found the "thesis" reminiscent of Spinoza, a shorthand way of equating the theory of self-organizing matter with atheistic materialism. The other relevant feature of Spinoza's philosophy, as it was understood in the Enlightenment, would have been the sensibility of matter, the only substance in the universe.[38] Mention of Spinoza raised just the kind of risk that Maupertuis implicitly courted but openly abjured. "I am happy that the Erlangen thesis pleased you: I judge it as you do, except that I do not find in it any Spinozism at all. Our friend Buffon never replied to me about the one I sent to him, maybe it is the shadow of Spinozism that displeased him. As for me, I tell you that I do not believe that there has been anything yet written on the production of organized bodies that compares with this thesis."[39] At some point, he must have had a small edition printed with both French and Latin versions of the text, still anonymously, and still not widely distributed; no copies of this edition, however, are known today.[40]

By this time, Maupertuis had gone public as the author of *Vénus physique* by including it in the 1752 edition of his collected works. His maneuvers for managing the release of the potentially controversial new work on generation coincided with his attempt to recover from Voltaire's attack. Fréron, a sworn enemy of Voltaire and the author of the journal *Année littéraire,* was an obvious ally in this campaign.

37. "Vous ne me parlez point . . . de la petite brochure sur la formation des etres organisées. Je suis curieux d'en savoir votre avis. Je vous prie aussy de me faire savoir si Buffon est content de ce que je dis de luy, et si Reaumur me scait gré de ce que je ne dis pas?" (M to La Condamine, 26 December 1752, SM, fol. 105).

38. See Diderot, in *Encyclopédie,* 15:463–74 s.v. "Spinosa" and "Spinosiste." Modern Spinozists, according to Diderot, hold "que la matière est sensible, ce qu'ils démontrent par le développement de l'oeuf, corps inerte, qui par le seul instrument de la chaleur graduée passe à l'état sentant & vivant, & par l'accroissement de tout animal qui dans son principe n'est qu'un point . . . devient un grand corps sentant & vivant dans un grand espace. De-là ils concluent qu'il n'y a que de la matière, & qu'elle suffit pour tout expliquer" (ibid., 174).

39. "Je suis bien aise que la these d'erlangen vous ait plû: j'en avois porté le meme jugement que vous, excepté que je n'y trouve aucunement le Spinozisme. Notre amy Buffon ne m'a jamais rien repondu sur celle que je luy ay envoyée, peut estre est-ce cette ombre de Spinozisme qui luy a déplu. Moy je vous avoue que je ne croye point qu'on ait encore rien sur la production des etres organisés, qui approche de cette these" (M to La Condamine, 13 January 1753, SM, fol. 160).

40. Trublet gave his version of the publishing history in "Avertissement," *Essai sur les corps organisés,* Berlin [Paris], 1754.

Maupertuis wrote to him and also asked La Condamine to lobby the journalist for a favorable review of his *Oeuvres.*[41] Fréron agreed willingly. In his review, he mentioned Dr. Baumann while discussing *Vénus physique,* without betraying any suspicion that the two texts were the work of the same author. He noted that Baumann had resolved the difficult question, unanswered in *Vénus physique,* of how to conceive of the "principle of union" guiding the combination of parts of the fetus.

> I learn that in September 1751, M. Baumann defended at Erlangen in Germany a thesis in which he supposes that all particles of matter, and especially the organic particles revealed by the latest experiments, are animated by a sort of instinct more or less perfect, much like that ordinarily ascribed to animals, and that these same particles preserve the memory of their former situation that they tend to take up again. From this, he explains quite convincingly a great number of phenomena, and in particular those of generation.[42]

With this, Baumann gained a wider, if still obscure, visibility and legitimacy; only a few readers would have recognized the sleight of hand by which Fréron introduced Baumann to solve *Vénus physique*'s conceptual problems.[43]

Maupertuis arrived in Paris in May 1753 for a stay of several months; shortly thereafter the "Erlangen thesis" was circulating around Paris. From correspondence with Nicolas Trublet we can see how the book was discussed, as well as the widespread public interest in its provenance. Trublet was a cleric and man of letters, as well as a royal censor and an enemy of Voltaire, and a longtime associate of Maupertuis.[44] In October 1753, he reported a conversation about a new book he had not yet seen, called "something like *System of the Universe.*" Maupertuis's old friend Pierre Le Monnier, the astronomer who had accompanied him to Lapland, had mentioned to Trublet a book "in your style [*de votre façon*], which he had loaned to the duchesse d'Ayen and then to M. Quesnay with whom he had left it at Fontainebleau; . . . in this book it is said that matter can think, etc." La Condamine had cleared up some of Trublet's confusion by telling him that this was Baumann's Erlangen thesis and noting that Fréron had discussed it in his journal. Further-

41. M to La Condamine, 26 December 1752, SM, fol. 150. Maupertuis met with Fréron in Nancy in May 1753, on his way from Berlin to Paris (Fréron to d'Hémery, 15 May 1753, in Balcou, *Le dossier Fréron,* 55).

42. Fréron, "Lettre VII," *Lettres sur quelques écrits de ce tems,* 8:163). See also Fréron to M, 21 December 1752, AS Fonds Maupertuis.

43. Fréron probably realized that Maupertuis and Baumann were the same person, although he never said as much.

44. Jacquart, *L'Abbé Trublet.* Trublet came from an old Malouin family, and was related to Maupertuis by marriage; he frequented the circle around Houdar de la Motte in the 1720s and '30s.

more, La Condamine reported rumors that Maupertuis had had something to do with it. Maupertuis was a popular subject of literary gossip at this time because of his recent battle with Voltaire in Prussia.[45] As we learn from Trublet's comment, a single copy of the little book made the rounds of the overlapping circles of court ladies (the duchesse d'Ayen) and men of letters (Quesnay, Le Monnier), fueling talk about its authorship as well as its ties to materialism and Spinozism. Trublet was fishing for direct confirmation from the suspected author, but Maupertuis adamantly refused to give it; he also claimed to have no copy of the book to give to Trublet. He had evidently given away a few copies of what he knew to be a risky text, to see what people would say. Certainly, the book's scarcity and its anonymity contributed to its notoriety, but there is no indication that Maupertuis was planning to publish it more widely.[46]

In November, Trublet told Maupertuis, who had left Paris for Saint-Malo, about a new book by Diderot that cited Baumann at some length. "This is the work that M. Le Monnier attributed to you, according to what M. l'Abbé de Pontbriant told me," Trublet noted.[47] In effect, Diderot's discussion of "a little Latin work" published in Erlangen and "brought to France by M. de M. . . . in 1753" linked the Baumann thesis to Maupertuis's name, albeit in conventionally elliptical form. Diderot's book, *Thoughts on the Interpretation of Nature,* was itself controversial and brought Maupertuis's potentially scandalous claims about thinking matter into public view, at a time when very few people had read them directly.[48] This exposure may well have forced Maupertuis's hand and pushed him toward publishing a more accessible French edition.

45. Trublet to M, 31 October 1753, AS Fonds Maupertuis. On La Condamine's hints: "Il me rappela que Fréron en avoit parlé dans ses feuilles à l'occasion de votre *Vénus physique* et me trouva tout d'un coup l'endroit; mais il m'avoua qu'on avoit dit que vous aviez quelque part à cette these" (ibid.). Fréron reported rumors and anecdotes about Voltaire's feud with Maupertuis in his letters to l'Hémery, lieutenant of Paris police, in Balcou, *Le dossier Fréron.*

46. He had followed the same strategy with *Réflexions sur les origines des langues,* circulating it privately before a normal edition appeared (M to JB II, 2 May 1740, BEB). On this work, see Beeson, *Maupertuis,* 153–61.

47. Trublet to M, 11 November 1753, AS Fonds Maupertuis.

48. Diderot, *Pensées sur l'interprétation de la nature,* mention of Baumann in Para. XII, 36. Diderot published the book anonymously in late 1753 with a tacit permission (Journal d'Hémery, BN f. fr. 22158, 6 Dec. 1753); see complex bibliographical history in Varloot, "Introduction," to Diderot, *Oeuvres,* 9:15–20. Trublet's reference to the book in November is the earliest possible date for publication, although it was probably still in the hands of the printer at that time. Fréron mentioned rumors about a book by Diderot "that will be entitled *Nouveau systeme de l'univers*" (Fréron to d'Hémery, 26 November 1753, in Balcou, *Dossier Fréron,* 121). Maupertuis read it shortly after it appeared (see M to Trublet, 25 December 1753, SM). It was reprinted in a larger edition in January 1754.

By December 1753, Trublet had finally gotten his hands on a copy of the Baumann thesis, "but in the original, that is to say in the French, as I submit that the Latin is only a translation and that the French is by you. It is one of the most beautiful things you have written. . . Durand [the Paris printer] must reprint this excellent work."[49] By this time two versions were circulating, in extremely limited numbers, one in Latin and one with the French text alongside the Latin.[50] Although Trublet had now seen through the pseudonym, Maupertuis was reluctant to give up the pretense and continued, rather perversely, to refer to Baumann as a third party:

> I do not think that M. Baumann has grounds to be pleased by the exposition that M. Diderot makes of his ideas, nor especially of the consequences he draws from them. . . I have read this thesis with great care and it seems to me that one could not without injustice see anything in it that is contrary to orthodoxy. . . . I think it would be easy for him to respond to M. Diderot. . . . After your judgment of the Erlangen thesis, I would like to be able to accept the compliments you make me about it, but I can assure you that you are mistaken in your conjecture, and that it appeared first in Latin, and that I sent it to M. Buffon more than a year before the French translation was printed, and I can also assure you that the Latin and the French are by the same author, whom I know and whom I have seen in Berlin. I cannot give you copies; I had only two which I gave away in Paris.[51]

The disavowal followed the same pattern he had used with Bernoulli in discussions of the *Examen désintéressée* a decade earlier. Trublet reminded his coy friend that he had also denied authorship of the *Nègre blanc,* even when everyone had recognized his style. "I don't even think that you want me to believe you. Certainly none of your friends believe you."[52]

Trublet took it upon himself, apparently with Maupertuis's acquiescence, to shepherd the French version through the printing process. He wrote a preface reviewing the publication history and reminded readers about Fréron's review and Diderot's contentious remarks.[53] Maupertuis quibbled about how visible his name should be (the book was formally anonymous), but eventually he let Trublet drop

49. Trublet to M 16 December 1753, AS Fonds Maupertuis.

50. See Trublet's "Avertissement," to [Maupertuis], *Essai sur les corps organisés* (1754), and "Avertissement" to Maupertuis, *Systeme de la nature,* in *Oeuvres,* vol. 2. Trublet had seen both Latin and French editions; Diderot quoted from the Latin.

51. M to Trublet, 25 December 1753, SM, fol. 107.

52. Trublet to M, 1 January 1754, AS Fonds Maupertuis.

53. Trublet did not sign the preface. Some sources give Trublet as the translator of the Baumann text, but his letters clearly indicate that he was publishing Maupertuis's original French text.

an elliptical hint about his identity in the preface. Trublet recognized that Maupertuis's authorship would give the book added interest: "If I consented to neither name nor indicate you, I would be harming the book, and besides . . . it would be pointless since a thousand people know it is your work." Trublet showed the text to Malesherbes, director of the book trade, and obtained a tacit permission to publish.[54] Printing commenced in March, "in small format, just like that of the *Lettres de M. de Maupertuis* [1752]." A thousand copies were printed, and the book appeared as *Essai sur la formation des corps organisés* (*Essay on the formation of organized bodies*), with a fictitious Berlin imprint.[55] Just two years later, Maupertuis included the *Essai* in his four-volume collected works, under a bland title reminiscent of Baumann's thesis, *Système de la nature.*

This episode illustrates the dynamics of publishing at a time when censorship was in force and authors might be locked up for writing imprudent books. Diderot himself had been imprisoned in Vincennes only recently, and Maupertuis had no desire to follow in his footsteps.[56] His caution about attribution was in part governed by this kind of fear, but rumors about his identity also inflated his image as an *esprit fort,* an image he cultivated up to a point. Many books occupied a borderline clandestine territory and circulated quite freely without being impounded, and talk of danger could make a book more appealing to readers. Maupertuis encouraged interest in his book, which he knew would provoke accusations of materialism and irreligion, by introducing it surreptitiously, disguised under a foreign name. He later claimed that he had adopted this ploy so he would not have to reply to criticisms.[57] At the same time, Maupertuis had an official reputation to protect, as president of the Berlin Academy. But he cannot have been altogether displeased by Diderot's notice of his book, and may even have expected it, since he had talked with Diderot upon his arrival in Paris in 1753.[58]

54. "M. De Malesherbes examined it himself, as well as the preface. No censors were involved, and he consents to the printing" (Trublet to M, 24 January 1754, AS Fonds Maupertuis). On Malesherbes's use of tacit permissions for potentially controversial books, see Shaw, *Problems and Policies of Malesherbes;* Pottinger, *The French Book Trade;* Bollème et al., *Livre et société dans la France du XVIIIe Siècle.*

55. "Petit brochure in 12° imprimé sans permission. On croit que c'est M. de Maupertuis qui en est l'auteur" (Journal of L'Hémery, 4 April 1754, BN, f.fr. 22158, fol. 24). L'Hémery apparently did not know of Malesherbes's tacit permission.

56. Another of Maupertuis's friends, La Beaumelle, had just been released from the Bastille at this time. "La Beaumelle est à la bastille depuis un mois: quelques-unes de ses Pensées et la note sur la mort de la famille de Louis XIV lui ont attiré cette disgrace, selon ce qu'on dit dans la ville" (M to Prades, 25 May 1753, in Koser, *Briefwechsel,* 294).

57. "Avertissement," *Système de la nature,* in *Oeuvres,* vol. 2.

58. Fréron reported to d'Hémery that Maupertuis had brokered an informal patronage arrangement between Diderot and the Prussian prince Henry, whereby the prince would give the

## *Système de la nature:* Desire, Aversion, Memory

What exactly made *Système de la nature* a contentious and potentially risky text? First and foremost, the author argued that an explanation of the origin and reproduction of organisms required that matter be allowed more properties than impenetrability, extension, and inertia. Although gravity and chemical affinities could account for many kinds of phenomena, organization required properties "of another order than those we call physical." Chemistry retained the same heuristic role it had played as the counterpoint to mechanical reductionism in *Vénus physique.* But now he argued that even affinities could not account for the complexity of plants and animals. "We must have recourse to some principle of intelligence, to something similar to what we call desire, aversion, memory."[59] This "principle of intelligence" resides in matter, down to its smallest parts. In effect, the elements of matter have special organic *rapports* or affinities, analogous to psychic entities: desires, aversions, perceptions, habits, and memories.

Desire and aversion are analogous to the affinities that unite chemical substances, directing elements to their places; memory links organic elements to the comparable particles in the parent organism. Each material element "retains a kind of memory (*souvenir*) of its previous situation and will resume it whenever it can, in order to form the same part in the fetus."[60] Since memory is not infallible, its status as a driving force for organization incorporated the possibility of mistakes or variations. The same process, or the same set of forces, produces individual variations and congenital defects. In these latter, the original cause of excesses or deficiencies could be strictly accidental, but their effects may then be perpetuated through normal generation. Once reproduction established a trait over time, "the particles become accustomed to their locations, which makes them place themselves similarly [in succeeding generations]."[61]

Instinct, perception, intelligence, desire, aversion, memory—Maupertuis used these terms more or less interchangeably, depending on the particular phenomenon under examination. At times, these terms functioned more metaphorically

---

philosopher a generous pension in return for entering into a philosophical correspondence (Fréron to d'Hémery, 26 May 1753, in Balcou, *Dossier Fréron,* 59).

59. *Système de la nature,* in *Oeuvres* 2:155.

60. "Mais chacun extrait de la partie semblable à celle qu'il doit former, conserve une espece de souvenir de son ancienne situation; et l'ira reprendre toutes les fois qu'il le pourra, pour former dans le foetus la même partie" (ibid., 158).

61. "L'habitude de la situation des parties dans le premier individu les fait se replacer de la même manière" (ibid., 160–61).

than literally; hence, the slippage among them. But they all pointed to relations among elements and a directedness that drives combination and development. Once the elements combine, their individuality disappears into the new whole, and they cannot be recognized as individuals. "Thus it is that an army, viewed from a distance, appears to our eyes as a great animal; thus a swarm of bees, when they are united around the branch of a tree, appears to us as a body which has no resemblance with the individuals that formed it."[62] Chemistry works this way as well: the compound has new properties of its own, and its identity supersedes those of the individual elements. Nevertheless, the elemental identities are not completely lost, as they come into play again when a compound (or an organism) is broken up (or propagated).

In collapsing psychic and physical properties into the same category, Maupertuis raised the specter of thinking matter, no doubt reminding his readers of the clandestine and much-maligned work by Julien de la Mettrie, *L'homme machine* (*Machine Man*), published in 1748.[63] But where La Mettrie had sought to scandalize, by stating his materialist thesis in the baldest possible terms, Maupertuis tried instead to make his claim theologically and scientifically palatable, knowing full well that he was on contentious ground. He hoped, he told his reader, to avoid confrontation with religion, but the evidence forced him into a position that might appear to challenge orthodoxy. So he took pains to show that the conflict is only apparent. Extension and thought obviously differ, but if they are both considered as "properties" they could belong to the same subject. "If it is true that we find more repugnance in conceiving of extension and thought in the same subject than in conceiving of extension and mobility, this is because experience shows the one continually to our eyes, and only lets us know about the other by reasoning and inductions."[64] This argument might remind the attentive reader of the rhetorical structure of Maupertuis's much earlier argument for the attractive force of matter. Why should we assume that impact is more intelligible than action at a distance, or that impenetrability is a more evident property than "attraction"? Each is equally mysterious, and equally evident from the examination of the phenomena.[65]

Furthermore, why should sentient matter entail a challenge to religion? Even theologians agreed that animals have some kind of intelligence, or at least the ability to perceive and remember, so there could be no contradiction in ascribing such properties to the parts of animals.

62. Ibid., 170–71. The image was taken up by Diderot in *D'Alembert's Dream*, some years later.

63. La Mettrie, *L'homme machine*.

64. Maupertuis, *Système de la nature*, in *Oeuvres*, 2:151–52.

65. Maupertuis, *Discours sur les différentes figures des astres* (see chapter 3 above).

> If we admit without peril some principle of intelligence in large collections of matter, such as the bodies of animals, what greater peril will we find in attributing such a principle to the smallest particles of matter? . . . The danger, if it existed at all, would be just as great in allowing it in the body of an elephant or a monkey, as in allowing it in a grain of sand.[66]

The cynic might accuse this strategy of disingenuousness, resting as it does on locating "intelligence" in the animal body, exactly the point at issue. But it also draws on the orthodox claim that animals cannot have souls, so that admitting desires, memories, habits, and instincts in animals means admitting their material basis. It could be read, then, as a clever display of wit at the theologians' expense. But Maupertuis also genuinely wanted to avoid "danger," at least in the form of open confrontation with the religious authorities, without pulling back from the substance of his provocative claims. On the one hand, he showed his reader that these properties were necessary for the explanation of phenomena and, on the other, he reserved space for faith, the human soul, and divine creation.

The ultimate origin of matter and life remained mysterious, outside the domain of science. The philosopher could accept the Biblical account of creation and limit himself to describing the laws nature has followed since the initial divine fiat. "Once this world was formed, by what laws is it conserved? What means were destined by the creator to reproduce the individuals that perish? Here we have a clear field and we can propose our ideas."[67] Theological concerns cropped up throughout the book, betraying the author's anxieties on this score. In the beginning, he anticipated "the murmurs of those who mistake stubbornness in their beliefs for a pious zeal." Towards the end, he comes back to this concern: "In spite of all that I said in the beginning of this essay, I still fear that objections (*le murmure*) will be renewed against what I am proposing. Nevertheless, I have shown in what seems to me an incontestable way that there was no more peril in admitting some degree of intelligence in the parts of matter than there would be in allowing it in the animals we regard as the most perfect."[68] Far from being dangerous, a "system" that would illuminate the process of generation could only enhance our appreciation for God's wisdom. Maupertuis took care to distinguish his notion of intelligent

66. "Or si, dans le gros amas de matière, tels que sont les corps des animaux, l'on admet sans péril quelque principe d'intelligence, quel péril plus grand trouvera-t-on à l'attribuer aux plus petites parties de la matiere? . . . Le péril, s'il existoit, seroit aussi grand à l'admettre dans le corps d'un éléphant ou d'un singe, qu'à l'admettre dans un grain de sable" (*Système de la nature,* in *Oeuvres* 2:149).

67. "Ce Monde une fois formé, par quelles loix se conserve-t-il? quels sont les moyens que le Créateur a destinés pour reproduire les individus qui périssent? Ici nous avons le champ libre, et nous pouvons proposer nos idées" (ibid., 155).

68. Ibid., 147, 178–79.

matter from the atomism of Lucretius, "an impious philosopher" admired by "the libertines of our day."[69] Eternal, passive atoms encountering each other by chance can explain organization no better than theologically safe preformation theories. Grounding the intelligence of higher animals in a universal property of matter meant erasing the sharp divide between brute and living, organized and unorganized. If matter can have intelligence, or perception, in its smallest parts, then organisms of different kinds simply have different degrees of it. But this does not imply independence from God.

> God, in creating the World, endowed each particle of matter with this property, by which he wished all the individuals he had formed to reproduce themselves. And since intelligence is necessary for the formation of organized bodies, it seems greater and more worthy of the Divinity that they form themselves by the properties that [God] distributed all at once to the elements, than if these bodies were in every instance the immediate productions of his power.[70]

Here, he articulated the same rationalist theology that Leibniz had espoused in his debate with Clarke and that Maupertuis himself had developed along different lines in his *Essay on Cosmology.* A rational God created a world amenable to scientific investigation. In the final analysis, he argued, an explanation based on intelligent matter provides a viable alternative to theories that denigrate God's own intelligence. The book ended with the stark juxtaposition of three possibilities: either brute matter formed the universe as we know it through chance combination (Lucretius), or God made everything in the world out of pieces, the way architects construct buildings, or, most plausibly, "the elements themselves endowed with intelligence arrange themselves and unite to fulfill the aims of the Creator."[71] Whatever his detractors claimed, Maupertuis made a serious effort to wrap his materialism in a layer of rational theology more reminiscent of Leibniz than of Newton. In part, this was no doubt a defensive gesture to deflect accusations of heresy, but it also reflected his struggle to reconcile the evidence for active matter with an undogmatic religious faith.

In Maupertuis's view, an infinitely wise God created the first organisms from matter endowed with its own inherent forces, or perceptions. The man of science must infer the existence of the forces from the effects they produce. "The formation of the first individuals having been miraculous, those that followed are nothing more than the effects of properties [analogous to desire, aversion, memory]. When the elements appropriate for each body find themselves in sufficient

69. Ibid., 182.
70. Ibid., 182–83.
71. Ibid., 184.

quantities and at distances where they can exert their action, they will unite together to continually repair the losses of the universe."[72] Once again, Maupertuis harked back discreetly to a problem that had bothered Newton too. What keeps the universe from running down, as bodies collide continually, losing some of their motion? Maupertuis had taken the Newtonian side on the question of the existence of hard bodies (which cannot rebound elastically, thereby losing motion) against the Leibnizian claim that matter is continuous down to infinity and therefore fundamentally elastic. The organic forces operate in Maupertuis's universe by analogy to Newtonian active principles, although he did not make this analogy explicit by any means. Instead of being the result of God's immediate intervention, as Newton had claimed in the General Scholium to the second edition of the *Principia* (1713), Maupertuis suggested that God imbued matter with forces of organization and left them to act, as properties of matter.

However strongly he denied the heterodoxy of active matter, Maupertuis's prime impulse was to make sense of the conundrums of everyday experience and recent experiments, to illuminate all the questions that had troubled him when he argued with Buffon about family resemblances and when he read about Needham's infusions. Resemblances between offspring and their parents, the production of hybrids between species and new varieties within species, and the regularities observed even in monstrous births supplied evidence that preexistence theories could not address. "By the combination [*réunion*] of elementary perceptions we will easily explain facts inexplicable in any other system: why passions and talents become hereditary in men and animals.... How does the dog transmit to his descendants his hunting ability? These inconceivable, but ordinary, phenomena ... are explained with the greatest ease by our system."[73] The particulate theory of epigenesis had accounted for commonly observed hereditary phenomena, and now it could explain even the inheritance of nonphysical traits like mathematical or musical ability, or a dog's propensity for hunting. These intangible attributes could be analyzed in the same way as six-digitism or eye color, because material interactions produce them in the same way.

In spite of his persistent refusal to acknowledge Leibniz as either source or inspiration, Maupertuis's usage of "perception" recalled certain aspects of Leibniz's monadology. By now we can see the complex and fraught nature of his debt to both Newton and Leibniz, and his attempt to put his own stamp on a strange, and hardly rigorous, synthesis of the two. The very real resonances of some of his formulations with Leibniz's approach, whether to mechanics or metaphysics, amplified the force of his reaction to König's accusations. His aloof stance during the

72. Ibid., 158.
73. Ibid., 174–75.

monad controversy in the Berlin Academy a few years before may be partially explained by an ambivalence toward Leibnizian metaphysics, and an unwillingness to be perceived in any way as a disciple of his predecessor in Berlin.[74] He certainly rejected the Leibnizian theory of substance, according to which the only true substances are metaphysical self-contained unities (monads). But Leibniz defined monads or "formal atoms" in terms of their perceptions and "appetitions." These determine the individuality and activity of the monad and connect each one to the rest of the universe.[75] Each monad possesses its own "internal force," bestowed by God, which becomes the source of all its actions and transformations, its "feeling and appetite."[76] It is difficult not to see reflections of those perceptions and appetites in Maupertuis's account of active, organic matter. Leibniz also argued that mechanical causes ("figure and motion") cannot explain perception and appetition, or activity; the activity of simple substances must differ in kind from the levers and pulleys of mechanics. For Leibniz, this rejection of the Cartesian universe of inert particles led him to abstract the metaphysical, substantial, plane of existence from the corporeal, and he spent considerable effort in making sense of their unconnected connection, the "pre-established harmony" imposed on the creation by God.

Maupertuis incorporated certain elements of Leibnizian metaphysics into his theory of life, particularly in the language of perception and appetition. However, because he rejected several of Leibniz's most fundamental tenets, he argued for a radically different organicism; when mechanical properties proved insufficient to the explanatory task at hand, Maupertuis simply attached various appropriate active properties directly to matter, for a different kind of active matter.[77] Most crucially, Maupertuis rejected the notion of a separate metaphysical

74. Maupertuis often referred to Wolff and König as "sectateurs" or "disciples" of Leibniz to disparage their originality and make them into dogmatic and uncritical thinkers. He had the utmost respect for Leibniz himself. The anonymous German translator of *Système de la nature* noted the consonance of Maupertuis's theory of active matter with Leibniz's monadology. German readers were certainly more likely to see these connections than their French counterparts. See translator's preface, Maupertuis, *Versuch von der Bildung der Körper.*

75. See Leibniz, *Monadology* in *Selections,* 533–52. While many of Leibniz's philosophical works were only published in the 1760s and later, a substantial selection was available in the 1740s. Desmaiseaux, *Receuil des divers pièces* (1740) included key Leibniz texts. See Bongie, "Introduction to *Les Monades,*" 38–40. Maupertuis knew Leibniz's *Theodicy* and *Monadology.* Leibniz's response to Locke (*Nouveaux essais sur l'entendement humaine*) remained unpublished in Maupertuis's lifetime; there the notion of "*petites perceptions*" was articulated fully, but there is no connection to Maupertuis's use of perceptions or instincts as properties of organic matter. For Maupertuis's reflections on Leibniz's "system," see Maupertuis, "Lettre VI: Sur les systêmes," in *Lettres de M. de Maupertuis, Oeuvres* 2:257–61.

76. Leibniz, *New System of Nature* (1695), in *Selections,* quotes on 115, 108.

77. Ernst Cassirer saw an "objective kinship" between Leibniz's ideas and Maupertuis's

plane parallel to the physical world, even though he reserved a salient role for metaphysical principles in physics. The organic order does not reveal a simple metaphysical principle like the principle of least action—although it may be simply that this principle is beyond our reach in this context. Nature does, however, fulfill the goals of the creator. Instead of seeking the origin of order in divine creation, Maupertuis attributed to matter the capability of generating its own order.

## Heredity: A New Kind of Evidence

Even without a metaphysical economy principle for the organic world, Maupertuis tied his theory of generation to a methodological principle of economy. Everyone agrees, he said, that scientific inquiry ought to explain phenomena with "the least number of principles and the simplest possible principles." But if the simplest definition of matter imaginable does not account for phenomena, then the philosopher must modify the definition to include additional properties. "A philosophy which does not explain phenomena could never pass as simple; and that which admits properties shown by experience to be necessary is never too complex."[78] In justifying the addition of psychic attributes to the list of basic properties of matter, Maupertuis was concerned to spell out the epistemological and methodological ramifications of his theory. The inadequacy of "physical properties of matter alone" did not rule out the possibility of a "physics" that would accurately reflect and conceptualize phenomena evident to the senses. Revising the conception of matter also meant multiplying the types of empirical evidence that could support a physical theory of life. And in fact, Maupertuis looked at evidence that traditionally had had no place in either anatomy or descriptive natural history, the two branches of "physics" concerned with generation. No one denied, for example, that offspring often resembled both parents, and might even exhibit traits visible only in their grandparents. But the inheritance of visible traits did not concern anatomists dissecting fetuses or naturalists observing the behavior of insects.

---

philosophy of nature, and claimed that Maupertuis substituted physical points for Leibniz's metaphysical points, transferring the properties of monads to material particles and undercutting the foundations of Leibniz's system. Cassirer, *Philosophy of the Enlightenment,* 86. Roger noted that Leibniz provided a "point of departure" for Maupertuis's theory, since Maupertuis described elements as "psychic unities on the model of the monad," while attributing to matter what Leibniz reserved for substance. Roger does not cite Cassirer, although the two analyses are similar (Roger, *Sciences de la vie,* 479). Neither of these analyses address the ambivalence of Maupertuis toward Leibniz, which is related to the former's efforts to address questions raised by experimental and observational evidence, as well as to his position in Berlin.

78. "Une Philosophie qui n'explique point les phénomenes ne sauroit jamais passer pour simple: & celle qui admet des propriétés que l'expérience fait voir nécessaires, n'est jamais trop composée" (*Système de la nature,* in *Oeuvres* 2:153).

Before releasing *Essai sur la formation des corps organisés* in French, Maupertuis published another short essay on generation, in the *Lettres,* one of the books that drew Voltaire's fire in 1752.[79] This little book of essays projected a different authorial identity from the anonymous risk-taker of *Vénus physique.* Here, he wrote as an elder statesman of science pronouncing on a wide range of topics, from Francis Bacon to happiness to philosophical systems and scientific controversies. The essay, "On the Generation of Animals," referred back to the conjectures of *Vénus physique,* without the playfulness and speculative tone of the earlier book. Where *Système de la nature* focused on a philosophical justification for the active properties of matter, this "Letter" laid out empirical evidence in more detail. It summarized the experiments of Buffon, Needham, and Réaumur, and added examples of inheritance of traits, in dogs and people.

As a pet fancier, Maupertuis had long been aware of the results of controlled breeding in developing new varieties of dogs and birds, though animal breeders did not trouble with theoretical explanations for their results (fig. 23). He referred to breeders as "so to speak, creators of new species," who developed new kinds of animals as marketable curiosities.[80] These familiar phenomena supplied evidence for the mixing of traits, and breeders knew perfectly well that they could perpetuate specific traits by mating animals selectively. According to the model developed in *Vénus physique,* an unusual trait initially appears by chance, but it may then be passed on through the process of combination governed by affinity-like forces. In the absence of a definitive interpretation of microscopical evidence, he analyzed patterns of inherited traits. Repeating Réaumur's tantalizing remarks about crossing varieties of chickens, he found his former colleague's refusal to impart the results particularly galling. "I am surprised that this able naturalist, who has undoubtedly done these experiments, would not inform us of the result."[81]

To remedy this lack, Maupertuis brought his own evidence to light, tracing the occurrence of six-digitism in humans through several generations to show that both males and females could transmit the abnormality.[82] He recorded the genealogy of the Ruhe family of Berlin, with names of individuals, their spouses, and their children, and the occurrence of extra digits among them. The trait occurred

79. Maupertuis, *Lettres de M. de Maupertuis;* English translation of the "Lettre sur la génération des animaux" appended to Hoffheimer, "Maupertuis and the Eighteenth-Century Critique of Preexistence," 138–44.

80. Maupertuis, *Vénus physique,* 134.

81. Maupertuis, "Sur la génération des animaux," *Oeuvres,* 2:307.

82. The example is discussed in both "Sur la génération" and *Système de la nature.* Several cases of hereditary abnormalities like extra digits had been discussed in the debate about the origins of monsters in the Paris Academy (see Tort, *L'ordre et les monstres*).

Fig. 23. One of Maupertuis's favorite Siberian dogs. It traveled with him from Berlin to Paris in 1753, where its portrait was drawn for Buffon's *Histoire naturelle, générale et particulière,* vol. 5 (1753). Courtesy of History Division, Biomedical Library, UCLA.

frequently enough to show that it was transmitted bilaterally, and that marriage to a five-digited spouse affected the frequency of its appearance in the offspring. "By such repeated marriages, it would probably die out, and it would perpetuate itself through marriages where the trait was common to both sexes."[83] To generalize the conclusions drawn from this genealogy, he turned to a probability calculation. Hypothetically considering the sixth finger as a random occurrence, he wondered about the probability of such a variation reappearing by chance in the descendants. Starting with his own knowledge of two six-fingered men in a city of 100,000, and adding three more to make up for any he might have missed, he deduced that the chances of a six-digited child being born by chance to a six-digited parent were 20,000 to 1. The probability would decrease by the same factor for each generation, to the point where the chances of three consecutive generations of the same family producing such individuals at random would be impossibly high; "numbers so large that the certainty of the best-demonstrated things in physics does not approach these probabilities."[84] This novel form of calculation (especially novel

83. "Sur la génération des animaux," *Oeuvres,* 2:308.
84. Ibid., 310.

Fig. 24. Iceland dog used by Maupertuis for breeding experiments. He obtained several of these dogs from a Danish correspondent. Original drawing by Fritsch, the Berlin Academy artist, given to Buffon by Maupertuis. *Histoire naturelle, générale et particulière,* vol. 5 (1753). Courtesy of History Division, Biomedical Library, UCLA.

in its application to a population) assumed the random occurrence of the trait as a hypothesis and showed it to be so unlikely as to be untenable.[85]

In addition to the genealogy of the Ruhe family and anecdotal evidence of hereditary six-digitism in other families, Maupertuis carried out breeding experiments with his pet dogs. When he came across a rare combination of colors in the coat of a female Iceland dog, he attempted to perpetuate the trait (fig. 24).[86] In the bitch's fourth litter, a male puppy with identical markings was born. This dog eventually went on to father yet another with the distinctive coloring. Although he was not able to predict the frequency of the trait, the result corroborated the evidence from the Ruhe family's six-digitism, that the trait was passed through both male and female lines. These examples seem prosaic, and not particularly startling; but they opened up questions about the meaning of individual variations

85. The calculation bears some resemblance to that used by Daniel Bernoulli to determine the probability of the planetary orbits occupying nearly the same plane at random. See Gower, "Planets and Probability."

86. Formey recalled making his way fearfully through the menagerie at Maupertuis's home in Berlin. "Il étoit dangereux quelquefois de passer a travers le plupart de ces animaux, par lesquels on

for a science of life—especially the analytic value of examining populations over time. Particular abnormalities took on significance, in these arguments, only in relation to others of the same type, and individuals become elements of a population extending over time and space. Over many generations, they might ultimately define a new species. "These varieties, once they are confirmed by a sufficient number of generations where they appear in both sexes, found a species; and it is perhaps thus that species have multiplied."[87] Thinking in terms of populations rather than individuals, where variations occur with a theoretically measurable regularity, was highly unusual in this period.

Maupertuis had accounted for the albino Negro with the notion of reversion to an earlier form; he used the same notion in discussing phenomena familiar to the animal breeder. For example, he reflected on the origin of supernumerary digits in dogs, which ordinarily have five digits on the front feet and four on the hind. "Nevertheless, it is not rare to find dogs with a fifth digit on the hind feet, although it is usually detached from the bone and without articulation. Is this fifth digit of the hind feet an extra one, or is it nothing other than a digit lost from variety to variety in the whole species, and which tends to reappear from time to time?"[88] In a more speculative vein, *Système de la nature* suggested not only that new varieties could become established as normal, but that all species may have developed from two individuals. "Could we not explain in this manner how the multiplication of the most dissimilar species could have sprung from just two individuals? They would owe their origin to some fortuitous productions in which the elementary parts deviated from the order maintained in the parents. Each degree of error would have created a new species, and as a result of repeated deviations the infinite diversity of animals that we see today would have come about."[89]

Breeding also raised questions about agency: could manipulation by human breeders produce new species? How malleable were living forms? And how could human efforts to breed new animals illuminate natural processes? Maupertuis understood the perpetuation of naturally occurring variations, like albinism, by analogy to selective breeding practices: "Whether one takes this whiteness [of the albino's skin] for an illness or accident, it will never be anything but a hereditary variation that reinforces itself or erases itself over the course of generations."[90]

---

étoit attaqués. Je craignois surtout beaucoup les chiens Islandois." These were the dogs used in Maupertuis's breeding experiments (Formey, *Souvenirs,* 1:218–19). Many letters attest to Maupertuis's trading dogs with his correspondents: see Tressan to M, 20 July [1756] in LS, 334.

87. "Sur la génération des animaux," *Lettres, Oeuvres* 2:309.

88. Ibid., 311–12.

89. *Système de la nature, Oeuvres* 2:164.

90. Maupertuis, *Vénus physique,* 98.

Similarly for polydactyly: "By such repeated marriages, it would probably die out, and it would perpetuate itself through marriages where the trait was common to both sexes."[91] Varieties occurring by chance could be perpetuated through cultivation, just as particular traits are preserved in animals by systematically selecting for them. Over time, cultivation could cause varieties to solidify into a separate species that would be stable enough to perpetuate itself.

This argument represented a radical break, however speculative, with earlier accounts, particularly in the importance of large-scale time intervals.[92] Placing individuals in the temporal sequence of their forebears and descendants meant rethinking the relevance of evidence and arguments. A dynamic nature full of changing forms replaced the static and perfectly ordered world, created as such. This dynamism located order not in unchanging structures designed to perform specific functions, but in the laws governing the process of change. "To make natural history a real science, one would need to apply oneself to research that would make known to us, not the particular shape of this animal or that, but the general processes of Nature in her production and her conservation."[93] The "general processes" turned out to be dynamic relations among particles, among forces, and among bodies. Instead of reducing chemical and organic activity to collisions of inert parts, as Descartes and his followers had done, this dynamic approach reversed the hierarchy, making passivity apparent and activity fundamental. Everything in nature is more or less fluid, Maupertuis suggests; inert substances are just less fluid than organic matter:

> all the matter we now see on the surface of our Earth was once fluid, whether dissolved in water or melted by fire. Now, in this fluid state, the matter of our globe was in the same situation as the liquors in which the elements that produce animals swim: and metals, minerals, precious stones were much easier to form than the least organized insect. The least active particles of matter will have formed metals and stones [*marbres*]; the most active formed animals and man. The only difference between these productions is that some continue in a fluid state, and in others the hardening of the matter containing their elements does not permit new productions.[94]

91. "Sur la génération des animaux," *Oeuvres* 2:308.

92. Roger notes that Maupertuis "put the living individual back into the succession of beings in its species" (Roger, *Sciences de la vie,* 481).

93. "Pour faire de l'Histoire naturelle une véritable Science, il faudroit qu'on s'appliquât à des recherches qui nous fissent connoître, non la figure particuliere de tel ou tel animal, mais les procédés généraux de la Nature dans sa production et sa conservation" (*Lettre sur le progrès des sciences, Oeuvres,* 2:418).

94. Maupertuis, *Système de la nature, Oeuvres,* 2:169.

In the preexistence model of generation, passive seeds waited to be activated; this kind of epigenesis posited organic molecules imbued with their own capacity for organization. Nature's very fluidity opened up possibilities for outcomes of any process contingent on a host of factors. In fact, all of nature as we know it could be returned to the primordial state by a cataclysmic disaster: "After such a flood or fire, new unions of elements, new animals, new plants or rather entirely new things could reproduce themselves."[95]

In challenging preexistence theories, Maupertuis went beyond the anatomy and physiology of generation to a theory of organization (on the submicroscopic scale) *and* a theory of heredity (on the macroscopic scale). By introducing hereditary patterns, this science combined the speculative analogical approach to the very small with calculation and observation of large groups. If the population replaces the individual as the unit of analysis, the laws of nature take on meaning expressed in new kinds of laws. Given that these laws can incorporate change, the theory historicized the problem of organization. Immediate creation of organized forms by God, whether at the beginning of time or since then, would mean that they have no history other than growing or unfolding. But if "elements," guided by memories of their places in the parent organism, mix and arrange themselves, each individual has a place in the history of the species and in the whole population of its kind at a given time. If all goes normally, the original situation of parts would determine their future place; in this sense teleology is built in to the organic matter itself. Maupertuis insisted that ascribing intelligence or perception to matter only seems problematic if we assume that this active property is the same as our human intelligence. Instead, he recommended thinking of intelligence as a property with degrees, like hardness; then the difference between properties of the simplest parts of matter and those of the human mind would be only a difference in degree.[96]

## Maupertuis and Diderot

As we have seen, Diderot read *Système de la nature* in 1753, when it was still masquerading as a thesis by Dr. Baumann. In his *Pensées sur l'interprétation de la nature,* he articulated a definition of matter's sensibility in dialogue with Baumann's text. Diderot and Maupertuis crossed paths in Paris in 1753, although no record remains

95. Ibid., 170.

96. "Au fond toute la répugnance qu'on a à accorder à la matiere un principe d'intelligence, ne vient que de ce que l'on croit toujours que ce doit être une intelligence semblable à la notre: mais c'est de quoi il faut bien se donner de garde. Si l'on réfléchit sur l'intelligence humaine, on y découvre une infinité de degrés tous différens entr'eux, dont la totalité forme sa perfection" (ibid., 180–81).

of actual conversations between them on these questions.[97] Nevertheless, we can trace a literary conversation in the pages of their published texts. The printed exchange must have prompted discussion among readers; this is corroborated by the account of Diderot's challenge to Baumann in Grimm's *Correspondance littéraire,* a literary review and gossip sheet that circulated in manuscript.[98] Diderot's use of Baumann's then-unpublished work, along with the response Maupertuis published a few years later, illustrates nicely the complex interplay among irony, philosophical method, literary technique, and empirical evidence familiar to denizens of the Paris intellectual scene. The printing of books, initially in small numbers and then in more accessible editions, widened the circle of conversation to include more readers. Maupertuis had not decided, when he first circulated a few copies of the book, whether or not he would publish it more widely. Diderot's challenge to Baumann pushed it into the public eye by provoking conversations about it and prodding Trublet to publish the book.

Diderot's book was a relatively unstructured series of reflections on the complementary methods of experiment and reasoning (or "interpretation"), juxtaposed with all sorts of examples from physics. Maupertuis's version of epigenesis served as a sounding board for Diderot's own thoughts. After discussing the use of experiment in natural philosophy, he examined "the hypothesis of the doctor of Erlangen."[99] He undertook to push it "as far as it can go," ostensibly to see whether it could hold up, but actually to articulate a rather similar "conjecture" of his own. If Baumann had stuck to the problem of generation, without mentioning the soul, Diderot argued, he would not have fallen into "the most seductive kind of materialism, attributing to organic molecules desire, aversion, feeling, and thought."[100] Without questioning the explanation of generation as such, he zeroed in on Baumann's treatment of the soul as an amalgamation of the perceptions of the constituent parts of the body, quoting the Latin text to show how it drifted perilously close to materialism. "Here we are surprised that the author either did not notice the terrible consequences of his hypothesis, or if he did notice the consequences, that he did not abandon the hypothesis."[101]

To explore these consequences, Diderot, applying the critical method he had outlined in the earlier sections of his book, manufactured a dialogue in which he

97. Diderot had sent Maupertuis a copy of *Lettres sur les aveugles* in 1749 (Diderot to M, 12 June 1749, in Chouillet, "Trois lettres inédites de Diderot," 8–9).

98. Grimm, *Correspondance littéraire, philosophique et critique,* 2:351–52.

99. For a probing analysis of Diderot's book, see Anderson, *Diderot's Dream.* Anderson argues that Diderot is performing a kind of "experiment" on Maupertuis's book (15).

100. Diderot, *Pensées sur l'interprétation de la nature,* 84.

101. Ibid., 77, 82.

asked Baumann/Maupertuis whether the universe forms a unified whole or not. "If he replies that it does not form a whole, he will undermine with a single word the existence of God, by introducing disorder into nature." But if he agrees that it is a whole, made up of ordered parts coming together into a larger order, "he will have to avow that in consequence of this universal copulation, the world, like a great animal, has a soul; that, the world possibly being infinite, this soul of the world, I do not say is, but may be God."[102] Pushing the argument, as he had promised to do, he arrived at the same Spinozist conclusion that La Condamine had spotted in Maupertuis's text. Diderot claimed to be simply generalizing Baumann's hypothesis about active matter from the process of animal generation to all of nature.[103] One cannot help suspecting him of bad faith, or at least a clever perversity. Grimm admired the way Diderot "noticed that one had not drawn from this thesis the strongest possible position [*tout le parti possible*]; but as we must treat these matters with extreme circumspection, he adroitly takes the part of refuting the supposed doctor Baumann, under the pretext of the dangerous consequences of this opinion, but actually in order to push it as far as it could go."[104] Diderot's readers knew that he was hardly reluctant to question the existence of God, especially if they had read the *Letter on the Blind,* and Diderot himself surely knew that imputations of Spinozism would not be to Maupertuis's taste. But his comment that Baumann's little book "will give plenty of torture to our philosophers," was actually a mark of approbation. Unlike the systematics of Linnaeus or the compulsive observations of Réaumur (both criticized in other sections of *The Interpretation of Nature*), the "hypothesis" developed by Baumann is "fertile," "the fruit of profound meditation." And most importantly, for Diderot, his engagement with it produced new knowledge. He used the conversation with his pseudonymous interlocutor not just as a pretext to push an atheistic materialism (which he professed to find "terrifying" [*effrayante*]), but to put forward a model of active matter based on the quality of sensibility.[105]

Diderot proposed as an alternative to desire and aversion, "a sensibility a thousand times weaker than that which the All-Powerful gave to those animals the nearest [on the scale of being] to dead matter." He implied that this "muffled sen-

102. Diderot took "copulation" from Maupertuis's Latin, which he had just quoted on the previous page ("cum aliis copulationem"); in French, Maupertuis used "*union*" (quotations on ibid., 81–82).

103. According to Venturi, Diderot went well beyond the "organic" treated by Maupertuis to a "cosmic" and even "mythic" dimension (Venturi, *La jeunesse de Diderot,* 301–03).

104. Grimm, *Correspondance littéraire, philosophique et critique,* 2:351–52.

105. He developed this account further in his *Rêve de d'Alembert* (1769, but published only posthumously). On sensibility in Diderot's fiction, see also Vila, *Enlightenment and Pathology,* 152–81.

sibility [*sensibilité sourde*]" would avoid the "dangerous consequences" of ascribing perception or thought to matter. Organic particles would find their way to an equilibrium position "convenient to rest," driven by an "automatic restlessness [*inquiétude automate*]."[106] Diderot was concerned less with explaining particular phenomena than with justifying a dynamic picture of nature to mesh with his epistemology and his conception of enlightened philosophical method. The philosopher seeks new knowledge in the same way that molecules seek their places, by a kind of restless touching and retouching, trial and error, a "*tatonnement*" like that of a blind man's stick. The ideal method involves moving back and forth from sense impressions to reflection, from experiment to theory, in a kind of oscillatory exploration. "It is the work of the bee," he said, by which he meant to conjure the image of productive work as well as restless traveling back and forth.[107] Any position, whether material or philosophical, is always either in flux or in a state of dynamic equilibrium, and potentially unstable. Diderot's literary conversation with Baumann/Maupertuis exemplified the method of reading that he articulated later in his article "Encyclopédie," where he showed the reader how to engage with texts as if in a conversation, in a vibratory back and forth motion. In a parallel to this dynamic model of reading, Diderot imagined nature itself perpetually either in an oscillatory motion or on the verge of being so.[108] His encounter with Baumann's book put him into this state of restless motion, and led him to articulate the materially based "automatic" or habitual restlessness. He used his critique, which was really a veiled appreciation, to extend his model of dynamic matter from chemistry and cosmology to living organisms.

The dynamic tension between the texts, and their authors, may have contributed to Maupertuis's agreement to publish his book, since Diderot's comments appeared before *Essai sur la formation des corps organisés* had been printed in France. A direct response to Diderot only materialized several years later, in the second volume of Maupertuis's collected works (1756). In the preface to this version, now titled *Système de la nature,* he claimed to have published under a pseudonym because "an unknown foreign author would be less the butt of objections, or at least I would not have to reply to them." But since he had been recognized, and then attacked for irreligion, he felt it imperative to reply.[109] He appended a

106. Diderot, *Pensées sur l'interprétation de la nature,* 84–85.

107. "C'est le travail de l'abeille" (ibid., 34).

108. Diderot, "Encyclopédie," in *Encyclopédie,* 5:635–48. Anderson, *Diderot's Dream,* chapter 3. Anderson argues that Diderot envisioned the *Encyclopédie* as the transcript of a giant conversation controlled by the system of cross-references (105).

109. For his hesitations about publishing a response to Diderot, see M to Trublet, 30 November 1754, SM, fol. 108v; see also Maupertuis, "Avertissement," in *Oeuvres,* 2:137–38.

"Response to the Objections of M. Diderot" to his original text, adding to its length considerably. Praising Diderot as "an Author who honors our Nation so much, who has enlightened it by so many writings where wit and invention shine from all sides," he then imported the full text of Diderot's remarks, amounting to ten printed pages, marked off with quotation marks at the beginning of each line. The reader could see just how Diderot had framed his interrogation with the conversation laid out graphically on the page. Just as Diderot had questioned Baumann, Maupertuis "asks" Diderot if there is any hypothesis which might not be pushed to dangerous consequences, if one were so inclined. "If one were less persuaded of the religion of the author of the Interpretation of Nature, one might suspect that his design was not so much to destroy the hypothesis as to derive from it the very consequences he calls 'terrible.'"[110] In his defense, Maupertuis brought Baumann into the conversation, reproducing extended passages in italics, and referring to "le Docteur Baumann" in the third person (fig. 25). Maupertuis's counterattack rests on challenging Diderot's use of the word "whole" [*un tout*].

> Is the Universe a whole, yes or no? In the negative, M. Diderot does not define the term "whole" and leaves it with the vaguest meaning; in the affirmative, he gives it a definite meaning, the meaning that he chooses, to conduct Doctor Baumann to a unfortunate conclusion. The Doctor could complain of this trap that it seems has been set for him, but he prefers to undertake to show that he has not fallen into it.[111]

If "whole" simply means a "regular edifice" where every part has its place, then he can safely answer yes, without implying that the melding of elementary perceptions in animals must be generalized to the whole universe. Individual bodies can have their own coherent identities, or "unique perceptions", without being in turn merged with other souls. "This manner of reasoning, which M. Diderot calls the act of generalization, . . . is only a kind of analogy . . .; it is incapable of proving either the falsehood or the truth of a system."[112] His main concern was to show that the charge of Spinozism had nothing to do with his conjectures about the perceptions of the elementary parts of matter. This was the interpretation he saw Diderot attempting to push, and Maupertuis subtly suggested that Baumann was

110. Ibid., 186, 197.

111. "L'Univers est-il un *tout,* oui or non? Dans la négative, M. Diderot ne définit point le terme *tout,* & le laisse dans le sens le plus vague; dans l'affirmative, il lui donne un sens déterminé, le sens qu'il lui plait pour conduire le Docteur Baumann à une conclusion facheuse. Le Docteur pourroit se plaindre de ce piege qu'il semble qu'on ait voulu lui tendre, mais il aime mieux s'attacher à faire voir qu'il n'y est pas pris" (Maupertuis, *Oeuvres,* 2:205).

112. Ibid., 206.

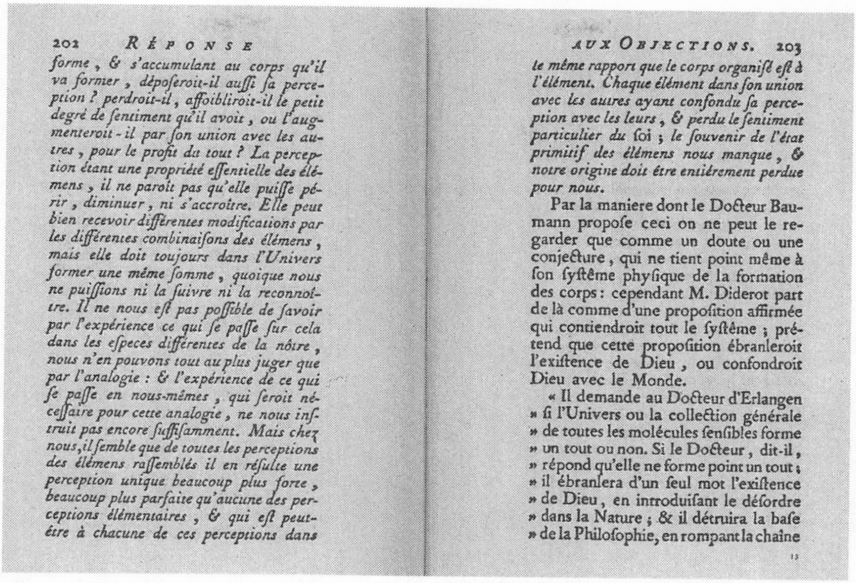

202 RÉPONSE

*forme, & s'accumulant au corps qu'il va former, dépoſeroit-il auſſi ſa perception? perdroit-il, affoibliroit-il le petit degré de ſentiment qu'il avoit, ou l'augmenteroit-il par ſon union avec les autres, pour le profit du tout? La perception étant une propriété eſſentielle des élémens, il ne paroît pas qu'elle puiſſe périr, diminuer, ni s'accroître. Elle peut bien recevoir différentes modifications par les différentes combinaiſons des élémens, mais elle doit toujours dans l'Univers former une même ſomme, quoique nous ne puiſſions ni la ſuivre ni la reconnoître. Il ne nous eſt pas poſſible de ſavoir par l'expérience ce qui ſe paſſe ſur cela dans les eſpeces différentes de la nôtre, nous n'en pouvons tout au plus juger que par l'analogie: & l'expérience de ce qui ſe paſſe en nous-mêmes, qui ſeroit néceſſaire pour cette analogie, ne nous inſtruit pas encore ſuffiſamment. Mais chez nous, il ſemble que de toutes les perceptions des élémens raſſemblés il en réſulte une perception unique beaucoup plus forte, beaucoup plus parfaite qu'aucune des perceptions élémentaires, & qui eſt peut-être à chacune de ces perceptions dans*

AUX OBJECTIONS. 203

*le même rapport que le corps organiſé eſt à l'élément. Chaque élément dans ſon union avec les autres ayant confondu ſa perception avec les leurs, & perdu le ſentiment particulier du ſoi; le ſouvenir de l'état primitif des élémens nous manque, & notre origine doit être entiérement perdue pour nous.*

Par la maniere dont le Docteur Baumann propoſe ceci on ne peut le regarder que comme un doute ou une conjecture, qui ne tient point même à ſon ſyſtême phyſique de la formation des corps: cependant M. Diderot part de là comme d'une propoſition affirmée qui contiendroit tout le ſyſtême; prétend que cette propoſition ébranleroit l'exiſtence de Dieu, ou confondroit Dieu avec le Monde.

« Il demande au Docteur d'Erlangen » ſi l'Univers ou la collection générale » de toutes les molécules ſenſibles forme » un tout ou non. Si le Docteur, dit-il, » répond qu'elle ne forme point un tout; » il ébranlera d'un ſeul mot l'exiſtence » de Dieu, en introduiſant le déſordre » dans la Nature; & il détruira la baſe » de la Philoſophie, en rompant la chaîne

13

Fig. 25. Two pages from Maupertuis's response to Diderot, showing different typefaces for different voices. Italics indicate quotation from Maupertuis's original (singled out by Diderot); normal type comments on quotation, referring to Baumann in third person; quotation marks indicate Diderot's own words. Maupertuis, "Reponse aux objections de M. Diderot," *Oeuvres* 2:202–03 (1756).

being used unfairly. "If we are to understand by a *whole* the God of Spinoza, M. Baumann assuredly will deny that the Universe is a whole; and no one will ever be able to hold that his system contains this idea."[113]

Maupertuis also played with the notion of the whole and used his response to Diderot to argue for a kind of skepticism that they both could have endorsed. "All our systems, even the most extensive," Maupertuis insisted, "only embrace a small part of the plan followed by the supreme Intelligence; we do not see either the relations of the parts among themselves, or their relation to the whole." Here again was evidence of the vast range in degrees of intelligence, from God to man to the smallest parts of nature. In a rather backhanded way, this conviction that the only possible perception of the whole universe would belong to the highest intelligence, namely God, supported Baumann's vision of intelligence as a property of all pieces

113. Ibid., 208.

of nature, varying in intensity. Diderot's rhetorical use of "the whole" brought Maupertuis to the familiar theme of the limitations of human knowledge. Baumann had not put forward a "system" like that of Spinoza, which aspired to a totalizing view of nature. Only the most alarmist readers could possibly follow Diderot in his Spinozist reading of the text in question. These were the people who "as soon as a philosophical proposition is presented to them, take it to the temple to judge it by the light of the lamp there."[114] They needed to learn how to read philosophy, and that is what Maupertuis set out to teach them in this exchange.

Finally, Maupertuis responded to Diderot's charge that "the most seductive materialism" followed necessarily from the "desire, aversion, sentiment, and thought" Baumann had attributed to organic molecules. He involved the reader in the conversation by formulating a series of rhetorical questions to show the spuriousness of Diderot's distinction between *sensibilité sourde* and perception. "He does not want perception to be able to belong to matter; and believes that sensation can belong to it: as if perception and sensation were different in kind. . . . Does M. Diderot seriously propose this distinction?"[115] The critique of Cartesian dualism that follows could just as well have been written by Diderot. Ultimately, Maupertuis's defense amounted to showing how close his position was to that of his interlocutor. "When he proposes to substitute *a sensation similar to an obscure and muted touch* for Dr. Baumann's *elementary perception,* it is a real game of words [*jeu de mots*] to win over or surprise the Reader; a sensation being a true perception. . . . M. Baumann did not say anything other than what M. Diderot wanted him to say."[116]

The reader of this exchange comes away with the impression that the authors have indeed engaged in an elaborate game of words, just as Maupertuis said. The game had a serious side, of course, but both men played with the flexibility of their conjectures and hypotheses, just as they played with style and genre. Evidently, they were each trying to show readers how to play such games, reading between the lines and stretching analogies to explore difficult questions. Diderot's whole book was in fact a defense of "conjecture" as a productive method; Maupertuis had used the same term for his interpretation of the evidence in *Vénus physique.* Maupertuis's reflections on the fluidity and contingency of natural forms and processes are not so far from Diderot's speculative ruminations on the "perpetual vicissitude" of nature: "I therefore ask," Diderot wrote, "if metals have always been and will always be as they are; if plants have always been and will always be

114. Ibid., 201.

115. "Il ne veut pas que la perception puisse appartenir à la matiere; & croit que la sensation peut lui appartenir: comme si la perception & la sensation étoient d'un genre différent. . . . Est-ce sérieusement que M. Diderot propose cette différence?" (ibid., 212).

116. Ibid., 214 (emphasis in original.)

as they are; . . . it would perhaps be forgiven in you, O skeptics, not to doubt that the world was created, but to doubt that it is such as it was, and as it will be.[117] Maupertuis willingly speculated, with the support of the latest empirical evidence, about matter's capacity for self-organization. But he drew the line at atheism. Knowing that Diderot had gone some distance in this direction, he decided he had to defend himself from imputations against his religious convictions. Even so, both Maupertuis and Diderot were aiming to destabilize the "spirit of systems" that d'Alembert had so eloquently reviled in the "Preliminary Discourse" to the *Encyclopédie*.[118]

As Maupertuis had written elsewhere, "Systems are inimical to the progress of the sciences."[119] Both Maupertuis and Diderot wished to distinguish the rehabilitation of conjecture or speculation from excessively rigid hypothesizing. Their alternative to system-building, of a type stereotypically associated in the eighteenth century with Descartes, was not simply descriptive empiricism, but rather a reasoned analysis of phenomena leading to probable conjectures. When they argued with each other, they did so respectfully, because neither one considered the other dogmatic. This exchange, then, was of a rather different character from Maupertuis's earlier disputes with competitors and challengers. Although he certainly wished to defend himself from charges of atheism, he had no interest in attacking Diderot personally or undermining the credibility of his suggestions about matter's sensibility. Diderot went farther than Maupertuis in his iconoclasm, and they did not agree on all details, but in many ways they were kindred spirits. Both men speculated freely, and drew their conjectures from phenomena, whether experiment or observation. And they both took pains to show that they were expert players of a literary game in which style and wit marked authors as worthy of being read.

In spite of the title chosen for the ultimate version of his theory of generation, Maupertuis did not construct a system of the reductive or dogmatic sort so maligned by his contemporaries. Given the unsystematic nature of his "system," in

117. "Je demande donc si les métaux ont toujours été et seront toujours tels qu'ils sont; si les plantes ont toujours été et seront toujours telles qu'elles sont; . . . un doute qu'on vous pardonnerait peut'etre, o sceptiques, ce n'est pas que le monde ait été créé, mais qu'il soit tel qu'il a été et qu'il sera" (Diderot, *Pensées sur l'interprétation de la nature,* 94).

118. D'Alembert, *Preliminary Discourse to the Encyclopedia;* Loveland shows that d'Alembert's distinction between "spirit of systems" and "systematic spirit," made into a canonical tenet of the Enlightenment by Cassirer and others, was not rigidly maintained, even by d'Alembert himself in other articles for the *Encyclopédie.* Castigation of "systems" did become commonplace at midcentury (Loveland, *Rhetoric and Natural History,* 101–14).

119. "Les systêmes sont de vrais malheurs pour le progrès des Sciences" (Maupertuis, "Sur les systêmes," *Lettres,* in *Oeuvres,* 2:257).

fact, the title takes on a tinge of irony. But it was also meant as a serious defense of a speculative, reflective, "fertile" mode of thinking and writing about nature. If systems inhibit progress in the sciences with the blinders of dogma, he suggested that doubts and queries, when carefully framed to reflect the current state of empirical knowledge, might well spur on inquiry in a productive direction.

# 11

## The Final Years

BY THE TURN OF THE NEW YEAR IN 1754, Maupertuis had been absent from Berlin for eight months, and Frederick was encouraging him to return to his post. "You have nothing to fear here from envy, and you should be reassured on the matter of reputation. Libels share the fate of ephemeral insects; they are bothersome for a time, but they perish sometimes even before they have been forgotten."[1] By February, the king was more insistent: "A certain degree of anarchy reigns in your academy since you have been gone, which bothers me very much. The chemists and the physicians are making a terrible noise; they are fighting about a position, everyone wishing to fill it with his own candidate."[2] In the event, the president did not make it back to Berlin until summertime, perhaps reluctant to face the turmoil in the Academy and the recurring problem of recruiting and retaining suitable "subjects." Frederick's literary and intellectual circle at Potsdam lacked the sparkle of earlier days, and the king must have seen that his spectacular clash with Voltaire could only deter other potential protégés from making such a commitment. In fact, Frederick had sent Maupertuis as an emissary to the illustrious d'Alembert, hoping to persuade him to forsake Paris for Prussia, but this, like so many other negotiations, had come to naught.[3] When he did finally return from France in the summer of 1754, his health temporarily restored, Maupertuis spent two more years in Berlin, before departing once again in June of 1756. Plagued by poor health, he attended only sixteen meetings of the Academy over this period, although he was still making crucial decisions about finances, publications, and personnel. He was alienated more than ever from the German intellectual community outside his immediate circle in the Academy and the court,

1. FII to M, 14 January 1754, in Koser, *Briefwechsel,* 297.

2. FII to M, 22 February 1754, in Koser, *Briefwechsel,* 298. For the dispute between the chemists Eller and Pott, see Euler to FII, [Dec. 1753], Euler, *Opera Omnia,* series 4A, 6:339. Their animosity continued for years; see Euler to M, 23 Oct. 1756, Euler, *Opera Omnia,* series 4A, 6:222.

3. "I saw d'Alembert yesterday and the day before, as the King ordered me to do, and as I believe that this would be the best acquisition that His Majesty could make, I forgot nothing of all that I thought most likely to make him wish to come to Berlin" (M to Abbé de Prades, 25 May 1753, in Koser, *Briefwechsel,* 294). On d'Alembert's relations with Frederick II, see Terrall, "Culture of Science."

as he saw several promising astronomers and physicists lured away from Berlin by the universities. In the case of Tobias Mayer, a professor at Göttingen who had agreed to direct the astronomical observatory of the Berlin Academy, Maupertuis interpreted the astronomer's failure to take up his post as the result of the animosity of the government in Hanover. "The Hanover minister, always trying to cross us, and who is continually giving us proofs of his envy and his bad wishes for the Academy, has retained Mayer."[4] As he complained to Bernoulli, who acted as go-between in negotiating with Swiss candidates, "Because of the low opinion of savants in Germany, there are very few savants in Germany who merit a high opinion, and for whom the sciences are more than a job."[5] Ultimately, his experience in Prussia did not live up to his high expectations for it, in spite of the king's unqualified support, and for this he blamed the Germans. "I have been so often deceived by several of these gentlemen who, after having gotten me to make them offers and propositions, have failed me, that I must be excused for not trusting [M. Huber, the new astronomer]. If I told you about a negotiation with M. Haller that lasted for six months . . . you would blush for someone who passes for a great man and even for an honorable man. . . . [A]ll these adventures bother me only because they have exposed me to the King's view as a man who is too inconsequential [*léger*] or too gullible."[6]

The northern climate conspired to discourage him as well, since his lung disease flared up every winter. In the summer of 1756, Maupertuis set off again, leaving the Academy in the care of Euler, for the comforting air of Saint-Malo.[7] Almost as soon as he left, Prussia and France were at war, in the opening battles of what was to become the Seven Years' War. For the next three years, he struggled not only with worsening health, but with the distressing situation of owing allegiance to two kings, formerly allies, but now bitter opponents. He agonized about whether to attempt to return to the potentially fatal climate of Berlin, as the citizen of an enemy power, or to find an honorable position in France. In his homeland, his ties to Frederick made him controversial at best; some influential figures in Louis XV's court viewed him as nothing more than a traitor. The possibilities in France fluctuated with the shifting political winds, as he called on old connections at court to speak for him, with only limited success. The larger events of the war superceded his private negotiations, even as they cut him off from Berlin. His wife had joined the Prussian royal household, where she was often beyond the reach of letters from France. Even communication with Frederick was impossible

4. M to FII, 19 Oct. 1754, in Koser, *Briefwechsel,* 300.

5. M to JBII, 14 December 1754, BEB.

6. M to JBII, 23 September 1755, BEB.

7. He went with Frederick's blessings (M to FII, 13 May 1756, in Koser, *Briefwechsel,* 320).

for long periods.[8] Maupertuis tried to maintain his cordial relations with the Prussian king, promising him that his absence from Berlin was only temporary, while simultaneously exploring other solutions to his dilemma that would allow him to remain in France, and perhaps even to bring his wife to join him.[9]

When Maupertuis's doctors in Brittany recommended a trip to Italy for the winter of 1757, Frederick did not insist on his return to Berlin; the Academy proceeded with its business as best it could under the leadership of Euler.[10] As it happened, Maupertuis never made it to Italy. He spent several months in Bordeaux, hosted by his old friend and sometime lover the duchesse d'Aiguillon. Although not strong enough to make the journey to Italy, the invalid continued to reflect on his position in the Republic of Letters, and, we may plausibly imagine, on his legacy to posterity. His letters from this period betray his awareness of his own mortality, even as he tried to keep all his earthly options open.[11] He continued to correspond with Formey and Euler in Berlin, and to follow the business of the Academy from a distance. In the summer of 1757, the Academy's finances were disrupted by war and the city was threatened with invasion. Although the Academy never had to evacuate the city, its premises fell into disrepair, and the academicians saw their pensions cut in half.[12] Euler's letters from this period contain a startling mixture of news about military campaigns, reports of war-induced austerity in Berlin (including two episodes of invasion or near-invasion), scientific news, and prosaic reports of the weekly proceedings at the Academy. The body continued to meet as usual, producing their *Mémoires* and dealing with deaths and appointments as best they could. Euler was pressed into service as examiner and translator of all Russian letters intercepted by Prussian forces.[13]

8. Frederick remarked in a letter from his camp outside the besieged Prague, "Apparemment que les Pompadouristes auront trouvé votre correspondance dangereuse" (FII to M, 29 May, 1757, in Koser, *Briefwechsel,* 323.)

9. Maupertuis conserved the rights to his property in France when he went to Prussia in 1745 (Tressan, "Eloge de Maupertuis," 319; Fouchy, "Eloge de Maupertuis," 272). The official approval for his departure from France is in the Archives nationales, Secretariat de la Maison du Roi, 1745; cited in Velluz, *Vie de Maupertuis,* 95.

10. FII to M, 29 May 1757, in Koser, *Briefwechsel,* 323.

11. See, e.g., M to La Beaumelle, 9 December 1757, in LS, 230: "Je vous avoue que je suis plus touché de quitter la France que je ne l'ai jamais été, parce que je vois que c'est pour la dernière fois."

12. "Cependant les revenus de notre Acad. y souffrent considerablement et je crois que pour l'année prochaine chacun sera oblige de se contenter de la moitié de sa pension; car nos capitaux placés sur la Landschaft et ailleurs n'y sauroient suppléer puisqu'on n'y a pas moyen d'en rien retirer" (Euler to M, 3 September 1757, in Euler, *Opera Omnia,* series 4A, 6:232). Euler also reported on the Academy's financial resources in letters of 15 January, February 1757, and 29 April 1757 (ibid. 226–30).

13. Euler to M, 14 Oct. 1758, in Euler, *Opera Omnia,* series 4A, 6:244.

The Berlin Academy also continued to announce its prize competitions. At Maupertuis's last meeting, in June of 1756, the mathematics class announced the topic for 1758: "Are the laws of statics and dynamics necessary or contingent?" The question was directly relevant to the work of both Maupertuis and Euler.[14] As it turned out, the prize was never awarded, for lack of worthy entries.[15] In the meantime, the absent president sent a manuscript written in his temporary abode in Bordeaux. Although he could not compete officially, this looked very much like his own response to the prize question.[16] In the *Essay on Cosmology,* he had raised the question of the necessity of the laws of mechanics without deciding it one way or the other. This new paper articulated Maupertuis's final synthesis of his earlier work on the laws of motion with reflections on epistemological certainty. It also clearly represented an effort to maintain an intellectual presence in the Berlin Academy, in spite of his physical distance and the tenuousness of his future. In reflecting on the relation of his own principle to the laws of motion attached to the names of the great men of previous generations, Maupertuis had ready to hand Descartes's *Principia philosophiae* and all available works of Leibniz, as well as those of Newton, Huygens, and Euler. Wolff was conspicuously (but not unexpectedly) absent from the citations. Merian read the first, metaphysical, part of Maupertuis's paper to the Academy's public meeting in January 1758 and continued with the physics sections in subsequent meetings.[17] As befit its author's standing, it was immediately sent to the printer for publication in the volume for 1756, then in press.[18]

The paper gave Maupertuis the opportunity to reexamine and restate the arguments of the *Essay on Cosmology* in light of "criticisms" and to close the book on the König affair without even mentioning it explicitly. The first section considered the proof for God's existence in connection with the necessity or contingence of the laws of motion and the related question of how to determine the certainty of different kinds of knowledge. He coined the term "replicability" to designate the the unambiguous truths of arithmetic and geometry. This property belongs above all to the idea of number, the simplest sort of abstraction inferred from the com-

14. Winter, *Registres,* 224 (3 June 1756).

15. In 1758, it was extended to 1760, but no winner was named then either. Winter, *Registres,* 239 (1 June 1758).

16. On the surviving submissions for the prize, see Clark, "Death of Metaphysics," 444–46. A manuscript of Maupertuis's paper, dated 30 August 1757 from Bordeaux, is in AS Fonds Maupertuis. The essay, published in *BAS* 1756, may well have discouraged contestants from sending contributions to Berlin.

17. Winter, *Registres,* 237–38 (26 January, 9 February, 23 February 1758).

18. Euler to M, 14 February 1758, in Euler, *Opera Omnia,* series 4A, 6:239, for plans to print the paper with the metaphysics section of the volume, because the mathematics section had already been completed.

parison of sensations, and also to extension. "I can add one extension to another equal extension and have as clear an idea of the double extension as I had of the original. . . . Extension, like number, can be increased and diminished at will . . ., a feature that does not belong to any other property of bodies [such as impenetrability]."[19] The replicability of mathematical objects means that mathematicians do not disagree about their results.

What of the other objects of scientific or philosophical investigation? Moral or metaphysical concepts have no replicability at all, he said, and cannot lay claim to any certainty. Between the absolutely certain class of mathematical truths and the nonreplicable claims of metaphysics lie physical objects, with some replicable attributes (like extension). "The speeds, for example, of bodies in motion, and the time they take to travel certain distances, bear such a natural relation to extension and numbers that these ideas become replicable."[20] Even forces can be replicable if their effects can be measured accurately. So dynamics can participate in the certainty of arithmetic and geometry, but only "if we always distinguish between what is replicable and what is not in each object; if we do not apply to an object in general that which pertains only to some of its parts."[21]

This epistemological background led into a discussion of the necessity or contingence of the laws of motion. If the laws of motion could be derived from extension, the laws themselves would also be replicable and thus necessarily true. But a priori derivations, like the ones attempted by Descartes, turned out to be erroneous, and experience had to supply verification for "hypotheses" such as Galileo's law of free fall and Newton's law of universal gravitation. "Instead of necessity, one sees in the establishment of these laws reasons for choice and preference."[22] This metaphysical stance, familiar from the *Essay on Cosmology,* also tied back into a strong statement of the limitations of human knowledge. If our ideas were eternal, divine archetypes, or if they were imprinted on the soul by God, unmediated by sense experience, we might have a science built on "the most solid of foundations." But if our knowledge comes from "the first impressions that objects made on our senses" and from the memory and comparison of these initial impressions by an active mind, knowledge could hardly be absolute, or necessary. "It would

19. Maupertuis, "Examen philosophique de la preuve de l'existence de Dieu employée dans *l'Essai de cosmologie,*" *BAS* 1756, 395. "Replicability" is not found in the dictionary of the Académie française (1694 or 1798 eds.). Maupertuis called it a "barbaric" word (ibid.).

20. Ibid., 399.

21. The *vis viva* controversy, Maupertuis claimed, arose because of this sort of confusion, one side claiming that force doubles when speed doubles, and the other claiming a quadruple force for a double speed (ibid., 400).

22. Ibid., 404.

only be a property belonging to our species: one extra sense in a superior species would make for it a new and more extensive science, to which we could never aspire: one sense less in the human species would have restricted our knowledge within bounds even more limited than they actually are."[23] Contingent laws of motion would then imply that our knowledge, far from being "a view of eternal truths," is rather "our own work, belonging only to our species."[24]

After examining the laws proposed by Descartes, Leibniz, Huygens, and Newton, to show how each is either mistaken or dependent on sense experience (and hence contingent), the paper culminates in a succinct list of "laws of motion" for elastic and inelastic collisions. The sixth and final law, the only one applicable to both types of bodies, is the principle of least action. "In refusing the supposed distinction of mathematical necessity to all these laws, one discovers in them another even more precious; namely, evidence of the choice of an intelligent and free being. They carry the imprint of the wisdom and power of their creator."[25] Finally, then, from an examination of epistemology, and the limits of human knowledge, Maupertuis once again situated his own principle, the most recently discovered of the laws of motion, in the context of the contingent science of mechanics and the demonstration of the existence of a wise and powerful deity. The very contingence of the principle makes it relevant to the theological discussion, and enhances its value.

At the end of his life, then, Maupertuis was still working to assert both the originality and the significance of the principle of least action. As well as defending the metaphysics of the *Essay on Cosmology,* he carefully distanced himself from Leibniz—without revisiting the question of his predecessor's use of extremum principles—through an examination of the laws governing collisions. Maupertuis pointed to the failings of the Leibnizian insistence on the elasticity of matter and the conservation of *vis viva,* as he returned to questions he had never fully resolved under the tutelage of Johann Bernoulli. The footnotes to this final paper show that he had carefully reread Leibniz, so he knew very well that his own theological position, stressing divine preference for maximizing efficiency, was not so far from that of Leibniz. By including his own principle as the most general of the laws governing mechanical collisions, he reiterated his claim to a place alongside the most illustrious mathematicians of the previous century. And he was reminding his colleagues in Berlin that he was still a player in the cosmopolitan republic of science.

At this time, Maupertuis had reason to be optimistic about the future of the principle of least action, in spite of the objections it had sustained, and in spite of

23. Ibid., 391.
24. Ibid., 392.
25. Ibid., 424.

his own precarious health and political uncertainties. König had been effectively silenced and had never published his projected work on Leibnizian physics. By 1757 he was dead, "his book on dynamics buried with him, if it ever existed."[26] More positive grounds for optimism had come unexpectedly from Turin, in the person of a young mathematician, Joseph-Louis Lagrange. Lagrange had discussed his new techniques for a variational calculus with Euler and had sent a paper on least action to the Berlin Academy shortly before Maupertuis's departure.[27] He was elected a foreign member soon thereafter and made it known that he would look favorably on a "sufficiently advantageous" offer of a pensioned position in Berlin.[28] He wrote to Maupertuis in the most effusive terms, seeking his support and approval. The young man explained that he was working on a paper that would "demonstrate with the greatest possible universality how your Principle always supplies with marvelous facility the solution to all cases that are most complicated and most difficult to resolve using ordinary methods, in dynamics as well as in hydrodynamics." Further, he assured Maupertuis that all attempts to undermine the generality of the principle were misguided. "As for myself, I count myself very fortunate to be able, after the learned Euler, to contribute in some manner to the universal application of such a principle, which brings such glory to its author, and which will always be esteemed as the most beautiful and important discovery of mechanics."[29] Maupertuis did not hear this kind of praise very often, and he responded by encouraging Lagrange in the strongest possible terms, while telling Euler to go ahead with arranging a pension for him immediately.[30] Unfortunately for all concerned, the war intervened, disrupting communications between Turin and Berlin and making new appointments to the Academy impos-

26. Merian to M, 10 September 1757, AS Fonds Maupertuis.

27. Euler reported to Lagrange that he had showed his paper to Maupertuis (Euler to Lagrange, 24 April 1756, Euler, *Opera Omnia,* series 4A, 5:389–90). It was presented to the Academy on 6 May (Winter, *Registres,* 227.)

28. Lagrange to Euler, 5 October 1756, Euler, *Opera Omnia,* series 4A, 5:403.

29. Lagrange to Maupertuis, 4 November 1756, AS Fonds Maupertuis.

30. Maupertuis argued that pensions released by the deaths of several members of the physics class could better be used to recruit Lagrange than to fill the physics spots: "vous scavés aussi bien que moi combien la classe de Phisique coute à l'Academie et combien il seroit à propos d'y eteindre quelques pensions au lieu de les perpetuer. Vous scavés combien la plupart de ces pensions ont été jusqu'ici en pure perte pour l'Academie . . . et combien il seroit plus avantageux pour l'Academie d'amasser une somme qui fut assés considerable pour nour procurer notre Géometre Piedmontois. Je ne crois donc point qu'il soit à propos de disposer de la pension de M. Leberkuhn a moins que le Roy, solicité peut-etre comme il le sera, n'en ordonne autrement" (M to Euler, 5 January 1757, AS Fonds Maupertuis). He wrote to Lagrange the same day (M to Lagrange, 5 January 1757, AS Fonds Maupertuis). The letter is published in Taton, "Sur quelques pièces de la correspondance de Lagrange." The letter from Lagrange remains unpublished.

sible. "In the present circumstances, no one dares make the proposal to the king," Euler reported.[31] Lagrange's paper was never published in Berlin, and he only came to the Academy much later, after Maupertuis's death and Euler's departure for Saint Petersburg in 1766.[32] Nevertheless, his enthusiastic commitment to a program that fit perfectly with Maupertuis's vision could only have encouraged the older man about the lasting impact of his work in mechanics, which only Euler had fully appreciated until then. It may well be that his exchange with Lagrange reinforced the focus of the "Examen philosophique," which culminated in a decisive formulation of the principle of least action. When he sat down to write this paper, which he very likely suspected would be his last work, it looked as though his program for dynamics would be carried on in Berlin by the ablest mathematicians of the day.[33]

## Contingency, Least Action, and Generation

Given Maupertuis's reluctance to construct any kind of totalizing system, his readers may well have wondered how to reconcile the principle of least action with the theory of generation. Building on the analysis of the "Examen philosophique," we can surmise that the properties of bodies that can be "reduced to replicability" differ fundamentally from organic properties. The latter seemed to be, at least when perceived by human minds, irreducible and nonreplicable by virtue of their complexity and their internal activity. Thus, when Maupertuis added active properties—whether described as desires and aversions or as instinct and intelligence—onto the mechanical attributes of matter, he did so only (he tells us) because the simpler, "replicable" properties could not adequately account for the evidence. Rhetorically, he presented his conjectures as the latest in a long history of philosophical attempts to understand nature and himself as the heir to a tradi-

31. Euler to M, 15 January 1757, in Euler, *Opera Omnia*, series 4A, 6:226.

32. The wartime hiatus in the journal's publication (from 1760 to 1764) does not explain the absence of Lagrange's paper from volumes 13 or 14 of the *BAS*; the manuscript may never have arrived in Berlin. Resuming correspondence with Euler in 1759, Lagrange announced the near-completion of a manuscript on the calculus of variations and the principle of least action, and asked for help in getting it published in Berlin. Euler insisted that the Berlin publishers were not dependable in wartime and disclosed that he had written up his own version of the calculus of variations. Lagrange's early papers on the calculus of variations and the extremum principle were published in the proceedings of the Turin Academy of Sciences, of which he was a founding member. On Lagrange's variational method, see Fraser, "J. L. Lagrange's Early Contributions to the Principles and Methods of Mechanics."

33. In 1758, Euler reported his own latest researches to Maupertuis, assuring him that "there is no doubt at all that your principle extends even a great deal further than anyone has extended it before" (Euler to M, 14 February 1758, in Euler *Opera Omnia*, series 4A, 5:239–40). For subsequent history of the principle of least action, see Terrall, "Metaphysics, Mathematics."

tion of explanation that had not yet fully succeeded. Thus, he gave his own work a place in the history of science, following from the failure of previous mechanical systems to grapple with organic phenomena—just as the principle of least action was a response to historical attempts to formulate laws of motion. Theories, speculations, laws, and empirical evidence are all human productions, with a history, constrained by the limitations of the human senses and intelligence. Organized bodies followed laws all right, but these laws were less accessible than mechanical principles to human understanding. Moreover, the fluctuations of local conditions and complex interlocking circumstances brought an element of contingency, or chance, into any given instance of organic generation.[34] Mathematics did not help with organic phenomena, except, revealingly, in making calculations of probability. Nevertheless, even without rigorous proof, knowledge about problems such as organization is more relevant to human existence than the necessary truths of mathematics. Geometrical demonstrations produce "truths that are to some extent indifferent to us."[35] In pursuit of certainty, mechanics idealizes its objects, abstracting them from empirical complexities. With the divinely chosen economy principle as its foundation, the mechanical world acts determinately to fulfill the extremum condition—hence the laws of physics, though contingent, are invariable and even universal. For the organic world, Maupertuis repeatedly asserted the contingency of the current order, arguing that things could have been otherwise. If the present order were disrupted, by a global catastrophe for example, a whole new order could come about without altering the primordial chemistry. Unpredictable organic forms can emerge simply from slight disruptions in the normal process of generation; if they are viable, these aberrations can then perpetuate themselves. Even language, political institutions, and organized religion—not to mention science—would have been different, given different initial conditions or different external constraints.[36]

By treating mechanics and living things separately, Maupertuis avoided an explicit discussion of the relation between the laws of physics and the laws of life, or organization. We saw in chapter 9 how he toyed with a way of connecting inert and organic matter genealogically when he speculated about how various types of matter had become differentiated over time from a primordial fluid. "Now, in this fluid state, the matter of our globe was just like the liquors in which the ele-

34. For an alternative interpretation of Maupertuis's organic forces as "psychobiological determinism," see Tonelli, *La pensée philosophique de Maupertuis,* 48–50.

35. Maupertuis, "Avant-propos," *Essay on Cosmology,* in *Oeuvres* 1:xx.

36. For the contingency of the sciences, see "Examen philosophique," 391; on language, see Maupertuis, *Reflexions philosophiques sur l'origine des langues,* in *Oeuvres* 1:259–85; on global catastrophes, see *Lettre sur la comète,* in *Oeuvres* 3:747–51; *Système de la nature,* in *Oeuvres* 2:170.

ments that produce animals swim: and metals, minerals, precious stones were much easier to form than the least organized insect. The least active particles of matter formed metals and stones [*marbres*]; the most active formed animals and man." With this speculative origin story, Maupertuis transformed the chaos of Genesis into a swirling mass of potential forms, analogous to the seminal fluids of animals. The organic particles destined to combine into organisms, simply manifested a more intense degree of that fundamental activity shared by all matter. Inert matter developed from the primeval active fluid when some portion of it solidified into particles too rigid to retain the capacity to give rise to "new productions."[37] In the hypothetical discussion of the origin of different kinds of matter, Maupertuis tentatively explored a way of making psychic properties (desire, aversion, memory) fundamental while maintaining the boundaries between mechanics and organization in the actual world. These boundaries were mirrored in the differences separating the science of mechanics (with its links to mathematics) from the emerging science of life, which had to address the radical contingency and flexibility of the organic world.

The complexity of living nature did not undermine the validity of the principle of least action, understood as a particular kind of final cause, for mechanics. Both strands in Maupertuis's work—metaphysical mechanics and the "sketch" of a theory of generation—supported a trenchant critique of natural theology, which read God's intentions in nature's design. Natural theology was closely tied to natural history, with an emphasis on describing particulars and avoiding questions about causes, so that this critique also challenged natural historical practice. The two efforts at redefinition, then, were related to the search for general principles, and a new kind of final cause. The principle of least action redefined the meaning of final cause in a Leibnizian direction; the theory of generation sought general processes based on forces, affinities, or instincts rather than particular external forms or behaviors. Both looked beyond the details to the general, as far as possible given the limits of the human mind.

## Final Travels of a Restless Soul

The contingency of political events—the shifting alliances that made Prussia and France enemies instead of allies—kept Lagrange in Turin and Maupertuis far from either Paris or Berlin. After his stay in Bordeaux, he traveled as far south as Toulouse, where he spent the winter of 1757–1758, occasionally attending the meetings of the local academy of sciences, and continuing to importune his friends

37. *Système de la nature,* in *Oeuvres* 2:169.

about his unsettled situation. He asked Bernoulli for help in getting news from Germany, where his wife was living with the court in Magdeburg. "Sixteen months ago, I left Berlin to come to the air of France in search of a remedy for the lung disease that kills me in Germany every winter. You know what has happened in Europe since my departure; and how two courts that by nature should remain united have become enemies. During all the past winter my health, which could not recover in my native air, was made even worse by all the news I received of the terrible state of our affairs. My illness and that of Europe have equally made the return to Berlin scarcely possible. . . ."[38]

When he finally gave up on the plan to travel to Italy, and started to head north for Berlin, he got only as far as Switzerland. After a few months in Neuchâtel, he suffered another relapse and retreated to Basel, where he spent the last months of his life in the home of Johann Bernoulli, the son of his old mentor. He died in July 1759. Shortly before his death, he sent a detailed justification of his actions since the beginning of the war to La Condamine, who had followed the progress of Maupertuis's last illness with intense concern.[39] Many of his French friends had viewed his refusal to break his ties to Frederick as morally questionable; even La Condamine had displayed a "coldness" in their most recent meeting in Paris. This "memorandum" to his old friend tells a story of secret dealings with various intermediaries at the French court and the vagaries of fortune that undermined his efforts to negotiate an honorable solution to the problem of dual allegiances, complicated by his matrimonial connection to Prussia. Maupertuis represented himself as the tragic victim of contingent circumstances that never seemed to work in his favor. "Read this memoir and I am sure that you will be touched by it, and that you will see the cause of my vacillations, for which you blamed me."[40]

Maupertuis began his saga in September 1756, when he was leaving Paris for Saint-Malo; France and Austria had formed their alliance, and Prussia had invaded Saxony, but was not yet at war with France. At that point, he spoke with his old friend and ally the comte d'Argenson, then secretary of state for Louis XV. They agreed that in case of war, Maupertuis would "leave the service of His Prussian Majesty" and return to the protection of the French king. "M. d'Argenson, who saw it coming sooner than I, soon wrote to me that he had spoken to the king of my resolution and my situation, that the king had been touched by it and led me to hope that they would put me in a position to live with my wife in Alsace or

38. M to JBII, 17 November 1757, BEB.

39. See La Condamine to JBII, 17 May 1759, 26 May 1759, 19 July 1759, 29 July 1759, 7 August 1759, BEB.

40. M to La Condamine, 23 April 1759, AS Fonds La Condamine.

some other border region . . . , or that they would place me in some foreign court or republic."[41] The response to this letter, informing d'Argenson that he was about to ask Frederick for his definitive release, arrived at Versailles the very day of the attempted assassination of Louis XV by Damiens. In the ensuing confusion, Maupertuis's dilemma was forgotten, and his most powerful patron, d'Argenson, lost his ministerial position. The new minister, the Abbé de Bernis, professed friendship but proved reluctant to do anything substantial.[42] When other letters to Versailles remained unanswered, Maupertuis wrote to Frederick asking for extended leave to go to Italy.[43] His attempts to deal with Bernis failed, even when conducted through an intermediary (probably the duchesse d'Aiguillon). "Finally, whether they valued me less, or whether political motives entered into my case, it was made known that they would do nothing for me before the end of the war. . . . I found myself in a quandary from which I could no longer see anything that could extricate me."[44] Eventually, outraged at discovering that orders had been given to put him under surveillance as a possible enemy spy, he wrote again to Bernis to ask if Louis XV would object to his returning to Prussia, since all other negotiations had stalled. With the explicit assurance that Louis would not hold this against him, Maupertuis decided to proceed on his journey—though it was to prove abortive because of his failing health. "Here was the conclusion of this grand negotiation, . . . a singular series of fatalities, and the Abbé Bernis, whose friendship I counted on, could never had done me more harm if he had been my worst enemy." In his final letter, Bernis was still assuring him that "His Majesty is persuaded that your feeling for him will never diminish . . . and that you will always remember that you were born a Frenchman."[45]

This was in fact the root of his distress, since that consciousness of being French now clashed with his continuing sense of obligation and attachment to the Prussian king. "In days gone by, I was Prussian at Versailles and French in Pots-

41. "Memoire pour M. de la Condamine," attached to M to La Condamine, 23 April 1759, AS Fonds La Condamine.

42. Maupertuis quotes Bernis as saying "I would wish that this position [as minister] would allow me to contribute to your peace [*repos*], your glory and your happiness" (ibid.). Only one letter from Bernis survives (AS Fonds Maupertuis)

43. M to FII, 1 May 1757, in Koser, *Briefwechsel*, 323.

44. "Mémoire pour M. de la Condamine," in M to La Condamine, 23 April 1759, AS Fonds La Condamine; Maupertuis's correspondence with the Abbé de Cicé in this period also documents the painful ambivalence of his position (AS Fonds Maupertuis). For the role of the duchesse d'Aiguillon, see Cicé to M, 16 September 1757, AS Fonds Maupertuis.

45. Quoted by Maupertuis in "Mémoire pour M. de la Condamine," in M to La Condamine, 23 April 1759, AS Fonds La Condamine.

dam, with as much ease in one place as the other. Today I can no longer be one or the other, either in one country or the other."[46] Furthermore, although he could have retreated to the countryside with his sister in Brittany, he had no fixed situation in France, and his income was rapidly diminishing. His French pension was suspended in 1758, and payment from Berlin came sporadically in nonnegotiable promissory notes.[47] Wartime living conditions in Berlin were sufficiently uncertain that it hardly seemed an appealing destination anyway. However, he remained on cordial personal terms with Frederick, who remarked at one point, "You are perhaps the only Frenchman in France who is thinking [favorably] of me."[48] Their correspondence shows that Maupertuis tried to keep all options open as long as possible, profusely congratulating the king on military successes and asking for permission for each of his own moves. In May 1758, writing from Toulouse, he asked Frederick for advice about the safest route back to Berlin. "It is with the greatest impatience that I bring back to Your Majesty's feet the remainder of the life I have consecrated to him."[49] He did not let on that he was also trying to arrange a way to rejoin his wife without returning to Berlin, which would have meant a definitive resignation from the Berlin Academy.[50] A few months later, he asked La Condamine what the duchesse d'Aiguillon advised. "She foresaw a tranquil and honorable position for me in France. . . . So she must tell me where I should wait for peace and what I should do, provided that His Prussian Majesty does not press me [to return], if the war continues."[51] This final period was clouded by an unaccustomed sense of powerlessness in the face of political events and by the knowledge that his reputation was being undermined by rumors about his loyalty. He no longer had the strength to play different interests off against each other as he had in earlier years. After his death, Bernoulli told Formey of his

46. M to JBII, 28 January 1758, BEB. To La Condamine, he wrote, "I have become too German for the French" (M to La Condamine, 1 January 1758, Hôtel Drouot, Auction Catalogue (1986): 183).

47. His sister encouraged him to remain in Saint-Malo and forget about his ties to Prussia (Cicé to M, 21 February 1759, AS Fonds Maupertuis). He explained his financial situation to Bernoulli (M to JBII, 4 September 1758, BEB). He had been receiving 4000 livres from the French crown; his pension from Frederick was paid from funds that were depleted by military expenses, rather than from the Academy, which still had some hard currency.

48. FII to M (from Dresden), 18 January 1757, in Koser, *Briefwechsel*, 322. For the impact of the war on life in Berlin, see, e.g., Formey to M, 19 November 1758, BJ; also Euler to M, 24 December 1757, 14 October 1758, 4 November 1758, 16 December 1758, 2 January 1759, in Euler, *Opera Omnia*, series 4A, 6:236–37, 242–51.

49. M to FII, 9 May 1758, in Koser, *Briefwechsel*, 324–25.

50. His wife, a Lutheran, did not want to come to France.

51. M to La Condamine, 5 November 1758, AS dossier Maupertuis.

friend's regrets: "He reproached himself and confessed to me, so to speak, as something he ought to ask God's pardon for, that he had sacrificed his fatherland, his friends, his relatives, and above all his aged father to ambition, and perhaps to spite."[52] When he accepted Frederick's patronage, his ambitions for glory alienated him from powerful supporters in France; these deathbed regrets reflect the unanticipated consequences of divided loyalties, as well as nostalgia for his personal connections in France.

Maupertuis had known that many of his old associates in Paris resented his continuing loyalty to Frederick and his failure to affirm his allegiance to France openly by resigning his presidency. The eulogy read by Grandjean de Fouchy to the Paris Academy publicly confirmed the tenuousness of Maupertuis's posthumous reputation in France. Breaking with the tradition of heroic panegyric that normally characterized academic eulogies, Fouchy castigated his subject as a traitor to the fatherland, brought down by an excessively active imagination and overweening ambition. Reporting on Maupertuis's move from Paris to Berlin, in response to the "seductive" offer from the Prussian king, Fouchy insisted, "It undoubtedly would have been better to continue giving to his king and to his fatherland [*patrie*] those services that were recognized, honored and rewarded there."[53] If only he had appreciated what he had in France, he would not have ended up in the questionable position of serving two masters.

Maupertuis was remembered generously in the Académie française, confirming his conviction that his literary *confrères* were less grudging and envious than some in the Academy of Sciences. (He had returned to meetings of the literary academy whenever he passed through Paris, whereas he never set foot in the Academy of Sciences after he left in 1745.)[54] His successor in the Académie française, Lefranc de Pompignan, eulogized him as both "man of letters and *philosophe*" and defended him on matters of literary style and religion alike. "What delight, what ravishing images in his *Vénus physique*! Those who know the author only as a savant dedicated to all that is austere and abstract in human knowledge will be surprised at the inexpressible charm that reigns in several passages of this work. One would

52. JBII to Formey, 12 September 1759, AS Fonds La Condamine. (René Moreau had died in 1746, shortly after Maupertuis moved to Berlin.)

53. Fouchy, "Eloge de M. de Maupertuis," *HAS* 1759, 271.

54. Académie française, *Registres*. After years of sporadic negotiations, he was finally reinstated as a veteran of the Academy of Sciences in 1756, just before his last visit to Paris, under the patronage of d'Argenson. See M to FII, 18 May 1756, in Koser, *Briefwechsel*, 320–21.

55. "Que d'agrément, que d'images ravissantes dans sa *Vénus physique*! Ceux qui n'en connoissent l'Auteur que comme un Savant livré à tout ce qu'il y d'austère & d'abstrait dans les connoissances humaines, seront étonnés du charme inexprimable qui règne dans plusieurs morceaux de cet Ouvrage. On croiroit quelquefois qu'il traduit Homère ou Milton" (Pompignan, *Discours prononcés dans*

sometimes think that he was translating Homer or Milton."[55] Where Fouchy had seen a dangerously overactive imagination, Pompignan praised the style that attracted readers outside the Academy. "He paints with so much warmth, with so much liveliness, that he transports us to the very places he describes. One scales with him the summits of Horrilakero; one follows him on the frozen waters of Tornea; one flies at his side on the fragile sleds of the Laplander."[56] Where Fouchy had faulted him for abandoning the Academy, Pompignan sympathized with the "difficult and painful" situation Maupertuis found himself in at the end of his life, as a result of the hostilities between his native and adopted countries. "He was born French, and he always felt that. His situation [*état*] bound him to Prussia: there he had his work, his fortune and finally a wife."[57] Even in death, however, Maupertuis was playing a part in a literary quarrel, since Pompignan used his eulogy to take potshots at "philosophie." Maupertuis, he noted, represented "true literature and sound philosophy," untainted by the materialism and irreligion running rampant in the work of some contemporaries. "Unbelievers will not be able to exploit M. de Maupertuis's sentiments. Whatever they say, whatever they write, his name will never enlarge the obituary list of *esprits-forts*."[58] In denying Maupertuis's materialism, Pompignan was stretching a point for his own purposes.

In the Berlin Academy, Maupertuis received a predictably heroic send-off, in spite of the war and his long absence from the Prussian capital. Delivering the eulogy in 1759, Formey drew on the heroic imagery that linked the mathematician adventurer with his immediate relatives and his more distant classical forebears as well:

> Saint Malo is a kind of republic of Argonauts; M. de Maupertuis's compatriots . . . bring back to their fatherland (*patrie*) riches which they have often devoted to the defense and health of that same country in the most glorious manner. M. de Maupertuis was the Jason of a different class of Argonauts. The treasures he sought in the

*l'Académie françoise, . . . 10 mars 1760* (Paris, 1760), 10). It was customary for new members to eulogize the previous holder of the chair to which they had been elected.

56. Ibid.

57. Ibid., 12–13.

58. Ibid., 17. Pompignan had Diderot and Voltaire in mind as *esprits forts.* The marquis d'Argens, who had known Maupertuis at Frederick's court, at the height of the conflict with Voltaire, objected to Pompignan's "canonization" of "Saint Maupertuis." D'Argens claimed that the devotion of the academy president's last years was a cynical ruse adopted to enlist the support of "the superstitious and fanatical" in his battle against Voltaire (D'Argens, *Histoire de l'esprit humain,* 11:200).

59. Formey, "Eloge de M. de Maupertuis," *BAS* 1759, 465. Formey even quoted a poem by Voltaire on the subject of scientific heroism, pointing out that the poet had excised Maupertuis's name from his verses in the aftermath of their quarrel. (ibid., 466).

world's extremities are the most precious of all that enrich the mind, and he shared them not only with his country (*patrie*), but with the whole human race.[59]

The corsair mythologized as Argonaut was a version of his identity that Maupertuis would have endorsed. It recalls his triumphant pose in his portrait, where he displayed the "treasures" retrieved from Lapland for his king, his nation, and himself. In proper elegiac style, Formey made that image universal, transforming his subject from adventurer into enlightened servant of humanity.

Even after his death, Maupertuis continued, for a time, to exert at least an imaginative hold on the Berlin Academy. Not long after learning of the president's demise, the Academy's botanist, Gleditsch, alone in the meeting room, saw an apparition: "Upon entering the room, he noticed M. de Maupertuis standing immobile, in the first corner to his left, with his eyes fixed upon him. It was about three hours after midday. The professor of natural history . . . went about his task without stopping at this phenomenon more than necessary to note it well. But he recounted this vision to his colleagues, and assured them that it had been as clear and perfect as if the person had been present."[60] Formey reported a slightly different version of this encounter, in which Gleditsch had been "terrified" by the vision, insisting on its reality to the amusement of his fellow academicians.[61] In Berlin, Maupertuis had been able, for a time, to exert an autocratic authority grounded in his personal connection to the king. The academicians recognized that they worked under his watchful eye, as evidenced by this unexpected appearance from beyond the grave. Not being superstitious, as befitted enlightened men of science, they went on about their business.

## Conclusion

At the end of his life, Maupertuis found himself buffeted by political and military events, and torn by competing personal loyalties. While his dilemma was undeniably poignant, even pathetic, it also reflected his lifelong penchant for strategic maneuvers aimed at enhancing his position. He had charmed and alienated powerful men and women in both regimes; he had seen his allies and patrons rise and fall; and he had fought many personal and intellectual battles. With numerous loyal friends, as well as numerous unrelenting enemies, he operated in an extensive network of connections that reached into many corners of the two kingdoms. These were not simply political connections, of course. Maupertuis's hunger for status and fame should not obscure his ardent commitment to his intellectual ambitions, not just for his own work but for the "sound philosophy"

60. Thiébault, *Frédéric le Grand, ou mes souvenirs.*

61. Formey to Algarotti, 12 February 1760, in Algarotti, *Opere,* 16:317.

touted by the editors of the *Encyclopédie* and their fellow travelers. Maupertuis took as his goal nothing less than general principles for understanding physics, life, cosmology and God. Behind the most self-aggrandizing stratagems lay a willingness to take risks, to speculate, to consider questions that might never be answered with certainty. His personal arrogance and vanity, and his quest for fame and reputation, were tied to his claims for the value of scientific investigation of nature. Along with many of his contemporaries, he promoted secular knowledge based on sensationalist epistemology, which led him to think about the historical development of both nature and society.[62]

The investigation of nature took place in a social world filled with overlapping and interpenetrating political realities and intellectual ideals. In France and Prussia, the absolutist state determined the rules and structures that governed the practice of science. But men of science also looked to the cosmopolitan Republic of Letters for legitimacy and rewards for their intellectual labors. Both of these contexts—absolute monarchy and the less structured Republic of Letters—informed the ideology and practice of science in the Enlightenment. In Maupertuis's case, his public persona, his scientific style, and many of his ideas were idiosyncratic, even eccentric. He cultivated a personal and intellectual flamboyance as a marker of his status, especially after he had established his mathematical credentials in Paris as a young man about town. But he played out his eccentricities at the institutional centers of science and culture, underwritten by the power and legitimacy of the crown, first in France and then in Prussia. His maneuvers in and around these institutions tell us a great deal about the status of science in the old regime, about the attention paid to scientific knowledge in the Republic of Letters, and about the role played by men of science in the literary sphere.

Men of science made their work useful to the state, and to absolutist rulers, but they also pursued knowledge in the service of the more idealized goals of human progress, rationality, and critical engagement. The repudiation of dogma, whether religious or philosophical, went along with a vision of the utility of science that integrated it into the progress of human society toward enlightenment. The intellectual enterprise continually wove in and out of the matrix of politics and social relations, leading to certain tensions and ambiguities, especially when the cosmopolitan ideal came up against political realities, as it did in a particularly harsh way at the end of Maupertuis's life. Of course, the Republic of Letters could coexist more or less peacefully with absolutism because it had no political or geographi-

62. From the perspective of the history of philosophy, Maupertuis belongs in the tradition of seventeenth-century skepticism; on this point, see Tonelli, *La pensée philosophique de Maupertuis.* Tonelli separates Maupertuis's "philosophy" from his "science," however, examining only intellectual connections, or "influences."

cal locus. Still, making an identity in science under these circumstances entailed speaking simultaneously in distinct but related voices. The voice of the loyal subject and servant of the state alternated with that of the unfettered mind in pursuit of disinterested truth. Everyone operating in this setting knew perfectly well that knowledge was serving particular interests, while professing disinterestedness themselves. Maupertuis's own position, after 1745, was complicated by his intellectually and politically ambiguous status as a Frenchman in Berlin, alienated from most German intellectuals, especially those working in the universities.

What did it mean to be a man of science in this period? Such a person was not yet a bureaucrat, nor a professional, as his nineteenth-century descendants would be, nor even an expert in the modern sense of the word. Nor were the mechanisms of legitimacy and validation strictly determined by the hierarchies of power localized in courts. Patronage relations still permeated all aspects of his life, however. Patronage mediated by institutions serving the absolutist state was gradually replacing personal patronage ties governed by the etiquette of princely courts.[63] In Maupertuis's lifetime, the two forms intermingled. In some ways, royal institutions like the Academy of Sciences reproduced the hierarchical power brokering of the older patronage system, and in other ways their corporate statutes and regulations allowed them a degree of separation from the maneuverings at court. Personal ties, whether to ministers, other aristocrats, kings, or senior academicians, very often came into play in negotiations for academic positions and moves up the institutional ladder. But the circumscribed dynamics of the patronage of the academies was distinct from, albeit tied to, court politics. The academician did not behave as a courtier, except on those exceptional occasions when he traveled to court to display his accomplishments and seek ministerial favors.[64]

We have seen how Maupertuis cultivated intellectual patrons at the beginning of his career, inside and outside the Paris Academy, and how he patronized others once he controlled the Berlin Academy. He also nurtured political connections to government ministers, well-placed courtiers, and, especially in Prussia, to the king himself. In crossing over into the Académie française, at a crucial point in his public career, he demonstrated the power of his connections to the social and po-

63. For the relation between court patronage and the evolving institutions of science in the seventeenth century, see Biagioli, "Scientific Revolution, Social Bricolage, and Etiquette." On patronage in the Jardin du Roi, and the transformations wrought by the French Revolution on these ties, see Spary, *Utopia's Garden;* Outram, *Georges Cuvier.* On the Enlightenment man of science, see Ferrone, "Man of Science," in *Enlightenment Portraits.*

64. On intersecting modes of patronage, see Terrall, "Emilie du Châtelet and the Gendering of Science."

litical elite to overcome solid opposition to his election. Maupertuis saw academies as vehicles for receiving patronage from the highest circles of government and as a framework for dispensing patronage himself. His obsession with marks of honor, such as titles and pensions, betrayed his desire to assert the noble status of his calling. This kind of status was not a necessary component of successful participation in academic or scientific life; many of Maupertuis's colleagues, from Clairaut to d'Alembert to Euler, had no such concerns, although they certainly benefited from the patronage system. Others, landed aristocrats by birth, had no need to derive noble status from their scientific positions, although they might still seek academic positions avidly. Nevertheless, a career in the royal institutions of science meant maneuvering in a space defined by hierarchies of power, as well as in the less structured space of the Republic of Letters.

Reputation, honor, and position redounded to the personal glory of the individual man, certainly. But they also validated the pursuit of the sciences as a form of life that was useful to the state and to the public. Public reputation was necessary for enhancing personal status, as well as for building a role for science in the state, a role that was by no means solidified in the mid-eighteenth century. The most successful academicians cultivated and exploited personal ties to powerful people outside the academies, to show ministers and monarchs that their interests lay in promoting, and paying for, disinterested science. The occupants of the exclusive pensioned positions in the academies were indeed free to choose their own intellectual directions, while simultaneously answering governmental demands for technical expertise.

The state and its ministers were not the only sources of honor and glory for science. Royal or aristocratic patronage did not preclude the enlightened sociability described by Diderot as the mark of the true philosopher (see chapter 1). As the audience for science, philosophy, and letters expanded, the Republic of Letters and the public came to play a legitimating role alongside ministers and kings. The status conferred by patronage cannot tell the whole story of the cultural location and meanings of science, even though the approval of the reading public was usually less tangible than the rewards of personal or institutional patronage. The relation between writer and public developed in the interstices of the many overlapping hierarchies of the old regime; hence, the fluidity of reputation derived from published works, and the many kinds of strategies that might lead to visibility and fame. All sorts of writers—journalists, novelists, playwrights, philosophers, chemists, mathematicians, travelers—referred to "the public" as the consumer and beneficiary of their works. It is notoriously difficult to draw a precise profile of the public addressed in scientific or literary works, since even fundamental information about print runs and book sales (not to mention reading habits) is very scarce.[65] Rhetorically, writers figured the public as an amorphous but sentient body capable of making demands and passing judgment, with interests and needs. Authors and publishers exaggerated the breadth and depth of this public, perhaps, but it was composed of people buying, reading, discussing, and circulating the periodicals and books that fed the appetite of the Republic of Letters across Europe. The reading public, considered as a group of individuals rather than an undifferentiated body, overlapped with academies and the court, so that an author's reputation in the world of letters might well translate into visibility at court and prominence in the academy. The audience for science in this period was variegated and heterogeneous, and the varied forms adopted for writing on scientific matters testify to this complexity. Whether written or performed, science might be variously amusing, useful, difficult, morally elevating, provocative, or subversive.

I have argued that Maupertuis's ambitions extended beyond the structured institutions of science, while remaining firmly grounded in those elite bodies. If his concern to maintain aristocratic status for his academic work, on the model of the kind of service that had earned his father a noble title, looked to the past, his turn to a wider public for his literary ventures was anything but conservative. Personal honor and ambition were only part of the story, however. What kind of science was Maupertuis promoting? It was not the entrepreneurial science of the instrument makers and public lecturers, flourishing in the shops and cafés of London and Paris in the same period.[66] This was a different kind of public science, aimed at the fashionable intellectual elites who wanted to read and discuss science and philosophy with a provocative edge. This edge might be supplied by Newtonianism (in the 1730s) or materialism (in the 1750s), just to take two examples among many. The former, translated into the language of Continental mathematics, eventually became incorporated into academic discourse, while the latter hovered on the fringes of the academies. The boundaries separating the official institutions from the less differentiated public were never impermeable; indeed, the learned pursuits of savants gained a measure of legitimacy by appealing to this readership.[67]

65. The extensive literature on the sociology of reading in the Enlightenment goes back to Mornet, "Les enseignements des bibliothèques privées," and up to the works of Chartier.

66. Stewart, *Rise of Public Science;* Golinski, *Science as Public Culture;* Walters, "Conversation pieces;" Geoffrey Sutton, *Science for a Polite Society.*

67. For Fontenelle's efforts to enlist the aristocratic audience, crucially including women, for science in the late seveteenth century, see Harth, *Cartesian Women.* See also Terrall, "Gendered Spaces, Gendered Audiences."

Much of the literature of the Enlightenment, most famously the *Encyclopédie* of Diderot and d'Alembert, addressed the question of how to read, as well as how to know. This was in part an artifact of the regime of censorship, where certain kinds of claims had to be masked. But it was also, as Diderot emphasized, a matter of teaching the public how to make connections, to see truths through the resonances of these connections, and to examine assumptions. Once his scientific reputation was well established through his mathematical contributions to the Paris Academy—where he wrote for a specialized readership with certain technical skills—Maupertuis took considerable trouble to craft many of his works for the same enlightened audience courted by the editors and publishers of the *Encyclopédie.* He took pains not only with his writing, but also with the details of publishing, including the presentation of books to hand-picked recipients as well as distribution to a wider public of anonymous readers. These were not enormous editions of popular works, however; they were usually short, readable books, often elegantly produced to appeal to readers' aesthetic sensibilities as well as their philosophical inclinations. Through the strategic use of print, authors like Maupertuis retailed an elite science and philosophy to a literary public. For Maupertuis, being a man of science was the means to reputation, and even glory; this glory was achieved through service to the crown, but also the promotion of "sound philosophy" and elegant style. The successful man of science, on this model, was also a man of letters.

# BIBLIOGRAPHY

## Manuscript Materials

*Archival sources are cited in full in the notes. The following are the major collections consulted:*


Académie des Sciences, Paris

Fonds Maupertuis

Fonds La Condamine

Dossiers for individual academicians

Pochettes de séances. (Manuscripts of some items submitted to Academy but not entered into the procès-verbaux.)

Procès-verbaux. (Records of attendance, elections, and other business; also manuscript copies of some papers presented at Academy meetings.)

Akademie der Wissenschaften, Berlin

Archives

Archives municipales de Saint-Malo

Maupertuis correspondence (18th-century copy), ms. ii. 24.

Bernoulli-Edition Basel. The manuscript papers of the Bernoulli family are held in the Universitätsbibliothek, Basel. The Bernoulli-Edition has transcribed much of the correspondence in preparing it for eventual publication. Transcripts of the following correspondences were consulted extensively:

Correspondance of Maupertuis with Daniel Bernoulli

Correspondence of Maupertuis with Johann I Bernoulli

Correspondance of Maupertuis with Johann II Bernoulli

Correspondence of Johann I Bernoulli with Gabriel Cramer

Correspondence of Johann I Bernoulli with J.-J. Dortous de Mairan

Correspondence of C.-M. La Condamine with Johann II Bernoulli

Biblioteka Jagiellonska, Cracow

Autograph collection

Varnhagen von Ense collection of correspondence (originally held by Preussische Staatsbibliothek, Berlin), including letters from Maupertuis to Formey

Bibliothèque de l'Institut, Paris

Autograph correspondence of academicians

Bibliothèque de l'Observatoire de Paris

Cassini papers

Correspondance of J.-N. Delisle

British Library
Correspondence of Pierre Desmaiseaux
Correspondence of Cromwell Mortimer
Hans Sloane Papers
Royal Society of London Archives
Staatsbibliothek, Berlin (Handschrift Abteilung)
Correspondence from J-H.-S. Formey to Maupertuis

## Works by Maupertuis

*Maupertuis's works are listed chronologically. For the volumes of MAS and BAS, the journal year is followed by the actual date of publication in brackets.*

"Sur la forme des instruments de musique," *MAS* 1724 [1726], 215–26.
"Sur une question de *maximis et minimis.*" *MAS* 1726 [1728] 84–94.
"Observations et expériences sur une espèce de salamandre." *MAS* 1727 [1729], 38–45
"Nouvelle manière de développer des courbes." *MAS* 1727 [1729], 340–9.
"Quadrature et rectification des figures formées par le roulement des polygones réguliers." *MAS* 1727 [1729], 204–13.
"Sur toutes les développées qu'une courbe peut avoir à l'infini." *MAS* 1728 [1730], 225–31.
"Sur quelques affections des courbes." *MAS* 1729 [1730], 277–82.
"La courbe *descensus aequabilis* dans un milieu résistant comme une puissance quelconque de la vitesse." *MAS* 1730 [1732], 233–42.
"Balistique arithmétique." *MAS* 1731 [1733], 297–98.
"Expériences sur les scorpions." *MAS* 1731 [1733], 223–29
"Problême astronomique." *MAS* 1731 [1733], 464–65.
"Sur la séparation des indéterminées dans les équations différentielles." *MAS* 1731 [1733], 103–09.
"De Figuris quas fluida rotata induere possunt problemata duo," *Philosophical Transactions,* no. 422 (1732): 240–56.
*Discours sur les différentes figures des astres avec une exposition des systèmes de MM. Descartes et Newton.* Paris, 1732. 2nd ed. Paris, 1742. English translation appended to J. Keill, *An Examination of Dr. Burnet's Theory of the Earth.* Oxford and London, 1734.
"Solution de deux problèmes de géométrie." *MAS* 1732 [1735], 442–45.
"Solution du même problème [épicycloides sphériques] et de quelques autres de cette espèce." *MAS* 1732 [1735], 255–59.
"Sur les courbes de poursuite." *MAS* 1732 [1735], 15–16.
"Sur les loix de l'attraction." *MAS* 1732 [1735], 343–62.
"Sur la figure de la terre et sur les moyens que l'astronomie et la géographie fournissent pour la déterminer." *MAS* 1733 [1735], 153–64.
"Sur la mouvement d'une bulle d'air qui s'élève dans une liqueur." *MAS* 1733 [1735], 255–59.
"Sur les figures des corps célestes." *MAS* 1734 [1736], 55–100.

"Lettre de Maupertuis à Mme de Vertillac [*sic*]," [1737]. *Mélanges publiés par la Société des Bibliophiles Français* 6 (1820): 3–10.

"Sur la figure de la terre." *MAS* 1735 [1738], 98–105.

"Méthode pour trouver la déclinaison des étoiles." *MAS* 1736 [1739], 375–90.

"Sur la figure de la terre." *MAS* 1736 [1739], 302–12.

"Observations sur la figure de la terre, déterminée par Messieurs de l'Académie des Sciences qui ont mesuré le degré du méridien au cercle polaire." *MAS* 1737 [1740], 389–466. Reprinted in *La figure de la terre* (1738).

[anon.,] *Anecdotes physiques et morales.* [n.p. 1738]. (Doubtful attribution.)

[anon.,] *Examen désintéressé des différens ouvrages qui ont été faits pour déterminer la figure de la terre.* Oldenbourg: Theobald Bachmuller [Paris], 1738 [1740]; 2nd ed. Amsterdam, 1741 [1743?]. Published with *Examen des trois dissertations que M. Desaguliers a publié sur la figure de la terre.*

*La figure de la terre déterminée par les observations de MM. de Maupertuis, Clairaut, Camus, Le Monnier, Outhier, Celsuis au cercle polaire.* Paris: Imprimerie royale, 1738. English translation, *The Figure of the Earth.* London: T. Cox, 1738. German translation by S. König. *Figur der Erden.* Zurich: Heidegger, 1741.

*Degré du méridien entre Paris et Amiens déterminé par la mesure de M. Picard, et pas les observations de Mrs. de Maupertuis, Clairaut, Camus, Le Monnier, de l'Académie royale des sciences.* Paris: Imprimerie royale, 1740.

[anon.,] *Elémens de géographie.* Paris, 1740; 2nd ed. Paris 1742. German translation, Zurich, 1743.

[anon.,] *Lettre d'un horloger anglais à un astronome de Pékin.* n.p., 1740 (edition of 4 copies).

"Loi du repos des corps." *MAS* 1740, 170–76.

*Discours sur la parallaxe de la lune pour perfectionner la théorie de la lune et celle de la terre.* Paris: Imprimerie royal, 1741; 2nd ed. Paris, 1755.

[anon.,] *Lettre sur la comète.* Paris, 1742 (2 editions). English translation in *An Essay towards a History of the Principal Comets.* London, 1769.

*Astronomie nautique: Ou élémens d'astronomie, tant pour un observatoire fixe, que pour une observatoire mobile.* Paris: Imprimerie royale, 1743; 2nd ed. Paris, 1751.

"Accord de différentes loix de la nature qui avoient jusqu'ici paru incompatibles." *MAS* 1744 [1748], 417–26.

[anon.,] *Dissertation physique à l'occasion du nègre blanc.* Leiden, 1744 (3 editions).

*Ouvrages divers.* Amsterdam, 1744.

"Traité de Loxodromie tracée sur la véritable surface de la mer." *MAS* 1744 [1748], 462–74.

*Vénus physique,* suivi de la "Lettre sur le progrès des sciences." Introductory essay, "L'Ordre du corps," by Patrick Tort. Paris: Aubier Montaigne, 1980. Original publication Paris, 1745.

"Les loix du mouvement et du repos, déduites d'un principe de métaphysique." *BAS* 1746 [1748], 267–94.

"Relation d'un voyage fait dans la Laponie septentrionale pour trouver un ancien monument." *BAS* 1747 [1749], 432–45.

[anon.,] *Réflexions philosophiques sur l'origine des langues et sur la signification des mots.* Paris, 1748.

*Essay de philosophie morale.* Berlin, 1749; London, 1750; Leiden, 1751.

*Essai de cosmologie.* Berlin, 1750; Leiden, 1751.

[Dr. Baumann, pseud.] *Dissertatio inauguralis metaphysica de universali naturae systemate.* Erlangen [Berlin], 1751. German translation, *Versuch von der Bildung der Körper, aus den Lateinischen des Herren von Maupertuis.* Leipzig, 1761.

*Lettres.* Dresden: Walther, 1752.

*Lettre sur le progrès des sciences.* Berlin: Etienne de Bourdeau, 1752; Paris, 1752.

"Des devoirs de l'académicien." *BAS* 1753 [1755], 511–21.

*Oeuvres.* 1 vol. Dresden: Walther, 1752; Lyon: Bruyset, 1752; Berlin: Etienne de Bourdeau, 1753.

*Discours académiques, lus dans l'Académie des Sciences de France, dans l'Académie françoise, et dans celle des Sciences et Belles-Lettres de Prusse.* Dresden, 1753.

*Maupertuisiana.* Hamburg [Leiden: Luzac], 1753.

"Dissertation sur les différens moyens dont les hommes se sont servis pour exprimer leurs idées." *BAS* 1754 [1756], 349–64.

[anon.,], *Essai sur la formation des corps organisés.* Berlin [Paris], 1754.

*Oeuvres.* 4 vols. Lyon, 1756; 2nd ed. 1768. Facsimile ed., with "L'Examen philosophique de la preuve de l'existence de dieu employé dans *L'essai de Cosmologie*" and introduction by Giorgio Tonelli. Hildesheim and New York: Olms, 1974.

"Examen philosophique de la preuve de l'existence de Dieu employée dans *L'essai de cosmologie.*" *BAS* 1756 [1758], 389–424.

*Versuch von der Bildung der Körper, aus den Lateinischen des Herren von Maupertuis, überseht von einem Freunde der Naturlehre.* Leipzig, 1761.

## Other Works

Aarsleff, Hans. "The Berlin Academy under Frederick the Great." *History of the Human Sciences* 2 (1989): 193–207.

Académie des Sciences (Paris). *Receuil des pièces qui ont remporté les prix de l'Académie.* 9 vols. Paris: Imprimerie royale, 1721–1777.

Académie française. *Les régistres de l'Académie françoise, 1672–1793.* Paris: Firmin-Didot, 1895.

*Actes de la journée Maupertuis.* Paris: Vrin, 1975.

Adams, Percy G. *Travelers and Travel Liars, 1660–1800.* Berkeley: University of California Press, 1962.

Aiton, Eric. *The Vortex Theory of Planetary Motions.* London: MacDonald, 1972.

Alembert, Jean LeRond d'. *Preliminary Discourse to the Encyclopedia of Diderot.* Translated by Richard Schwab. Indianapolis: Bobbs-Merrill, 1963. Originally published in 1758.

———. *Traité de dynamique.* Paris: David, 1743.

———. *Traité de dynamique.* 2nd ed. *Oeuvres philosophiques, historiques et littéraires de d'Alembert.* Paris: A. Belin, 1821.

Algarotti, Francesco. *Opere del conte Algarotti.* 17 vols. Venice: Carlo Palese, 1791–1794.

Anderson, Wilda C. *Diderot's Dream.* Baltimore: Johns Hopkins University Press, 1990.

Argens, Jean-Baptiste de Boyer, marquis d'. *Histoire de l'esprit humain ou mémoires secrets et universels de la république des lettres.* 13 vols. Berlin: Haude & Spener, 1765–1768.

Aucoc, Leon. *L'Institut de France et les anciennes académies.* Paris: Plon Nourrit, 1889.

Bachelard, Suzanne. *Les polémiques concernant le principe de la moindre action au XVIIIième siècle.* Paris: Palais de la découverte , 1961.

Badinter, Elisabeth. *Emilie, Emilie: L'ambition feminine au XVIIIe siècle.* Paris: Flammarion, 1983.

———. *Les passions intellectuelles: Désirs de gloire (1735–1751).* Paris: Fayard, 1999.

Baker, John. *Abraham Trembley of Geneva, Scientist and Philosopher, 1710–1784.* London: E. Arnold, 1952.

Baker, Keith, ed. *The Political Culture of the Old Regime.* Oxford: Pergamon Press, 1987.

Balcou, Jean, ed. *Le dossier Fréron: Correspondances et documents.* Geneva: Droz, 1975.

Bartholmess, Christian. *Histoire philosophique de l'Académie de Prusse depuis Leibniz jusqu'à Schelling.* 2 vols. Paris: Franck, 1850.

Basset des Rosiers, Gilles. *L'Anti-Vénus physique.* n. p. [Paris?], 1746.

———. *Critique de la lettre sur la comète, ou lettre d'un philosophe à une demoiselle âgée de 9 ans.* Paris, 1742.

Bataille, Marie-Louise. "Tournières." *Les peintres français du XVIIIe siècle: Histoire des vies et catalogue des œuvres,* 2 vols. Edited by M. Louis Dimier, 227–43. Paris and Brussells: G. van Oest, 1928–30.

Beaune, Jean-Claude and Jean Gayon, eds. *Buffon 88: Actes du Colloque international pour le bicentenaire de la mort de Buffon.* Paris: J. Vrin, 1992.

Beeson, David. "Lettre d'un horloger anglois à un astronome de Pékin." *Studies on Voltaire and the 18th Century* 230 (1985): 189–222.

———. *Maupertuis: An Intellectual Biography.* Oxford: Voltaire Foundation, 1992.

Bénézit, E., ed. *Dictionnaire critique et documentaire des peintres, sculpteurs, dessinateurs, et graveurs de tous les temps et de tous les pays.* Paris: Gründ, 1999.

Benguigui, Isaac, ed. *Théories électriques du XVIIIe siècle: Correspondance entre l'Abbé Nollet (1700–1770) et le physicien genèvois Jean Jallabert (1712–1768).* Geneva: Georg, 1984.

Bernoulli, Jean. *Discours sur les loix de la communication du mouvement* (1724). In idem, *Opera omnia.* 3:7–107. Lausanne and Geneva: Bousquet and Sociorum, 1742.

———. "Méthode pour trouver les tautochrones, dans des milieux résistants, comme le quarré de la vitesse," *MAS* 1730, 78–101.

———. "Nouvelles pensées sur le systême de M. Descartes." In idem, *Opera Omnia,* 3:131–73. Lausanne and Geneva: Bousquet and Sociorum, 1742.

———. *Opera Omnia.* 4 vols. Lausanne and Geneva: Bousquet and Sociorum, 1742.
Bertoloni Meli, Domenico. "Caroline, Leibniz, and Clarke." *Journal of the History of Ideas* 60 (1999): 469–86.
———. *Equivalence and Priority: Newton Versus Leibniz.* New York: Oxford University Press, 1993.
Besterman, Theodore, ed. *Les lettres de la Marquise du Châtelet.* 2 vols. Geneva: Voltaire Foundation, 1958.
——— et al., eds. *The Complete Works of Voltaire.* 135 vols. Geneva and Toronto: Institut et Musée Voltaire and University of Toronto Press, 1968–2001.
Beuchot, A. J., ed. *Oeuvres de Voltaire.* 72 vols. Paris: Lefevre, 1829.
Biagioli, Mario. *Galileo, Courtier: The Practice of Science in the Culture of Absolutism.* Chicago: University of Chicago Press, 1993.
———. "Scientific Revolution, Social Bricolage, and Etiquette." In *The Scientific Revolution in National Context,* edited by Roy Porter and Mikulaś Teich, 11–54. Cambridge: Cambridge University Press, 1992.
Blay, Michel. *La naissance de la mécanique analytique: La science du mouvement au tournant des XVIIième et XVIIIième siècles.* Paris: Presses Universitaires de France, 1992.
Bollème, Geneviève. *Livre et société dans la France du XVIIIième siècle.* Paris, The Hague: Mouton, 1965.
Bongie, Lawrence. "Introduction to *Les Monades,* by Etienne Bonnot de Condillac." *Studies on Voltaire and the 18th Century* 187 (1980): 11–197.
Boudri, J. Christiaan. *What Was Mechanical about Mechanics? The Concept of Force between Metaphysics and Mechanics from Newton to Lagrange.* Dordrecht: Kluwer, 2002.
Bouguer, Pierre. *La figure de la terre.* Paris: Jombert, 1749.
Bouhier, Jean. *Correspondance littéraire du président Bouhier,* edited by Henri Duranton. 21 vols. Saint-Etienne: Centre de Saint-Etienne, 1974.
Bowler, Peter. "Evolutionism in the Enlightenment." *History of Science* 12 (1974): 159–83.
Breger, Herbert. "Über den von Samuel König veröffentlichen Brief zum Prinzip der kleinsten Wirkung." In *Pierre Louis Moreau de Maupertuis: eine Bilanz nach 300 Jahren,* edited by H. Hecht, 363–81. Berlin: Verlag A. Spitz, 1999.
Brennan, Thomas. *Public Drinking and Popular Culture in 18th-Century Paris.* Princeton: Princeton University Press, 1988.
Briggs, Robin. "The Académie royale des sciences and the pursuit of utility." *Past and Present* 131 (1991): 38–88.
Brown, Harcourt. "From London to Lapland and Berlin." In idem, *Science and the Human Comedy.* Toronto: University of Toronto Press, 1976.
———. "From London to Lapland: Maupertuis, Johann Bernoulli and La Terre Applatie, 1728–1738." In *Literature and History in the Age of Ideas,* edited by Charles Williams, 69–94. Columbus: Ohio State University Press, 1975.

———. "Madame Geoffrin and Martin Folkes: Six New Letters." *Modern Language Quarterly* 1 (1940): 215–40.

———. "Maupertuis *Philosophe:* Enlightenment and the Berlin Academy." *Studies on Voltaire and the 18th Century* 24 (1963): 255–69.

Browne, Janet. *Charles Darwin: A Biography.* New York: Knopf, 1995.

Brunet, Pierre. *Etude historique sur le principe de la moindre action.* Paris: Hermann, 1938.

———. *L'Introduction des théories de Newton en France au XVIIIième siècle avant 1738.* Paris: Albert Blanchard, 1931.

———. *Maupertuis.* 2 vols. Paris: Albert Blanchard, 1929.

Buffon, Georges-Louis Leclerc de. *Correspondance générale,* edited by H. Nadault de Buffon. Geneva: Slatkine Reprints, 1971. Originally published in 1885.

———. "Découverte de la liqueur séminale dans les femelles vivipares et du réservoir qui la contient." *MAS* 1748, 211–28.

———. *Histoire naturelle, générale et particulière.* 15 vols. Paris: Imprimerie Royale, 1749–1767.

———. *Oeuvres philosophiques,* edited by Jean Piveteau. Paris: Presses universitaires de France, 1954.

Burney, Charles. *An Essay Towards a History of the Principal Comets That Have Appeared Since the Year 1742.* London: T. Becket, 1769.

Calinger, Ronald. "Frederick the Great and the Berlin Academy of Sciences." *Annals of Science* 24 (1968): 239–49.

———. "The Newtonian-Wolffian Controversy." *Journal of the History of Ideas* 30 (1969): 319–30.

Casini, Paolo. "D'Alembert, l'économie des principes et la métaphysique des sciences." In *Jean d'Alembert savant et philosophe: Portrait à plusieurs voix,* edited by M. Emery and P. Monzani, 135–65. Paris: Editions des Archives Contemporaines, 1989.

Cassini, Jacques. "De la carte de France, & de la perpendiculaire à la méridienne de Paris," *MAS* 1733, 389–405.

———. *De la grandeur et de la figure de la terre.* Paris: Imprimerie royale, 1720.

———. "Du mouvement véritable des cometes à l'égard du soleil et de la terre," *MAS* 1731. 299–346.

———. "Reponse aux remarques qui ont été faites dans le *Journal Historique de la République des Lettres* sur le traité *De la grandeur et de la figure de la terre,*" *MAS* 1732, 497–513.

Cassini de Thury, César-François. *La méridienne de l'Observatoire Royal de Paris.* Paris, Imprimerie royale, 1744.

Cassirer, Ernst. *The Philosophy of the Enlightenment.* Translated by F. Koelln and James Pettegrove. Princeton: Princeton University Press, 1951.

Celsius, Anders. *De observationibus pro figura telluris determinanda in Gallia habitis, disquisitio.* Uppsala: Typis Hojerianis, 1738.

Censer, Jack Richard. *The French Press in the Age of Enlightenment.* London and New York: Routledge, 1994.

Chartier, Roger. *The Cultural Origins of the French Revolution.* Durham, N.C.: Duke University Press, 1991.

———. *The Cultural Uses of Print in Early Modern France.* Translated by Lydia G. Cochrane. Princeton: University of Princeton Press, 1987.

———. *Forms and Meanings: Texts, Performances, and Audiences from Codex to Computer.* Philadelphia: University of Pennsylvania Press, 1995.

Châtelet, Gabrielle Emilie Le Tonnelier de Breteuil du. *Les lettres de la marquise du Châtelet,* edited by Theodore Besterman. 2 vols. Geneva: Institut et Musée Voltaire, 1958.

Chaussinand-Nogaret, Guy. *The French Nobility in the Eighteenth Century: From Feudalism to Enlightenment.* Translated by William Doyle. Cambridge: Cambridge University Press, 1985.

———, ed. *Histoire des élites en France du XVIième au XXième siècle: L'honneur, le mérite, l'argent.* Paris: Tallandier, 1991.

Chouillet, Anne-Marie. "Du nouveau sur *l'Encyclopédie:* Une lettre inédite de d'Alembert." *Recherches sur Diderot et sur l'Encyclopédie* 11 (1991): 18–31.

———. "Rôle de la presse périodique de langue française dans la diffusion des informations concernant les missions en Laponie ou sous l'équateur." In *La figure de la terre du XVIIIième siècle à l'ère spatiale,* edited by Henri Lacombe and Pierre Costabel, 171–90. Paris: Gauthier-Villars, 1988.

———. "Trois lettres inédites de Diderot." *Recherches sur Diderot et sur l'Encyclopédie* 11 (1991): 8–16.

Clairaut, Alexis-Claude. "Détermination géometrique de la perpendiculaire a la méridienne tracé par M. Cassini; avec plusieurs méthodes d'en tirer la grandeur & la figure de la terre, " *MAS* 1733: 406–416.

———. "Sur les explications cartésienne et newtonienne de la réfraction de la lumière," *MAS* 1739, 259–75.

Clarke, Samuel. *A Collection of Papers which Passed between the Late Learned Mr. Leibnitz and Dr. Clarke.* London, 1717.

Clark, William. "The Death of Metaphysics in Enlightened Prussia." In *The Sciences in Enlightened Europe,* edited by William Clark, Jan Golinski, and Simon Schaffer, 423–73. Chicago: University of Chicago Press, 1999.

———, Jan Golinski, and Simon Schaffer, eds. *The Sciences in Enlightened Europe.* Chicago: University of Chicago Press, 1999.

Collé, Charles. *Journal et mémoires de Charles Collé sur les hommes de lettres, les ouvrages dramatiques et les évènements les plus mémorables du règne de Louis XV (1748–1772),* edited by H. Bonhomme. Paris: Firmin Didot, 1868.

Costabel, Pierre. "L'affaire Maupertuis-König et les questions de fait." *Arithmos-arrythmos, Skizzen der Wissenschaftsgeschichte.* Munich, 1979.

———. "Science positive et forme de la terre au début du XVIIIième siècle." *La figure de la terre du XVIIIième siècle à l'ère spatiale*, edited by Henri Lacombe and Pierre Costabel, 97–114. Paris: Gauthier-Villars, 1988.

Craveri, Benedetta. *Madame du Deffand et son monde.* Paris: Seuil, 1987.

Cunat, Charles. *Saint-Malo illustré par ses marins.* Rennes, 1857.

Dahlgren, E.-W. "Jérôme de Pontchartrain et les armateurs de Saint-Malo." *Revue historique* 88 (1904): 225–63.

Danzel, Theodor Wilhelm. *Gottsched und seine Zeit: Auszüge aus seinem Briefwechsel.* Leipzig: Verlag der deutschen Buchhandlung, 1848.

Darnton, Robert. *The Business of Enlightenment: A Publishing History of the Encyclopédie, 1775–1800.* Cambridge: Harvard University Press, 1979.

Daston, Lorraine. "The Ideal and Reality of the Republic of Letters in the Enlightenment." *Science in Context* 4 (1991): 367–86.

———, and Katharine Park. *Wonders and the Order of Nature, 1150–1750.* Cambridge: MIT Press, 1998.

Dawson, Virginia P. *Nature's Enigma: The Problem of the Polyp in the Letters of Bonnet, Trembley, and Réaumur.* Philadelphia: American Philosophical Society, 1987.

Dear, Peter. "A Mechanical Microcosm: Bodily Passions, Good Manners, and Cartesian Mechanism." In *Science Incarnate: Historical Embodiments of Natural Knowledge*, edited by Christopher Lawrence and Steven Shapin, 51–82. Chicago: University of Chicago Press, 1998.

DeJean, Joan E. *Ancients against Moderns: Culture Wars and the Making of a Fin de Siècle.* Chicago: University of Chicago Press, 1997.

Delambre, J. B. J. *Grandeur et figure de la terre*, edited by G. Bigourdan. Paris: Gauthier-Villars, 1912.

Delorme, Suzanne. "Tableau chronologique de la vie et des oeuvres de Fontenelle." *Revue d'histoire des sciences* 10 (1957): 288–309.

Des Essarts, Nicolas-Toussaint Le Moyne. *Causes célèbres, curieuses et intéressantes de toutes les cours souveraines du royaume, avec les jugmens qui les ont décidées.* 16 vols. Paris, 1774.

Desaguliers, J. T. "A Dissertation Concerning the Figure of the Earth." *Philosophical Transactions* (1725). No. 386 (201–22), No. 387 (239–55), No. 388 (277–304).

Desfontaines, Pierre-François Guyot. *Jugemens sur quelques ouvrages nouveaux.* 11 vols. Paris, 1744–1745.

———. *Observations sur les écrits modernes.* 34 vols. Paris: 1735–1743.

Desmaiseaux, Pierre, ed. *Receuil de diverses pièces sur la philosophie, la religion naturelle, l'histoire, les mathématiques, &c., par Leibniz, Clarke, Newton, and autres auteurs célèbres.* Amsterdam, 1740.

Diderot, Denis. *Correspondance.* 16 vols, edited by Georges Roth. Paris: Editions de Minuit, 1955–1959.

———. *Pensées sur l'interprétation de la nature*, edited by Jean Varloot. *Oeuvres complètes*, Vol. 9. Paris: Hermann, 1981. Originally published in Paris in 1754.

———, and Jean Le Rond d' Alembert, eds. *Encyclopédie; ou Dictionnaire raisonné des sciences, des arts et des metiers.* 17 vols. Paris: Briasson, 1751–1765.

*Dissertation qui a remporté le prix proposé par l'Académie royale des sciences et belles lettres sur le systeme des monades avec les pieces qui ont concouru.* Berlin: Haude and Spener, 1748.

Dorn, Walter. "Prussian Bureaucracy in the Eighteenth Century." In *Frederick the Great: A Profile,* edited by Peter Paret, 57–78. New York: Hill and Wang, 1972.

Droysen, Hans, Fernand Caussy, and Gustav Berthold Volz, eds. *Nachträge zu dem Briefwechsel Friedrichs des Grossen mit Maupertuis und Voltaire, nebst verwandten Stucken.* Leipzig: S. Hirzel, 1917.

Duchesneau, François. *La physiologie des lumières: Empirisme, modèles et théories.* The Hague: Martinus Nijhoff, 1982.

Duclos, Charles Pinot. "Mémoires sur la vie de Duclos, écrits par lui-même." In *Mémoires biographiques et littéraires,* edited by M. de Lescure, 1–37. Paris: Firmin-Didot, 1881.

Duguay-Trouin, René. *Mémoires de Monsieur Du Guay-Trouin.* Amsterdam: Pierre Mortier, 1741.

Duranton, Henri, ed. *Corréspondance littéraire du président Bouhier.* Saint-Etienne: Centre de Saint-Etienne, 1974.

Ehrman, Esther. *Mme. du Châtelet.* Leamington Spa: Berg, 1986.

Euler, Leonhard. "De curvis elasticis." Translated by W. A. Oldfather, C. A. Ellis, D. M. Brown. *Isis* 20 (1933): 72–160.

———. *Dissertation sur le principe de la moindre action avec l'examen des objections de M. le Prof. Koenig faites contre ce principe.* Berlin: Haude and Spener, 1753.

———. "Essay d'une démonstration métaphysique du principe général de l'équilibre." *BAS* 1751: 246–54.

———. *Gedancken von den Elementen der Cörper.* Berlin: Haude and Spener, 1746. Reprinted in *Opera Omnia,* series 3, vol. 2.

———. "Harmonie entre les principes généraux de repos et de mouvement de M. de Maupertuis." *BAS* 1751, 169–98.

———. *Lettres à une princesse d'Allemagne* in *Opera Omnia,* series 3, vol. 11. Originally published in 1768.

———. *Methodus inveniendi lineas curvas maximi minimive proprietate gaudentes sive solutio problematis isoperimetrici latissimo sensu accepti.* Lausanne, 1744 reprinted in *Opera Omnia,* series 1, vol. 24.

———. *Opera Omnia,* series 2, vol. 5, *Commentationes mechanicae.* Edited by J. O. Fleckenstein. Lausanne: Orell Füssli, 1957.

———. *Opera Omnia,* series 4A, vol. 5, *Correspondance avec Clairaut, d'Alembert et Lagrange.* Edited by A. P. Juškević and René Taton. Basel: Birkhäuser, 1980.

———. *Opera Omnia,* Series 4A, vol. 6, *Correspondance avec Maupertuis et Frédéric II.* Edited by P. Costabel, E. Winter, A.T. Grigorijan, and A. P. Juškević . Basel: Birkhäuser, 1986.

———. "Recherches sur les plus grands et les plus petits qui se trouvent dans les actions des forces." *BAS* 1748, 149–88.

———. "Réflexions sur quelques loix générales de la nature qui s'observent dans les effets des forces quelconques." *BAS* 1748, 189–218,

———. "Sur le principe de la moindre action." *BAS* 1751, 199–218.

Farge, Arlette. *Subversive Words: Public Opinion in Eighteenth-Century France.* Translated by Rosemary Morris. Cambridge: Polity Press, 1994.

Fellman, E. "Johann Samuel König." In *Dictionary of Scientific Biography.* Vol. 7. Edited by Charles Gillispie, 441–44.

Ferrone, Vincent. "Man of Science." In *Enlightenment Portraits,* edited by Michel Vovelle and translated by Lydia G. Cochrane, 190–225. Chicago: University of Chicago Press, 1997.

Fontenelle, Bernard le Bovier de. "Eloge de M. le Chevalier Newton." *Mémoires de l'Académie Royales des Sciences* (Paris) 1727: 151–72.

———. *Entretiens sur la pluralité des mondes.* Paris: C. Blageart, 1686.

———. "La figure de la terre." *HAS* 1735, 51–55.

———. *Oeuvres complètes.* 3 vols. Paris: A. Belin, 1818.

———. "Sur l'attraction Newtonienne," *HAS* 1732, 85– 93.

———. "Sur les monstres," *HAS* 1740, 37–50.

———. "Sur les mouvemens en tourbillon," *HAS* 1728, 103.

———. "Sur les soudéveloppées," *HAS* 1728, 58–63.

Fontius, Martin, Rolf Geissler, and Jens Häseler, eds. *Correspondance passive de Formey: Antoine-Claude Briasson et Nicolas-Charles-Joseph Trublet.* Paris and Geneva: Champion and Slatkine, 1996.

Formey, Jean-Henri-Samuel. *La belle Wolfienne.* The Hague: Veuve de Charles le Vier, 1741.

———. "Eloge de M. de Maupertuis." *BAS* 1760, 464–512.

———. *Histoire de l'Académie Royale des Sciences de Berlin.* Berlin: Haude and Spener, 1750.

———. *Mélanges philosophiques.* 2 vols. Leiden, 1754.

———. *Recherches sur les élémens de la matière.* Leiden, 1747.

———. *Souvenirs d'un citoyen.* 2 vols. Berlin: Lagarde, 1789.

Fosca, François. *Histoire des cafés de Paris.* Paris: Firmin-Didot, 1934.

Fouchy, Jean-Paul Grandjean de. "Eloge de M. de Maupertuis." *HAS* 1759, 2.

Fraser, Craig. "The Calculus as Algebraic Analysis: Some Observations on Mathematical Analysis in the 18th Century." *Archive for the History of Exact Sciences* 39 (1989): 317–35.

———. "D'Alembert's Principle: The Original Formulation and Application in Jean d'Alembert's *Traité de Dynamique* (1743)." *Centaurus* 28 (1985): 31–61, 145–69.

———. "J. L. Lagrange's Early Contributions to the Principles and Methods of Mechanics." *Archive for the History of Exact Sciences* 28 (1983): 197–241.

Frederick of Prussia. *Oeuvres de Frédéric le Grand.* Edited by J.D.E. Preuss. 31 vols. Berlin: Imprimerie royale, 1846–1857.

———. "Testament politique" (1752). *Le mémorial des siècles (XVIIIième siècle): Frédéric II roi de Prusse.* Introduction by Pierre Gaxotte. Paris, 1967.

Fréron, Elie. *Lettres sur quelques écrits de ce tems.* 13 vols. London: Duchesne, 1749–1754.

Funck-Brentano, Frantz. *Les nouvellistes.* Paris: Hachette, 1905.

Gallois, Léon. "L'Académie des sciences et les origines de la carte de Cassini." *Annales de géographie* 100 (1909): 193–204, 289–307.

Gandt, François de. "1744: Maupertuis et d'Alembert entre mécanique et métaphysique." In *Pierre Louis Moreau de Maupertuis: Eine Bilanz nach 300 Jahren,* edited by Hartmut Hecht, 277–91. Berlin: Verlag A. Spitz, 1999.

Gasking, Elizabeth B. *Investigations into Generation, 1651–1828.* London: Hutchinson, 1967.

Gastelier, Jacques-Elie. *Lettres sur les affaires du temps: 1738–1741.* Edited by Henri Duranton. Paris and Geneva: Champion and Slatkine, 1993.

Gaukroger, Stephen. *Descartes: An Intellectual Biography.* Oxford: Oxford University Press, 1995.

Genuth, Sara Schechner. *Comets, Popular Culture and the Birth of Modern Cosmology.* Princeton: Princeton University Press, 1997.

Geoffroy, Etienne François. "Table des différens rapports observés en chymie entre différentes substances." *MAS* 1718, 202–12.

Gillispie, Charles. *Science and Polity in France at the End of the Old Regime.* Princeton: Princeton University Press, 1980.

Glass, Bentley, ed. *Forerunners of Darwin: 1745–1859.* Baltimore: Johns Hopkins Press, 1959.

Goldgar, Anne. *Impolite Learning: Conduct and Community in the Republic of Letters, 1680–1750.* New Haven: Yale University Press, 1995.

Goldstine, Herman Heine, ed. *Die Streitschriften von Jacob und Johann Bernoulli: Variationsrechnung.* Basel: Birkhauser, 1991.

Goodman, Dena. "Enlightenment Salons: The Convergence of Female and Philosophic Ambitions." *Eighteenth-Century Studies* 22 (1989): 329–50.

———. "Governing the Republic of Letters: The Politics of Culture in the French Enlightenment." *History of European Ideas* 13 (1991): 183–99.

———. "Public Sphere and Private Life: Toward a Synthesis of Current Historiographical Approaches to the Old Regime." *History and Theory* 31 (1992): 1–20.

———. *The Republic of Letters: A Cultural History of the French Enlightenment.* Ithaca: Cornell University Press, 1994.

Gordon, Daniel. *Citizens without Sovereignty: Equality and Sociability in French Thought, 1670–1789.* Princeton: Princeton University Press, 1994.

———. " 'Public Opinion' and the Civilizing Process in France: The Example of Morellet." *Eighteenth-Century Studies* 22 (1989): 302–28.

Gossman, L. "Berkeley, Hume and Maupertuis." *French Studies* 14 (1960): 304–24.

Gower, Barry. "Planets and Probability: Daniel Bernoulli on the Inclinations of the Planetary Orbits." *Studies in the History and Philosophy of Science* 18 (1987): 441–54.

Grabiner, Judith. *The Origins of Cauchy's Rigorous Calculus.* Cambridge: MIT Press, 1981.

Graffigny, Françoise de. *Correspondance de Madame de Graffigny,* edited by English Showalter, P. Allan, and J. A. Dainard. 6 vols. Oxford: Voltaire Foundation, 1985–2000.

Grau, Conrad. "Maupertuis in Berlin." In *Pierre Louis Moreau de Maupertuis: Eine Bilanz nach 300 Jahren,* eEdited by Hartmut Hecht, 35–55. Berlin: Verlag A. Spitz, 1999.

Greenberg, John. "Degrees of Longitude and the Earth's Shape: The Diffusion of a Scientific Idea in Paris in the 1730s." *Annals of Science* 41 (1984): 151–58.

———. "Geodesy in Paris in the 1730s and the Paduan Connection." *Historical Studies in the Physical Sciences* 13 (1983): 239–60.

———. "Mathematical Physics in Eighteenth-Century France." *Isis* 77 (1986).

———. *The Problem of the Earth's Shape from Newton to Clairaut: The Rise of Mathematical Science in Eighteenth-Century Paris and the Fall of "Normal" Science.* Cambridge: Cambridge University Press, 1995.

Grimm, Friedrich Melchior. *Correspondance, littéraire, philosophique et critique par Grimm, Diderot, Raynal, Meister, etc.* 16 vols. Edited by Maurice Tourneux. Paris: Garnier frères, 1877–1882.

Guerlac, Henry. *Newton on the Continent.* Ithaca and London: Cornell University Press, 1981.

Guéroult, Marcel. *Dynamique et métaphysique leibniziennes.* Paris: Les Belles Lettres, 1934.

Habermas, Jürgen. *The Structural Transformation of the Public Sphere: An Inquiry into a Category of Bourgeois Society.* Cambridge: Harvard University Press, 1989.

Hagner, Michael. "Enlightened Monsters." In *The Sciences in Enlightened Europe,* edited by William Clark, Jan Golinski, and Simon Schaffer, 175–217. Chicago: University of Chicago Press, 1999.

Hahn, Roger. *The Anatomy of a Scientific Institution: The Paris Academy of Sciences, 1666–1803.* Berkeley: University of California Press, 1971.

Hall, A. Rupert. *Philosophers at War: The Quarrel between Newton and Leibniz.* Cambridge and New York: Cambridge University Press, 1980.

Hankins, Thomas. "Eighteenth-Century Attempts to Resolve the *Vis Viva* Controversy." *Isis* 56 (1965): 281–97.

———. *Jean d'Alembert: Science and the Enlightenment.* Oxford: Clarendon Press, 1970.

———. *Science and the Enlightenment.* Cambridge: Cambridge University Press, 1985.

Harman, Peter M. "Dynamics and Intelligibility: Bernoulli and Maclaurin." In *Metaphysics and Philosophy of Science in the Seventeenth and Eighteenth Centuries,* edited by R. S. Woolhouse, 213–25. Dordrecht: Kluwer, 1988.

Harnack, Adolf. *Geschichte der Königlich Preussischen Akademie der Wissenschaften zu Berlin.* 3 vols. in 4. Berlin: Reichsdruckerei, 1900.

Harth, Erica. *Cartesian Women: Versions and Subversions of Rational Discourse in the Old Regime.* Ithaca: Cornell University Press, 1992.

———. *Ideology and Culture in Seventeenth-Century France.* Ithaca: Cornell University Press, 1983.

Hecht, Hartmut, ed. *Pierre Louis Moreau de Maupertuis: Eine Bilanz nach 300 Jahren.* Berlin: Verlag A. Spitz, 1999.

———. "Pierre Louis Moreau de Maupertuis et la notion de nature au siècle des Lumières." *Cahiers de la revue de théologie et de philosophie* 18 (1996): 205–10.

Heilbron, John. *Elements of Early Modern Physics.* Berkeley: University of California Press, 1982.

Hervé, G. "Les correspondantes de Maupertuis." *La revue de Paris* 5 (1911): 751–58.

Hine, Ellen McNiven. "Dortous de Mairan and Eighteenth-Century 'Systems Theory.'" *Gesnerus* 52 (1995): 54–65.

———. "Dortous de Mairan, the 'Cartonian.'" *Studies on Voltaire and the Eighteenth Century* 266 (1989): 163–79.

Hoffheimer, Michael H. "Maupertuis and the Eighteenth-Century Critique of Pre-existence." *Journal of the History of Biology* 15 (1982): 119–44.

Home, R. W. "Out of a Newtonian Straitjacket: Alternative Approaches to Eighteenth-Century Physical Science." In *Studies in the Eighteenth Century* 4, edited by R. F. Brissenden and J. C. Eade, 235–49. Canberra: Australian National University Press, 1976.

Houdar de la Motte, Antoine. *Œuvres de Monsieur Houdar de la Motte.* 10 vols. Paris: Prault, 1754.

Ibrahim, Annie. "Matière inerte et matière vivante: La théorie de la perception chez Maupertuis." *Dix-huitième siècle* 24 (1992): 95–103.

Iliffe, Rob. "'Aplatisseur du monde et de Cassini': Maupertuis, precision measurement, and the shape of the earth in the 1730s." *History of Science* 31 (1993): 335–75.

Iltis, Carolyn. "Madame du Châtelet's Metaphysics and Mechanics." *Studies in the History and Philosophy of Science* 8 (1977): 29–48.

Irailh, Simon-Augustin. *Querelles littéraires ou mémoires pour servir à l'histoire des révolutions dans la République des Lettres.* 3 vols. Paris: Durand, 1761.

Jacob, Margaret C. *Living the Enlightenment: Freemasonry and Politics in Eighteenth-Century Europe.* New York: Oxford University Press, 1991.

Jacquart, Jean. *L'Abbé Trublet, critique et moraliste, 1697–1770: Un témoin de la vie littéraire et mondaine au XVIIIième siècle.* Paris: A. Picard, 1926.

Janik, Linda Gardiner. "Searching for the Metaphysics of Science: The Structure and Composition of Mme. du Châtelet's *Institutions de Physique,* 1737–1740." *Studies on Voltaire and the 18th Century* 201 (1982): 85–113.

Jurin, James. *The Correspondence of James Jurin (1684–1750): Physician and Secretary to the Royal Society.* Edited by Andrea Rusnock. Amsterdam: Rodopi, 1996.

Kaulfuss-Diesch, Carl. "Maupertuisiana." *Zentralblatt für die Bibliothekswesen* 39 (1922): 525–46.

Kerviler, René. *La Bretagne à l'Académie française au XVIIIe siècle.* Paris: V. Palme, 1889.

Kim, Mi Gyung. *Affinity, That Elusive Dream: A Genealogy of the Chemical Revolution.* MIT Press: Cambridge, 2002.

Kneser, A. *Das prinzip der kleinsten Wirkung von Leibniz bis zur Gegenwart.* Leipzig and Berlin: Teubner, 1929.

Koerner, Lisbet. *Linnaeus: Nature and Nation.* Cambridge: Harvard University Press, 1999.

König, J. S. "De universali principio aequilibrii et motus." *Nova Acta Eruditorum* (March 1751): 125–35, 162–76.

Konvitz, J. *Cartography in France, 1660–1848: Science, Engineering and Statecraft.* Chicago: University of Chicago Press, 1987.

Koser, R., ed. *Briefwechsel Friedrichs des Grossen mit Grumbkow und Maupertuis.* Leipzig: S. Hirzel, 1898.

La Barre de Beaumarchais, Antoine de. *Amusemens littéraires: ou, Correspondance politique, historique, philosophique, critque, and galante.* 3 vols. The Hague: Van Duren, 1740.

La Beaumelle, Laurent Angliviel. *Vie de Maupertuis.* Paris: Ledoyen, 1856.

La Condamine, Charles-Marie de. *Journal du voyage fait par ordre du roi à l'équateur.* Paris: Imprimerie Royale, 1751.

———. *Mesure des trois premiers degrés du méridien dans l'hémisphère australe.* Paris: Imprimerie Royale, 1751.

———. "Sur une nouvelle espèce de végétation métallique," *MAS* 1731, 466–82.

Lacombe, Henri, and Costabel Pierre, eds. *La figure de la terre du XVIIIième siècle à l'ère spatiale.* Paris: Gauthier-Villars, 1988.

Lafuente, Antonio, and Antonio Delgado. *La Geometrización de la Tierra: Observaciones y Resultados de la Expedición Geodésica Hispano-Francea al Virreinato del Perú (1735–1744).* Madrid: Instituto Arnau de Vilanova, 1984.

Lafuente, Antonio, and José L. Peset. "La question de la figure de la terre. L'agonie d'un débat scientifique au XVIIIième siècle." *Revue d'histoire des sciences* 37 (1984): 235–54.

Lalande, Joseph Jerome de. *Bibliographie astronomique.* Paris: l'Imprimerie de la République, 1803.

La Mettrie, Julien Offroy de. *L'homme machine: A Study in the Origins of an Idea.* Edited by Aram Vartanian. Princeton: Princeton University Press, 1960. Originally published in 1748.

Le Blanc, Jean-Bernard. *Lettres d'un françois.* 3 vols. The Hague: J. Neaulme, 1745.

Le Monnier, Pierre-Charles. *Histoire céleste, ou Recueil de toutes les observations astronomiques faites par ordre du roy* . Paris: Briasson, 1741.

Le Sueur, Achille. *Maupertuis et ses correspondants.* Montreuil-sur-Mer: Notre-Dame des Prés, 1896. Reprinted Geneva: Slatkine, 1971.

Leibniz, Gottfried Wilhelm. *Selections.* Edited by Philip P. Wiener. New York: Charles Scribner, 1951.

Lespagnol, André. *Messieurs de Saint-Malo: Une élite négociante au temps de Louis XIV.* 2 vols. Rennes: Presses Universitaires de Rennes, 1997.

Lesser, Friedrich-Christian. *Théologie des insectes.* 2 vols. Translated by P. Lyonnet. Paris, 1745.

*Lettres concernant le jugement de l'Académie, à l'occasion de la lettre attribuée à Leibniz par M. le Professeur Koenig.* Berlin: Etienne de Bourdeaux, 1752; 2nd ed. with *Apologie de M. de Maupertuis,* Paris: Durand & Pissot, 1753.

Lignac, Joseph-Adrien Lelarge de. *Lettres à un amériquain sur l'histoire naturelle, générale et particulière de Monsieur de Buffon.* 5 vols. Hamburg: 1751–1752.

López-Beltrán, Carlos. "Forging Heredity: From Metaphor to Cause, a Reification Story." *Studies in the History and Philosophy of Science* 25 (1994): 211–35.

Lougee, Carolyn. *"Le paradis des femmes": Women, Salons and Social Stratification in Seventeenth-Century France.* Princeton: Princeton University Press, 1976.

Lough, John. *Essays on the Encyclopédie of Diderot and d'Alembert.* London and New York: Oxford University Press, 1968.

Loveland, Jeff. *Rhetoric and Natural History: Buffon in Polemical and Literary Context.* Oxford: Voltaire Foundation, 2001.

Luynes, Charles Philippe d'Albert de. *Mémoires du duc de Luynes sur la cour de Louis XV (1735–1758).* 17 vols. Paris: Firmin Didot, 1860–1965.

Maindron, Ernest. *L'Académie des sciences.* Paris: F. Alcan, 1888.

———. *Les fondations de prix à l'Académie des sciences.* Paris: Gauthier-Villars, 1881.

Mairan, Jean-Jacques Dortous de. "Animaux coupés & partagés en plusieurs parties," *HAS* 1741, 33–35.

———. "Dissertation sur l'estimation & la mesure des forces motrices des corps." *MAS* 1728, 1–49.

———. "Recherches géometriques sur la diminution des degrés terrestres en allant de l'équateur vers les pôles." *MAS* 1720, 231–277.

———. "Suite des recherches physico-mathématiques sur la réflexion des corps." *MAS* 1723, 343–86.

Martin, J.-P. *La figure de la terre: récit de l'expédition française en Laponie suédoise (1736–1737).* Cherbourg: Isoète, 1987.

Mazzolini, Renato G., and Shirley A. Roe, eds. *Science against the Unbelievers: the Correspondence of Bonnet and Needham, 1760–1780.* Oxford: Voltaire Foundation, 1986.

McClellan, James. *Science Reorganized: The Scientific Societies in the Eighteenth Century.* New York: Columbia University Press, 1973.

Merian, J. B. "Eloge de M. Formey," *BAS* 1797, 49–82.

———. *Mémoires pour servir à l'histoire du jugement de l'Académie.* Berlin, 1753.

Mervaud, Christiane. *Voltaire et Frédéric II: Une dramaturgie des lumiéres, 1736–1778.* Oxford: Voltaire Foundation, 1985.

Meyer, Jean. *La noblesse bretonne au XVIIIième siècle.* Paris: S.E.V.P.E.N., 1966.

Molières, Joseph Privat de. "Les loix astronomiques des vitesses des planetes dans leurs orbes, expliquées méchaniquement dans le systême du plein," *MAS* 1733, 301–12.

Monod-Cassidy, Hélène. *Un voyageur-philosophe au XVIIIe siècle: L'Abbé Jean-Bernard Le Blanc.* Cambridge: Harvard University Press, 1941.

Montesquieu, Charles de Secondat de. *Correspondance de Montesquieu.* 2 vols. Edited by François Gébelin. Paris: E. Champion, 1914.

Montucla, Jean Etienne. *Histoire des mathématiques, dans laquelle on rend compte de leurs progrès depuis leur origine jusq'à nos jours.* Paris: C. A. Jombert, 1758.

Morel, Anne. "La guerre de course à Saint-Malo de 1681–1715." *Mémoires de la Société d'histoire et d'archéologie de Bretagne* 37 (1957): 5–103.

Mornet, Daniel. "Les enseignements des bibliothèques privées (1750–1780)." *Revue d'histoire littéraire de la France* 17 (1910): 449–96.

Müller, Jakob. *Die ungegründete und idealistische Monadologie.* Frankfurt am Main, 1745.

Needham, John Turberville. *An Account of Some Microscopical Discoveries.* London: F. Needham, 1745.

———. *Nouvelles découvertes faites avec le microscope . . . avec un mémoire sur les polypes à bouquet . . . par A. Trembley.* Leiden: Luzac 1747.

———. *Nouvelles observations microscopiques; avec des expériences intéressantes sur la composition et la décomposition des corps organisés.* Paris: Ganeau, 1750.

———. *Observations upon the Generation, Composition, and Decomposition of Animal and Vegetable Substances.* London, 1749.

———. "A Summary of Some Late Observations upon the Generation, Composition, and Decomposition of Animal and Vegetable Substances." *Philosophical Transactions* no. 490 (1748): 615–66.

Nordenmark, N. V. E. *Anders Celsius: Professor i Uppsala 1701–44.* Uppsala: Almqvist and Wiksell, 1936.

Nordmann, Claude. "Expédition de Maupertuis et Celsius en Laponie." *Cahiers d'histoire mondiale* 10 (1966): 74–97.

Olmsted, John W. "The Scientific Expedition of Jean Richer to Cayenne (1672–1673)." *Isis* 34 (1942): 117–28.

Ostoya, Paul. "Maupertuis et la biologie." *Revue d'histoire des sciences et de leurs applications* 7 (1954): 60–80.

Outhier, Reginald. *Journal d'un voyage au nord en 1736 et 1737.* Edited by André Balland. Paris: Seuil, 1994. Originally published in Paris in 1744.

Outram, Dorinda. *The Enlightenment.* Cambridge: Cambridge University Press, 1995.

———. *Georges Cuvier: Vocation, Science, and Authority in Post-Revolutionary France.* Manchester: Manchester University Press, 1984.

———. "Politics and Vocation: French Science, 1793–1830." *British Journal for the History of Science* 13 (1980): 27–43.

Panza, Marco. "De la nature épargnante aux forces généreuses: Le principe de moindre action entre mathématiques et métaphysique. Maupertuis et Euler, 1740–1751." *Revue d'histoire des sciences* 48 (1995): 435–520.

Paret, Peter. *Frederick the Great: A Profile* . New York: Hill and Wang, 1972.

Passeron, Irene. "La forme de la Terre est-elle une preuve de la vérité du système newtonien?" In *Terre à decouvrir, terres à parcourir: Exploration et connaissance du monde XIIe–XIXe siècles,* edited by Danielle Lecoq and Antoine Chambard, 129–45. Paris: L'Harmattan, 1998.

———. "Maupertuis, passeur d'intelligibilité. De la cycloïde à l'ellipsoïde aplati en passant par le 'newtonianisme': années parisiennes." In *Pierre Louis Moreau de Maupertuis: Eine Bilanz nach 300 Jahren,* edited by Hartmut Hecht, 17–33. Berlin: Verlag A. Spitz, 1999.

Paul, Charles B. *Science and Immortality: The Eloges of the Paris Academy of Sciences (1699–1791).* Berkeley & Los Angeles: University of California Press, 1980.

Pelletier, Monique. *La carte de Cassini: L'extraordinaire aventure de la carte de France.* Paris: Presses de l'Ecole nationale des ponts et chaussées, 1990.

———. "Cartographie et pouvoir sous le règnes de Louis XIV et Louis XV." In *Terre à decouvrir, terres à parcourir: Exploration et connaissance du monde XIIe–XIXe siècles,* edited by Danielle Lecoq and Antoine Chambard, 113–27. Paris: L'Harmattan, 1998.

Penisson, Pierre. "Maupertuis philosophe géographe." *Corpus* (1998): 45–58.

Pluche, Noël-Antoine. *Le spectacle de la nature.* 8 vols. Paris, 1732–1750.

Pomeau, René, and Christiane Mervaud. *De la cour au jardin, 1750–1759.* Oxford: Voltaire Foundation, 1991.

Pomian, Krzysztof. *Collectors and Curiosities: Paris and Venice 1500–1800.* Cambridge, U.K., and Cambridge, MA: Polity Press; Basil Blackwell, 1990.

Pons, George. "Les années berlinoises de Maupertuis ou Maupertuis vu par les allemands de son temps." *Annales de Bretagne et des Pays de l'Ouest* 84 (1976): 681–94.

Pottinger, David T. *The French Book Trade in the Ancien Regime, 1500–1791.* Cambridge: Harvard University Press, 1958.

Procope-Couteau, Michel. *L'Art de faire des garçons.* Montpellier, [1748].

Proust, Jacques. *Diderot et l'Encyclopédie.* Paris: A. Colin, 1967.

Pulte, Helmut. *Das Prinzip der kleinsten Wirkung und die Kraftkonzeptionen der rationalen Mechanik: Eine Untersuchung zur Grundlegungsproblematik bei Leonhard Euler, Pierre Louis Moreau de Maupertuis und Joseph Louis Lagrange.* Stuttgart: Franz Steiner Verlag, 1989.

Radelet-de Grave, Patricia. "La 'Diatribe du Docteur Akakia, Médicin du pape.'" *Revue des questions scientifiques* 168 (1998): 209–49.

Ramati, Ayval. "Harmony at a Distance: Leibniz's Scientific Academies." *Isis* 87 (1996): 430–52.

Réaumur, René-Antoine Ferchault de. *L'art de faire éclorre et d'élever en toute saison des oiseaux domestiques de toutes espèces.* Paris: Imprimerie royale, 1749. 2nd ed. 1751.

———. The Art of Hatching and Bringing Up Domestick Fowls of All Kinds at Any Time of the Year. London: C. Davis, 1750.

———. *Mémoires pour servir à l'histoire des insectes.* 6 vols. Paris: L'Imprimerie royale, 1734–1742.

Rigaud, Stephen Peter, ed. *Correspondence of Scientific Men of the Seventeenth Century.* 2 vols. Hildesheim: Olms, 1965.

Rives, D. B., ed. *Lettres inédits du chancelier d'Aguesseau.* 2 vols. Paris: Imprimerie royale chez C. J. Trouvé, 1823.

Roe, Shirley. "Buffon and Needham." In *Buffon 88: actes du Colloque international pour le bicentenaire de la mort de Buffon,* edited by Jean-Claude Beaune and Jean Gayon, 439–50. Paris and Lyon: Vrin, 1992.

———. "John Turberville Needham and the Generation of Living Organisms." *Isis* 74 (1983): 159–84.

———. *Matter, Life, and Generation: Eighteenth-Century Embryology and the Haller-Wolff Debate.* Cambridge: Cambridge University Press, 1981.

Roger, Jacques. *Buffon: A Life in Natural History.* Translated by Sarah Bonnefoi. Ithaca: Cornell University Press, 1997. Originally published in 1989.

———. *Les sciences de la vie dans la pensée française du XVIIIième siècle.* 2nd ed. Paris: Armand Colin, 1971.

Rosenberg, Hans. *Bureaucracy, Aristocracy and Autocracy: The Prussian Experience, 1660–1815.* Cambridge: Cambridge University Press, 1958.

Sabra, A. I. *Theories of Light From Descartes to Newton.* 2nd ed. Cambridge: Cambridge University Press, 1981.

Sadoun-Goupil, Michelle. *Du flou au clair? Histoire de l'affinité chimique: De Cardan à Prigogine.* Paris: C.T.H.S., 1991.

Schaeper, Thomas J. *The French Council of Commerce, 1700–1715: A Study of Mercantilism after Colbert.* Columbus: Ohio State University Press, 1983.

Schaffer, Simon. "Authorized Prophets: Comets and Astronomers After 1759." *Studies in Eighteenth-Century Culture* 17 (1987): 45–74.

Schieder, Theodor. *Friedrich der Grosse: Ein Königtum der Widerspruche.* Frankfurt am Main: Propylaen Verlag, 1983.

Schiller, Joseph. "Queries, Answers, and Unsolved Problems in Eighteenth-Century Biology." *History of Science* 12 (1974): 184–99.

Sewell, William. "*Etat, Corps,* and *Ordre:* Some Notes on the Social Vocabulary of the French Old Regime." In *Sozialgeschichte heute: Festschrift für Hans Rosenberg,* edited by Hans-Ulrich Wehler, 49–68. Gottingen: Vandenhoeck & Ruprecht, 1974.

Sgard, Jean, Michel Gilot, and Françoise Weil. *Dictionnaire des journalistes, 1600–1789.* Grenoble: Presses Universitaires de Grenoble, 1976.

Shank, John Bennett. "Before Voltaire: Newtonianism and the Origins of the Enlightenment in France, 1687–1734." Ph.D. diss., Stanford University, 2000.

Shapin, Steven. "Personal Development and Intellectual Biography: The Case of Robert Boyle." *British Journal for the History of Science* 26 (1993): 335–46.

———. *The Social History of Truth.* Chicago: University of Chicago Press, 1994.

Shapin, Steven and Simon Schaffer. *Leviathan and the Air-Pump: Hobbes, Boyle, and the Experimental Life.* Princeton, N.J.: Princeton University Press, 1985.

Shaw, Edward. *Problems and Policies of Malesherbes as Directeur de la Librairie in France* (1750–1763). Albany, NY: State University of New York, 1966.

Shea, William R. "The Unfinished Revolution: Johann Bernoulli (1667–1748) and the Debate Between the Cartesians and the Newtonians." *Revolutions in Science.* Edited by William R. Shea, 70–92. Canton, MA: Science History Publications, 1988.

Sloan, Philip. "Organic Molecules Revisited." *Buffon 88: actes du Colloque international pour le bicentenaire de la mort de Buffon.* Edited by Jean-Claude Beaune and Jean Gayon. Paris and Lyon: Vrin, 1992.

Smith, Jay M. *The Culture of Merit: Nobility, Royal Service, and the Making of Absolute Monarchy in France, 1600–1789.* Ann Arbor: University of Michigan Press, 1996.

Société des bibliophiles français, Paris, ed. "Lettre de Maupertuis à Mme de Vertillac [sic]." *Mélanges publiés par la Société des bibliophiles français.* Vol. 6, 1829. Reprint, Geneva: Slatkine Reprints, 1970.

Sorrenson, Richard. "George Graham, Visible Technician." *British Journal for the History of Science* 32 (1999): 203–21.

———. "Scientific Instrument Makers at the Royal Society of London, 1720–1780." Ph.D. diss., Princeton University, 1993.

Spary, E. C. "Codes of Passion: Natural History Specimens as a Polite Language in Late Eighteenth-Century France." In *Wissenschaft als kulturelle Praxis, 1750–1900,* edited by Hans Erich Bödeker, Peter Hanns Reill, and Jürgen Schlumbohm, 105–35. Göttingen: Vanderhoek and Ruprecht, 1999.

———. "Enlightened Natures." In *The Sciences in Enlightened Europe,* edited by William Clark, Jan Golinski, and Simon Schaffer, 272–304. Chicago: University of Chicago Press, 1999.

———. *Utopia's Garden: French Natural History from Old Regime to Revolution.* Chicago: University of Chicago Press, 2000.

Stewart, Larry R. *The Rise of Public Science: Rhetoric, Technology, and Natural Philosophy in Newtonian Britain, 1660–1750.* New York: Cambridge University Press, 1992.

Stroup, Alice. *A Company of Scientists: Botany, Patronage, and Community at the Seventeenth-Century Parisian Royal Academy of Sciences.* Berkeley: University of California Press, 1990.

———. *Royal Funding of the Parisian Académie royale des Sciences during the 1690s.* Philadelphia: American Philosophical Society, 1987.

Struik, Dirk. *A Source Book in Mathematics, 1600–1800.* Cambridge: Harvard University Press, 1969.

Sutton, Geoffrey. *Science for a Polite Society: Gender, Culture and the Demonstration of Enlightenment.* Boulder, Colo.: Westview Press, 1995.

Symcox, Geoffrey. *The Crisis of French Sea Power, 1688–1697: From the guerre d'escadre to the guerre de course.* The Hague: M. Nijhoff, 1974.

Taton, René. "L'expédition géodesique de Laponie (avril 1736–août 1737)." In *La figure de la terre du XVIIIième siecle à l'ère spatiale,* edited by Henri Laombe and Pierre Costabel, 115–38. Paris: Gauthier-Villars, 1988.

———. "Sur quelques pièces de la correspondance de Lagrange pour les années 1756–1758." *Bolletinno di Storia delle Scienze Matematiche* 8 (1988): 3–19.

Terrall, Mary. "The Culture of Science in Frederick the Great's Berlin." *History of Science* 28 (1990): 333–64.

———. "Emilie du Châtelet and the Gendering of Science." *History of Science* 33 (1995): 283–310.

———. "Gendered Spaces, Gendered Audiences: Inside and Outside the Paris Academy of Sciences." *Configurations* 2 (1995): 207–32.

———. "Metaphysics, Mathematics, and the Gendering of Science in Eighteenth-Century France." In *The Sciences in Enlightened Europe,* edited by William Clark, Jan Golinski, and Simon Schaffer, 246–71. Chicago: University of Chicago Press, 1999.

———. "Representing the Earth's Shape: The Polemics Surrounding Maupertuis's Expedition to Lapland." *Isis* 83 (1992): 218–37.

———. "Salon, Academy and Boudoir: Generation and Desire in Maupertuis's Science of Life." *Isis* 87 (1996): 217–29.

Thiébault, Dieudonné. *Frédéric le Grand, ou mes souvenirs de vingt ans de séjour à Berlin.* 5 vols. Paris: Bossange, 1827.

Tonelli, Giorgio. "La necessité des lois de la nature au XVIIIième siècle et chez Kant en 1762." *Revue d'histoire des sciences et de leurs applications* 12 (1959): 225–41.

———. *La pensée philosophique de Maupertuis: Son milieu et ses sources.* Hildesheim: G. Olms, 1987.

Torlais, Jean. *Réaumur: Un esprit encyclopédique en dehors de l'Encyclopédie.* Paris: Desclée de Brouwer, 1937.

Tort, Patrick. *L'ordre et les monstres.* Paris: Le Sycomore, 1980.

Trembley, Abraham. *Mémoires pour servir à l'histoire d'un genre de polypes d'eau douce.* Leiden, 1744.

Tressan, L. E. "Eloge de M. Moreau de Maupertuis." In *Oeuvres posthumes du comte de Tressan.* 2 vols. Paris: Desray, 1791.

Trystram, Florence. *Le procès des étoiles.* Paris: Payot, 1993.

Tuilier, André. *Histoire de l'Université de Paris et de la Sorbonne.* 2 vols. Paris: Nouvelle Librairie de France, 1994.

Tweedie, Charles. *James Stirling, a Sketch of His Life and Works along with His Scientific Correspondence.* Oxford: Clarendon Press, 1922.

Vaillot, René. *Madame du Châtelet.* Paris: A. Michel, 1978.

Vartanian, Aram. "Diderot and Maupertuis." *Revue internationale de philosophie* 38 (1984): 46–66.

———. "Trembley's Polyp, La Mettrie, and Eighteenth-Century French Materialism." *Journal of the History of Ideas* 11 (1950): 259–86.

Velluz, Leon. *Maupertuis.* Paris: Hachette, 1969.

Venturi, Franco. *Jeunesse de Diderot (1713–1753).* Translated by Juliette Bertrand. Paris: Albert Skira, 1939.

Vila, Anne. *Enlightenment and Pathology.* Baltimore: Johns Hopkins University Press, 1998.

Voltaire. *Elémens de la Philosophie de Newton.* In *Oeuvres Complètes de Voltaire.* Edited by R. L. Walters and W. H. Barber, vol. 15. Oxford: Voltaire Foundation, 1992.

———. *The Complete Works of Voltaire.* 135 vols. Edited by Theodore Besterman et al. Geneva and Toronto: Institut et Musée Voltaire and University of Toronto Press, 1968–2001.

———. *Histoire du Docteur Akakia et du natif de St-Malo.* Edited by Jacques Tuffet. Paris: A.G. Nizet, 1967.

———. *Lettres philosophiques, ou Lettres anglaises.* Edited by Raymond Naves. Paris: Garnier frères, 1967.

———. *Oeuvres de Voltaire.* 72 vols. Edited by A. J. Beuchot. Paris: Lefevre, 1829.

Walters, Alice. "Conversation Pieces: Science and Politeness in Eighteenth-Century England." *History of Science* 35 (1997): 121–54.

Walters, R. L., and W. H. Barber. "Introduction to the *Elements de la Philosophie de Newton.*" In *Oeuvres complètes de Voltaire,* 15:98–118. Oxford: Voltaire Foundation, 1992.

Westman, Robert. "The Astronomer's Role in the Sixteenth Century: A Preliminary Study." *History of Science* 18 (1980): 105–47.

Winter, Eduard. *Die Registres der Berliner Akademie der Wissenschaften, 1746–1766.* Berlin: Akademie-Verlag, 1957.

Wolf, Charles. *Histoire de l'Observatoire de Paris de sa fondation à 1793.* Paris: Gauthier-Villars, 1902.

# INDEX

Made in the USA
Lexington, KY
17 March 2012